# EAI/Springer Innovations in Communication and Computing

**Series Editor**

Imrich Chlamtac, European Alliance for Innovation, Ghent, Belgium

The impact of information technologies is creating a new world yet not fully understood. The extent and speed of economic, life style and social changes already perceived in everyday life is hard to estimate without understanding the technological driving forces behind it. This series presents contributed volumes featuring the latest research and development in the various information engineering technologies that play a key role in this process. The range of topics, focusing primarily on communications and computing engineering include, but are not limited to, wireless networks; mobile communication; design and learning; gaming; interaction; e-health and pervasive healthcare; energy management; smart grids; internet of things; cognitive radio networks; computation; cloud computing; ubiquitous connectivity, and in mode general smart living, smart cities, Internet of Things and more. The series publishes a combination of expanded papers selected from hosted and sponsored European Alliance for Innovation (EAI) conferences that present cutting edge, global research as well as provide new perspectives on traditional related engineering fields. This content, complemented with open calls for contribution of book titles and individual chapters, together maintain Springer's and EAI's high standards of academic excellence. The audience for the books consists of researchers, industry professionals, advanced level students as well as practitioners in related fields of activity include information and communication specialists, security experts, economists, urban planners, doctors, and in general representatives in all those walks of life affected ad contributing to the information revolution.

Indexing: This series is indexed in Scopus, Ei Compendex, and zbMATH.

**About EAI** - EAI is a grassroots member organization initiated through cooperation between businesses, public, private and government organizations to address the global challenges of Europe's future competitiveness and link the European Research community with its counterparts around the globe. EAI reaches out to hundreds of thousands of individual subscribers on all continents and collaborates with an institutional member base including Fortune 500 companies, government organizations, and educational institutions, provide a free research and innovation platform. Through its open free membership model EAI promotes a new research and innovation culture based on collaboration, connectivity and recognition of excellence by community.

Behçet Uğur Töreyin • Hatice Köse •
Nizamettin Aydin • Seifedine Nimer Kadry •
Ömer Melih Gül
Editors

# 8th EAI International Conference on Robotic Sensor Networks

## EAI ROSENET 2024

Springer

EAI
RESEARCH MEETS INNOVATION

*Editors*

Behçet Uğur Töreyin
Istanbul Technical University
Istanbul, Türkiye

Hatice Köse
Istanbul Technical University
Istanbul, Türkiye

Nizamettin Aydin
Istanbul Technical University
Istanbul, Türkiye

Seifedine Nimer Kadry
Noroff College
Oslo, Norway

Ömer Melih Gül
Istanbul Technical University
Istanbul, Türkiye

International Conference on Robotics and Networks rosenet rosenet2024 ROSENET 8 Crete Greece 2024 09 03 2024 09 05 https://rosenet.eai-conferences.org/2024/

ISSN 2522-8595 ISSN 2522-8609 (electronic)
EAI/Springer Innovations in Communication and Computing
ISBN 978-3-031-92142-1 ISBN 978-3-031-92143-8 (eBook)
https://doi.org/10.1007/978-3-031-92143-8



This Springer imprint is published by the registered company Springer Nature Switzerland AG
The registered company address is: Gewerbestrasse 11, 6330 Cham, Switzerland

# Contents

# About the Editors

**Behcet Uğur Töreyin** received a BS degree from the Middle East Technical University, Ankara, Turkey in 2001 and the MS and PhD degrees from Bilkent University, Ankara, in 2003 and 2009, respectively, all in electrical and electronics engineering. He is now a Full Professor with the Informatics Institute at Istanbul Technical University. His research interests broadly lie in signal processing and pattern recognition with applications to image/video analysis and communication systems. His research is focused on developing algorithms to analyze and compress signals from a multitude of sensors such as visible/infrared/hyperspectral cameras, microphones, passive infrared sensors, vibration sensors, and spectrum sensors for wireless communications. For further information, visit the Signal Processing for Computational Intelligence Group website at URL: http://spacing.itu.edu.tr.

**Hatice Köse** (Member, IEEE) received the PhD degree from the Computer Engineering Department, Bogaziçi University, Turkey. She has been a Full Professor with the Faculty of Computer and Informatics Engineering, Istanbul Technical University, Turkey, coordinating the GameLaboratory and Cognitive Social Robotics Laboratory since 2010. From 2006 to 2010, she was a Research Fellow with the University of Hertfordshire. Her current research interests include gesture communication (involving sign language) and imitation-based interaction games with social humanoid robots for the education and rehabilitation of children with hearing impairment and children with ASD. She is leading several national projects and taking part in several Horizon 2020 projects, Erasmus+, and Cost actions on socially assistive robots, sign language tutoring robots, and human–robot interaction.

**Nizamettin Aydın** (Senior Member, IEEE) received the BSc and MSc degrees in electronics and communication engineering from Yildiz Technical University, Türkiye, 1984 and 1987, respectively, and the PhD degree in medical physics from the University of Leicester, UK, in 1994. He was with the Department of Clinical Neurosciences, Kings College London, and the Division of Clinical Neuroscience, St George's Hospital Medical School, as a Research Fellow, from 1998 to 2001.

He was a Senior Research Fellow with the Institute for Integrated Micro and Nano Systems, University of Edinburgh, from 2001 to 2004. In 2004, he was appointed as the Head of the Computer Engineering Department, and in 2006 he founded the Software Engineering Department, both at Bahcesehir University, Türkiye. From 2009 to 2023, he was a Professor with the Computer Engineering Department, Yildiz Technical University, and was the Head of the Department, from 2011 to 2023. He is now with the Computer Engineering Department at Istanbul Technical University, Türkiye. His research interests and contributions cover a wide range of subjects, including biomedical signal/image processing, bioinformatics, system-on-chip, AI, data science, and computer science. He was awarded the IEE (now IET) Institute Premium Award from 2000 to 2001. He serves as the IEEE Türkiye Section Chair.

**Seifedine Nimer Kadry** has a bachelor's degree in 1999 from Lebanese University, MS degree in 2002 from Reims University (France) and EPFL (Lausanne), PhD in 2007 from Blaise Pascal University (France), HDR degree in 2017 from Rouen University. He is a Full Professor of Data Science at Lebanese American University, Lebanon, and Noroff University College, Norway. At present, his research focuses on Data Science, education using technology, system prognostics, stochastic systems, and applied mathematics. He is an ABET program evaluator for both computing and Engineering Technology. He is a Fellow of IET, IETE, and IACSIT. He was a distinguished speaker of the IEEE Computer Society.

**Ömer Melih Gül** (SIEEE'17, MIEEE'21, SMIEEE'24) received BSc, MSc, and PhD degrees from the Department of Electrical and Electronics Engineering at Middle East Technical University (METU), Ankara, Turkey, in 2012, 2014, and 2020, respectively, by also working as a research assistant at the same department. His research interests include machine learning applications, wireless security, networking, scheduling, IoT, UAV, robotics, and blockchain. He has co-authored over 50 papers and 4 book chapters. He was awarded third place in the 2019 Lance Stafford Larson Outstanding Student Paper Award by the IEEE Computer Society. He was also awarded third place in the poster competition at 2021 IEEE Rising Stars Global Conference. In 2022, he worked as a postdoctoral fellow at the School of Electrical Engineering and Computer Science at the University of Ottawa, Canada. He is a recipient of the best paper award at 48th Wireless World Research Forum (WWRF) in 2022. In 2023, he worked as an Assistant Professor in the Department of Computer Engineering at Bahcesehir University, Istanbul, Turkey, where he supervised 4 MSc theses and co-supervised 1 thesis. Since March 2024, he has been working as an associate professor in the Informatics Institute at Istanbul Technical University (ITU), Istanbul, Turkey, where he is supervising 1 PhD and 4 MSc students. He serves as an Editor for (Springer) *Telecommunication Systems*, (Springer) *Wireless Networks*, (Springer) *Cluster Computing*, (Elsevier) *Sustainable Computing: Informatics and Systems*, and also the *International Journal of Interactive Multimedia and Artificial Intelligence*. As cochair, he organized CIEAI workshop at IEEE Fog and Mobile Edge Computing (FMEC) 2023. Moreover, he became the Publicity Chair in the IEEE iThings 2024. He organized two EAI International Conferences on Robotic Sensor Networks (ROSENET 2023, ROSENET 2024) as general chair. On the IEEE volunteering side, he held several prestigious IEEE volunteer roles including Vice-Chair of Students and Young Professionals Committee, Region 8 (EMEA) Coordinator, and Member-at-Large at MGA Board of the IEEE Computer Society. Currently, he has been Vice-Chair of the Geographic Activities Committee at the IEEE Computer Society. He was awarded the 2022 IEEE MGA Young Professionals Achievement Award by IEEE MGA. In addition, he was awarded the Outstanding Contribution Certificate in 2021 and the Continuous Service Certificate in 2024 by the IEEE Computer Society.

# Part I
# Main Track

# Computer-Aided Alzheimer's Disease Diagnosis from Magnetic Resonance Images Using Cycle Generative Adversarial Networks and Deep Transfer Learning

**Seda Coşkun Eliküçük and Volkan Müjdat Tiryaki**

## 1 Introduction

Alzheimer's disease is the most prevalent type of dementia and a worldwide disease causing the death of neurons in the human brain [1]. Cognitive decline and memory impairment are some of the disease symptoms. Alzheimer's disease has wide-ranging effects on basic skills such as reading, speaking, and writing due to the loss of neurons in the brain. Alzheimer's disease was first diagnosed in a 51-year-old woman in 1907 by Dr. Alois Alzheimer. As the disease progressed, the patient experienced a variety of complex symptoms and significant memory loss. It has been noted that she was able to correctly name some objects shown to her, but forgot these objects after a short time. Additionally, comprehension difficulties, hallucinations, and reading problems have also been identified. The disease took four and a half years and the patient died at the end of this period [2].

The development of artificial intelligence (AI) algorithms has created revolutionary effects in medicine. AI techniques in areas such as medical imaging, disease diagnosis, treatment planning, and patient monitoring have enabled healthcare professionals to provide faster, more accurate, and personalized services. Development of new AI methods could enable medical doctors to make more accurate diagnoses and to develop better treatment options.

Early diagnosis of the Alzheimer's disease is crucial for treatment and disease management [3]. Magnetic Resonance Imaging (MRI) is a noninvasive method that

S. Coşkun Eliküçük
Diyarbakır Bağlar District National Education Directorate, Diyarbakır, Turkey
e-mail: seda.coskun.elikucuk@roi.k12.tr

V. M. Tiryaki (✉)
Informatics Institute, Istanbul Technical University, İstanbul, Turkey
e-mail: tiryakiv@itu.edu.tr

 
B. U. Töreyin et al. (eds.), *8th EAI International Conference on Robotic Sensor Networks*, EAI/Springer Innovations in Communication and Computing,
https://doi.org/10.1007/978-3-031-92143-8_1

can be used to screen and monitor Alzheimer's disease. AI methods have been successfully conducted in medical image analysis; however, the performance of computer-aided Alzheimer's diagnosis may be limited due to lack of larger datasets. Deep learning (DL)-based computer-aided diagnosis systems may enable significant performance improvements in medical diagnosis. Computer-aided diagnosis systems for MR images have been developed [4–6]. However, only a few studies exist in the literature involving generative AI for computer-aided Alzheimer's disease diagnosis [7, 8].

DL has gained popularity with the development of new methods in the last decade. With the application of this technology in medical image analysis, significant advances have been made in the screening and early diagnosis of diseases. The use of AI algorithms in the analysis of medical imaging such as MR and computed tomography increases diagnostic accuracy and strengthens clinical decision support systems [9]. Additionally, generative AI techniques such as generative adversarial networks (GAN) play an important role in overcoming the challenges of limited datasets. The potential of GANs in healthcare has been proven by achieving successful results in various medical imaging applications [10]. Using an open source database containing patients with dementia, a machine learning method has been developed for the early diagnosis of Alzheimer's disease [11]. DL-based morphometric analysis was used to diagnose of Alzheimer's disease [12].

GAN-based methods have significant potential in medical image analysis. GANs have a model in which two artificial neural networks are trained adversely to increase the reality of the synthetic images. While the generator network creates realistic images, the discriminator network recognizes whether the generated image is real or fake. This competition allows both networks to perform better gradually. The main reason for using GANs in medicine is that they offer the opportunity to increase the volume of medical datasets, which is timely and costly. The data volume increase is especially valuable for rare diseases or certain patient groups. Successful applications have been achieved with GANs for improving MR images [13], completing missing data [14], or converting various image modes [15].

CycleGAN can be used to perform unsupervised transformation from image to image. CycleGAN has the ability to transform one set of images into another, minimizing the need for labeled data. The collection brain MR images of the same person with and without Alzheimer's symptoms require a lot of time and effort. In this study, the advantages of CycleGAN-assisted computer-aided Alzheimer's disease diagnosis methods were investigated.

The steps of literature research, preparation of the dataset, creation of MR images with generative artificial intelligence, training of deep learning models, performance measurement, and evaluation of the results were discussed in detail. The importance of the effective use of generative AI techniques in the early diagnosis of Alzheimer's disease was emphasized and a new perspective in this field was presented.

# 2 Materials and Methods

The present study involves generation of human brain MR images by artificial intelligence. MR image collection, computational environment, generative model training, automated diagnosis model training, and performance evaluation methods are explained in the following sections.

## *2.1 MR Image Dataset*

Alzheimer's disease diagnosis has been implemented by using MR images with and without Alzheimer's disease symptoms. This study has been conducted with a two-class dataset consisting of MR images without Alzheimer's symptoms and with Alzheimer's symptoms. In the first phase of the study, a transfer learning method was applied to the original dataset for the automatic diagnosis of Alzheimer's. Subsequently, synthetic data augmentation was performed using GANs, and the dataset was extended. The model performances were re-evaluated by applying deep transfer learning on the extended dataset.

The aim of this study is to investigate the use of synthetic MR images on the performance of computer-aided Alzheimer's disease diagnosis. In this study, the preparation of the dataset, training of the models, and evaluation of the results are addressed step by step, and methods for the effective use of GANs and other techniques in Alzheimer's diagnosis processes are presented in detail.

The MR image dataset used in this study is the "Alzheimer's Dataset" uploaded by Sarvesh Dubey on the Kaggle platform [16]. The dataset contains four different types of MR images: those without Alzheimer's symptoms and those three different levels of Alzheimer's symptoms. In this study, MR images without Alzheimer's symptoms and those with some Alzheimer's symptoms have been used. The size of the MR images is $128 \times 128$ pixels, and the dataset includes a total of 5440 images. At the time of writing this document, the dataset was inaccessible from the Kaggle website.

The MR images have been randomly split into 80% for training and 20% for validation. The images in the test folder of the original dataset have been used for testing purposes. The number of MR images in the training, validation, and test datasets used in this study is provided in Table 1. The data is unbalanced (59%–41%) since the number of MR images without Alzheimer's symptoms is greater than the ones with Alzheimer's symptoms.

**Table 1** Original Alzheimer's disease dataset training, validation, and test samples

| Dataset | Number of MR images with Alzheimer's symptoms | Number of MR images without Alzheimer's symptoms | Total |
|---|---|---|---|
| Training | 2048 | 1434 | 3482 |
| Validation | 512 | 358 | 870 |
| Test | 640 | 448 | 1088 |
| Total | 3200 | 2240 | 5440 |

## *2.2 Computational Environment*

Deep learning training data has been stored in Google Drive. High performance computing required during CycleGAN training and deep transfer learning was performed using Google Colab Pro. Computing acceleration has been achieved by NVIDIA Tesla T4 GPU and using 25 GB RAM. Model training and data preprocessing was done using NumPy, Pandas, TensorFlow, OpenCV, SciPy, Seaborn, Keras, and Matplotlib libraries [17–24].

## *2.3 MR Image Generation by CycleGAN*

CycleGAN was used to generate images with Alzheimer's symptoms from images without Alzheimer's symptoms and images without Alzheimer's symptoms were obtained from images with Alzheimer's symptoms. Two images were obtained in each cycle, and a total of 200 MRI images were obtained, 100 with Alzheimer's disease symptoms and 100 without Alzheimer's disease symptoms. The obtained images were randomly added to the training and validation data as 80% and 20%. The codes written by Mahnaz Arjmand were used to implement the CycleGAN model [25]. Fig. 1 shows the CycleGAN architecture adapted to Alzheimer's disease images [26, 27]. The aim of this adaptation is to produce synthetic MR images that can mimic the neurodegenerative changes which are characteristics of Alzheimer's disease. The figure shows the operation of the generator and discriminator networks for bidirectional image generation.

In the first stage of CycleGAN, images are loaded and processed from data generators. Due to the limited memory of Colab platform and high RAM requirement of the GAN model, the MR images were resized to $64 \times 64$ pixels.

In the training process of CycleGAN, two generators and two discriminators work together. The composite model evaluates the images produced by one generator through a discriminator while checking the reality of images reconstructed by the other generator. This model optimizes both the quality of image generation and cycle loss by calculating adversarial losses and cycle losses.

The discriminator model of CycleGAN includes a model capable of distinguishing between two different image sets. In the model, a series of convolution and

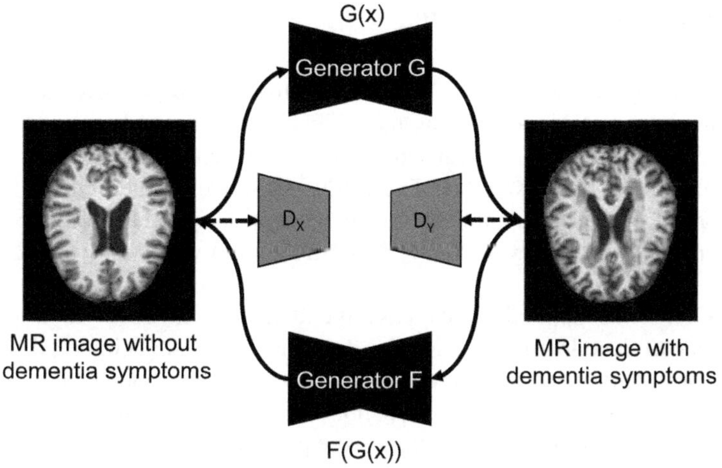


**Fig. 1** Bidirectional MR image generation pipeline using CycleGAN for modeling the changes in the human brain occurred during Alzheimer's disease. On the left, an MRI image labeled "MR image without dementia symptoms" is connected to "Generator G," which outputs "MR image with dementia symptoms". "MR image with dementia symptoms" on the right is connected to "Generator F" which generates "MR image without dementia symptoms". The chart represents CycleGAN process for transforming MRI images. Dx and Dy discriminators are trained to classify real and fake MR images without and with dementia symptoms, respectively [26, 27]

activation functions were performed with increasing filter numbers in each layer. There is a neuronal output in the last layer of the model. This model was used to distinguish real and fake images.

CycleGAN generator models enable the generation of target images from source images. The model consists of a series of convolutional layers and residual blocks. The model initially starts with a large filter and then processes the data with smaller filters sequentially. Then, it is returned to the initial size with decoder blocks. In the last layer, the predicted image was obtained using the tangent hyperbolic activation function.

During the training phase, the discriminator and generator models were updated respectively at each epoch. The adversarial loss and cycle losses of the CycleGAN have been optimized. Real and synthetic images were used to evaluate the model performance. Additionally, model results were visualized at regular intervals and the models were recorded.

Model performance was evaluated during and after the training process. Visualized outputs were obtained, showing real and generated images side by side. These operations were performed to determine the differences between real and fake images and to evaluate the performance of the network.

## 2.4 Deep Transfer Learning

Transfer learning is a technique of speeding up the training process and improving performance on a new task by using the weights of a previously trained model. In this study, transfer learning was carried out for the Alzheimer's disease classification problem using ResNet50 weights. The ResNet50 model trained on the ImageNet was imported [28, 29]. The number of artificial neurons in the fully connected layer was determined empirically as $512 \times 256$. MR images were resized to $224 \times 224$ pixels.

CycleGAN was applied to increase the data size in order to improve the performance of deep learning models for the detection of Alzheimer's disease. The transfer learning has been applied on the original and the extended dataset with and without data augmentation. Vertical flip, horizontal flip, wrapping, zoom, width and horizontal shifts, and rotation operations were performed during data augmentation.

Transfer learning of the ResNet50 model was carried out in three stages. In the first stage, the pre-trained weights of the ResNet50 model were frozen and only the fully connected layers were trained with a learning rate of $10^{-3}$. In the second stage, 33% of the layer weights in the fully connected layers were trained with a learning rate of $10^{-4}$. In the last stage, all layers were unfrozen and the model was trained with a learning rate of $10^{-5}$ [30]. The performance evaluation of each model has been carried out using various binary classification metrics [7].

## 2.5 Performance Evaluation

Binary classification model metrics were selected to evaluate the computer-aided Alzheimer's disease diagnosis performance. Precision, accuracy, sensitivity, specificity, F1 score, area under the receiver operating curve (AUC-ROC), and Matthews correlation coefficient (MCC) [31] score metrics were used to evaluate the model's performance on validation and test data. The metrics were defined as:

$$Pre = \frac{TP}{TP + FP} \tag{1}$$

$$Rec = \frac{TP}{TP + FN} \tag{2}$$

$$Spe = \frac{TN}{TN + FP} \tag{3}$$

$$Acc = \frac{TP + TN}{TP + TN + FP + FN} \tag{4}$$

$$F1\ score = 2 \times \frac{Pre \times Rec}{Pre + Rec} \tag{5}$$

$$MCC = \frac{(TP \times TN) - (FP \times FN)}{\sqrt{(TP + FP)(TP + FN)(TN + FP)(TN + FN)}} \tag{6}$$

where true positive (TP) represents the accurate classification of an MR image with Alzheimer's disease symptoms, while true negative (TN) accurately represents the classification of an MR image without Alzheimer's disease symptoms. A false positive (FP) represents a sample that is actually an MR image without Alzheimer's disease symptoms being incorrectly classified as an MR image with Alzheimer's disease symptoms, while a false negative (FN) refers to a sample that is actually an MR image with Alzheimer's disease symptoms being incorrectly classified as an MR image without Alzheimer's disease symptoms. MCC value measures the performance between $+1$ and $-1$.

## 3 Results

The automated diagnosis of Alzheimer's disease from MR images has been examined in the following image generation and image classification sections.

### *3.1 MR Image Generation with CycleGAN*

The original dataset has been extended by adding the generated MR images via CycleGAN. The number of MR images in the extended dataset has been shown in Table 2.

Representative original MR images and the ones generated by CycleGAN were shown in Fig. 2.

The MR images given in Fig. 2 show that generative AI can produce similar images to the original ones. The differences between the real and the fake images

**Table 2** Alzheimer's disease extended dataset training, validation, and test images

| Dataset | Number of images with Alzheimer's symptoms | Number of images without Alzheimer's symptoms | Total |
|---|---|---|---|
| Training | 2128 | 1514 | 3642 |
| Validation | 532 | 378 | 910 |
| Test | 640 | 448 | 1088 |
| Total | 3200 | 2240 | 5640 |

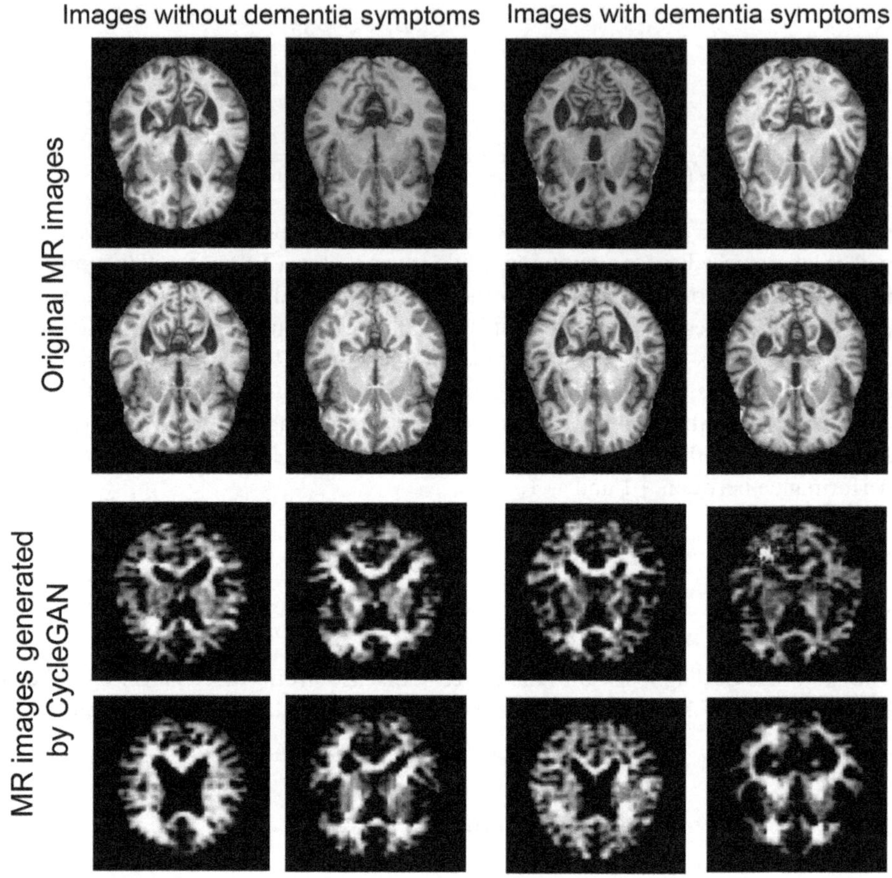


**Fig. 2** Comparison of brain MR images. The top two rows show original MR images, with the left column labeled "Images without dementia symptoms" and the right column labeled "Images with dementia symptoms." The bottom two rows display MR images generated by CycleGAN, following the same labeling. The images illustrate differences in brain structure between those with and without dementia symptoms

were distinguishable. More realistic images may be generated by more data and better representative models.

### 3.2 *Performance Evaluation of Computer-Aided Diagnosis of Alzheimer's Disease*

Binary classification of MR images with and without Alzheimer's disease symptoms was implemented to automate the diagnosis. The first deep transfer learning model

**Table 3** Performance comparison between the original and the extended dataset on the test data when data augmentation is not used

| Metric | Original dataset | Extended dataset |
|---|---|---|
| Precision | 0.6905 | 0.7101 |
| Recall | 0.6473 | 0.6451 |
| Specificity | 0.7969 | 0.8156 |
| Accuracy | 0.7353 | 0.7454 |
| F1 score | 0.6682 | 0.6760 |
| MCC | 0.4490 | 0.4658 |
| AUC-ROC | 0.8085 | 0.8245 |

**Table 4** Performance comparison between the original and the extended dataset on the test data when data augmentation is used

| Metric | Original dataset | Extended dataset |
|---|---|---|
| Precision | 0.6837 | 0.6845 |
| Recall | 0.7478 | 0.7411 |
| Specificity | 0.7581 | 0.7609 |
| Accuracy | 0.7537 | 0.7528 |
| F1 score | 0.7143 | 0.7117 |
| MCC | 0.5001 | 0.4971 |
| AUC-ROC | 0.8393 | 0.8429 |

was trained using the original MR images, where the number of images were given in Table 1. The transfer learning approach was also implemented with the dataset which was extended by the 200 MR images generated by CycleGAN. The performance evaluation comparison between original and extended dataset without data augmentation is given in Table 3.

Performance improvements have been observed in the extended dataset compared to the original dataset in all metrics except recall. Further experiments have been performed to see whether performance improvement will be seen when data augmentation is used. The results are given in Table 4.

The performance comparison between the original and the extended dataset has shown that addition of generated MR images improved the precision, specificity, and AUC-ROC performance slightly when data augmentation is applied.

## 4 Discussion

The use of generative AI in medicine has attracted a lot of interest in the last decade. Generative AI can be used to extend medical imaging data. The advantages of using CycleGAN in the field of computer-aided Alzheimer's diagnosis were investigated. The use of generative AI allowed performance improvements in Alzheimer's disease diagnosis. The other advantage of using generative AI in medical imaging is that the synthetic images do not contain any patient information.

The purpose of using MR images with Alzheimer's disease symptoms in this study is to improve the diagnostic performance of the disease which is a crucial problem. Evaluating the potential of AI techniques in this field may lead to more effective and widespread use in the future. The integration of synthetic MR image generation via CycleGAN to augment limited datasets is a novel contribution that addresses a significant challenge in medical imaging.

Computer-aided Alzheimer's disease diagnosis performance improvements have been achieved via CycleGAN-generated MR images. However, when data augmentation was used, the benefits of using generative AI have diminished. This could be due to the low quality of the CycleGAN-generated images. In fact, the differences between the real and synthetic MR images were clear for the authors. This study shows how GAN-based synthetic data generation can be used against data scarcity in medicine. However, the investigations reveal that further research and development is required. Advances in this field can make significant contributions to improving the quality of life of patients by providing higher accuracy and reliability in the early diagnosis of Alzheimer's disease. In future works, we will investigate higher resolution MR image generation for modeling Alzheimer's disease characteristics. We will also investigate the effects of ensuring the data balance by the generated data.

The use of AI and generative AI could lead to significant improvements in healthcare. Recent research suggests that the robotics applications offer huge impacts in medicine. Combination of research from robotics and generative AI such as the present study may revolutionize the healthcare services.

## References

1. Nawaz, H., Maqsood, M., Afzal, S., Aadil, F., Mehmood, I., Rho, S.: A deep feature-based real-time system for Alzheimer disease stage detection. Multimed. Tools Appl. **80**(28), 35789–35807 (2021)
2. Alzheimer, A., Stelzmann, R.A., Schnitzlein, H.N., Murtagh, F.R.: The English translation of Alzheimer's 1907 journal article: "Über eine eigenartige erkrankung der hirnrinde". Clin. Anat. **8**(6), 429–431 (1995)
3. Chetelat, G.A., Baron, J.C.: Early diagnosis of Alzheimer's disease: contribution of structural neuroimaging. NeuroImage. **18**(2), 525–541 (2003)
4. Arimura, H., Magome, T., Yamashita, Y., Yamamoto, D.: Computer-aided diagnosis systems for brain diseases in magnetic resonance images. Algorithms. **2**(3), 925–952 (2009)
5. Özdemir, C., Dogan, Y.: Advancing early diagnosis of Alzheimer's disease with next-generation deep learning methods. Biomed. Signal Process. Control. **96**, 106614 (2024)
6. Türk, Ö.: MR görüntülerinden Alzheimer tespitinde boyut azaltma ve derin öğrenme yaklaşımlarının karşılaştırılması. Dicle Üniversitesi Mühendislik Dergisi. **13**(3), 485–491 (2022)
7. Coşkun Eliküçük, S.: Manyetik rezonans görüntüleri kullanılarak çevrimsel çekişmeli üretici ağ destekli Alzheimer teşhisi. MSc Thesis, Siirt University Graduate School of Natural and Applied Sciences (2024)
8. Yao, Z., Wang, Z., Xie, W., et al.: Applications of generative artificial intelligence in brain MRI image analysis for brain disease diagnosis. NPT. **1**(2), 65–77 (2024). https://doi.org/10.15212/npt-2024-0007
9. Zhang, Z., Seeram, E.: The use of artificial intelligence in computed tomography image reconstruction—a literature review. J. Med. Imaging Radiation Sci. **51**(4), 671–677 (2020)
10. Ahmad, W., Ali, H., Shah, Z., Azmat, S.: A new generative adversarial network for medical images super resolution. Sci. Rep. **12**, 9533 (2022)

11. Sertkaya, M.E., Ergen, B.: Alzheimer hastalığının erken teşhisinin çoklu değişken kullanarak tespiti. Avrupa Bilim ve Teknoloji Dergisi. **35**, 306–314 (2022)
12. Çelebi, S.: Alzheimer teşhisi için derin öğrenme tabanlı morfometrik analiz. Iğdır Üniversitesi Fen Bilimleri Enstitüsü Dergisi. **13**(3), 1454–1467 (2023)
13. Zhou, X., Qiu, S., Joshi, P.S., Xue, C., Killiany, R.J., Mian, A.Z., Chin, S.P., Au, R., Kolachalama, V.B.: Enhancing magnetic resonance imaging-driven Alzheimer's disease classification performance using generative adversarial learning. Alzheimer's Res Ther. **13**(1), 92 (2021)
14. Isola, P., Zhu, J.Y., Zhou, T., Efros, A.A.: Image-to-image translation with conditional adversarial networks. In: Proceedings of the IEEE Conference on Computer Vision and Pattern Recognition, Honolulu, USA, pp. 1125–1134 (2017)
15. Yu, J., Lin, Z., Yang, J., Shen, X., Lu, X., Huang, T.S.: Generative in painting with contextual attention. In: Proceedings of the IEEE Conference on Computer Vision and Pattern Recognition, pp. 5505–5514 (2018)
16. Dubey, S.: Alzheimer's dataset (4 class of images). Kaggle (2023) https://www.kaggle.com/datasets/tourist55/alzheimers-dataset-4-class-of-images. Last Accessed 2024/08/28
17. Abadi, M., Barham, P., Chen, J., Chen, Z., Davis, A., Dean, J., Devin, M., Ghemawat, S., Irving, G., Isard, M., Kudlur, M., Levenberg, J., Monga, R., Moore, S., Murray, D.G., Steiner, B., Tucker, P., Vasudevan, V., Warden, P., Wicke, M., Yu, Y., Zheng, X.: TensorFlow: a system for large-scale machine learning. In: 12th USENIX Symposium on Operating Systems Design and Implementation, OSDI 16, pp. 265–283. Savannah, USA (2016)
18. Bradski, G.: The OpenCV library. Dr. Dobb's J. Software Tools. **120**, 122–125 (2000)
19. Chollet, F.: Keras: The Python Deep Learning library (2015). https://keras.io. Last Accessed 2024/08/26
20. Harris, C.R., Millman, K.J., van der Walt, S.J., Gommers, R., Virtanen, P., Cournapeau, D., Wieser, E., Taylor, J., Berg, S., Smith, N.J., Kern, R., Picus, M., Hoyer, S., van Kerkwijk, M.H., Brett, M., Haldane, A., Del Río, J.F., Wiebe, M., Peterson, P., Gérard-Marchant, P., Sheppard, K., Reddy, T., Weckesser, W., Abbasi, H., Gohlke, C., Oliphant, T.E.: Array programming with NumPy. Nature. **585**(7825), 357–362 (2020)
21. Hunter, J.D.: Matplotlib: a 2D graphics environment. Comput. Sci. Eng. **9**(3), 90–95 (2007)
22. McKinney, W.: Data structures for statistical computing in Python. In: Proceedings of the 9th Python in Science Conference, Austin, USA, pp. 51–56 (2010)
23. Virtanen, P., Gommers, R., Oliphant, T.E., Haberland, M., Reddy, T., Cournapeau, D., Burovski, E., Peterson, P., Weckesser, W., Bright, J., van der Walt, S.J., Brett, M., Wilson, J., Millman, K.J., Mayorov, N., Nelson, A.R.J., Jones, E., Kern, R., Larson, E., Carey, C.J., Polat, İ., Feng, Y., Moore, E.W., VanderPlas, J., Laxalde, D., Perktold, J., Cimrman, R., Henriksen, I., Quintero, E.A., Harris, C.R., Archibald, A.M., Ribeiro, A.H., Pedregosa, F., van Mulbregt, P.: SciPy 1.0: Fundamental algorithms for scientific computing in Python. Nat. Methods. **17**(3), 261–272 (2020)
24. Waskom, M.L.: Seaborn: Statistical data visualization. J. Open Sour. Software. **6**(60), 3021 (2021)
25. Arjmand, M.: Converting images to cartoon form using cycle-GAN. Kaggle (2023) https://www.kaggle.com/code/mahnazarjmand/converting-images-to-cartoon-form-using-cycle-gan/notebook#Conclusion. Last Accessed 2024/08/11
26. Garrucho, L., Kushibar, K., Osuala, R., Diaz, O., Catanese, A., Del Riego, J., Lekadir, K.: High-resolution synthesis of high-density breast mammograms: application to improved fairness in deep learning based mass detection. Front. Oncol. **12**, 1044496 (2023)
27. Zhu, J.Y., Park, T., Isola, P., Efros, A.: A unpaired image-to-image translation using cycle-consistent adversarial networks. In: IEEE International Conference on Computer Vision (ICCV). Venice, Italy, pp. 1–10 (2017)
28. Deng, J., Dong, W., Socher, R., Li, L.J., Li, K., Fei-Fei, L.: Imagenet: a large-scale hierarchical image database. In: 2009 IEEE Conference on Computer Vision and Pattern Recognition, Miami, USA, pp. 248–255 (2009)

29. He, K., Zhang, X., Ren, S., Sun, J.: Deep residual learning for image recognition. In: Proceedings of the IEEE Conference on Computer Vision and Pattern Recognition, pp. 770–778 (2016)
30. Shen, L., Margolies, L.R., Rothstein, J.H.: Deep learning to improve breast cancer detection on screening mammography. Sci. Rep. **9**(1), 12495 (2019)
31. Matthews, B.W.: Comparison of the predicted and observed secondary structure of T4 phage lysozyme. Biochim. Biophys. Acta Protein Struct. **405**(2), 442–451 (1975)

# Enhancing Credit Risk Assessment with Federated Learning Through a Comparative Study

**Mustafa Aktaş, Ruşen Akkuş Halepmollası, and Behçet Uğur Töreyin**

## 1 Introduction

Digital transformation has dramatically advanced traditional banking systems, creating a critical impact on modern banking practices and the financial sector [28]. Although financial consumers increasingly prefer digital systems and credit cards for daily transactions [3], the easy access to various tools and techniques for conducting fraudulent credit card transactions raises concerns regarding the acceptance of digital systems [4]. Also, the increasing complexity of the financial market and the growing diversity of customer demands create significant challenges for service providers [7].

Financial risk refers to the likelihood of a company's financial outcomes deviating from expected business objectives. These variances, which are frequently the result of unforeseen and uncontrollable circumstances, may arise during a variety of financial processes, resulting in the organization experiencing economic losses or generating earnings that are lower than anticipated. In this context, one of the most pressing risks faced in the financial sector is credit card default, which occurs when borrowers fail to meet their repayment obligations. The challenge of accurately predicting credit card defaults is compounded by the rapidly evolving digital landscape, which has made it easier for users to accumulate debt. Meanwhile,

M. Aktaş (✉) · R. A. Halepmollası
Istanbul Technical University, Istanbul, Turkey

TÜBİTAK Informatics and Information Security Research Center, Kocaeli, Turkey
e-mail: aktasm22@itu.edu.tr; aktas.mustafa@tubitak.gov.tr; halepmollasi@itu.edu.tr; rusen.halepmollasi@tubitak.gov.tr

B. U. Töreyin
Istanbul Technical University, Istanbul, Turkey
e-mail: toreyin@itu.edu.tr



B. U. Töreyin et al. (eds.), *8th EAI International Conference on Robotic Sensor Networks*, EAI/Springer Innovations in Communication and Computing,
https://doi.org/10.1007/978-3-031-92143-8_2

in the financial sector, building predictive models presents significant challenges due to data privacy and security issues as traditional centralized machine learning (ML) methods require data to be aggregated at a central location [8]. Federated learning (FL) is proposed to address those issues. FL is a collaborative ML technique that enables disconnected entities to train models without sharing their raw data, addressing privacy issues [21].

In this chapter, considering the aforementioned issues, we present a risk prediction model using FL approach that allows for collaborative model training without the need for direct data sharing to predict credit card default assessment based on users' credit card debts and payments. To this end, we utilized a publicly available default of credit card clients (DCCC) dataset [32] and applied data preprocessing steps such as data scaling and encoding. We also employed sampling techniques, namely the synthetic minority oversampling technique (SMOTE) and random undersampling (RUS). To reduce the complexity and improve model performance, we used two feature reduction methods, which are principal component analysis (PCA) and k-best feature selection. We utilized five ML algorithms, including logistic regression (LR), multilayer perceptron (MLP), support vector machine (SVM), XGBoost (XGB), and random forest (RF). Furthermore, to investigate the impact of FL, we compared two distinct aggregation methods through our experiments. Our contributions are as follows:

- We propose a novel FL approach for credit card default prediction that maintains data privacy, while ensuring it competes with centralized ML methods.
- We perform an extensive comparison of five ML algorithms (LR, MLP, SVM, XGB, and RF) under both centralized and federated approaches in a total of 135 experiments.
- We introduce the F1-score weighted aggregation (FedF1) method that achieves comparable or superior results to the traditional federated averaging (FedAvg) method.
- The code implementation of our experiments is publicly available for FL research reproducibility and exploration.

The structure of this chapter is organized as follows: Sect. 2 summarizes previous works on FL and ML approaches in finance. In Sect. 3, we present the system architecture. Section 4 describes the dataset and our methodology including data preprocessing, dimensional reduction methods, built models, and train/test configuration phases. In Sect. 5, the results of the study are reported and discussed. Section 6 concludes the paper and offers future works.

## 2 Related Work

The research community has widely studied financial risk from different perspectives. In this section, we discuss previous studies related to central ML (CML) approach and FL approach in finance, respectively.

## 2.1 Central Machine Learning in Finance

There are several studies in which ML methods are applied for customer segmentation [26], financial risk prediction [30], fraud detection [13], and price anomaly detection [5, 24], addressing the increasing data volumes and complexity in financial operations. For instance, Wang et al. [30] conducted a comprehensive study on the application of ML techniques for financial risk prediction within the context of the US financial markets. Authors focused on predicting main types of financial risks such as interest rate risk and cash risk, using LR and RF. They stated that ML methods are effective in accurately assessing financial risks and emphasized their expanding role in risk management strategies.

ML techniques, especially ensemble learning approaches including RF and DT, are increasingly employed in financial fraud detection. Guo et al. [16] demonstrated the efficacy of RF algorithm in identifying fraud in financial transactions, comparing this algorithm with XGB and DTs. The study utilized the well-known "Credit Card Fraud Detection" dataset [13] from Kaggle, which contains 284,807 transactions, of which only 492 are fraud. Initially, they achieved an out-of-bag (OOB) score of 0.924 and an AUC of 0.967 using default parameters. With hyperparameter tuning, they improved those metrics to 0.933 for OOB and 0.978 for AUC and emphasized the importance of hyperparameter optimization in ML applications [16].

Moreover, feature selection is a crucial process in credit card default assessment, as the selected variables significantly impact the model outputs. Zhou et al. [33] developed the "FDAF-score" feature selection model, which combines Fisher discriminant analysis (FDA) and F-score methodologies. Thus, authors aimed to ensure that feature combinations have maximum discrimination ability, contrasting with single feature selection methods. They evaluated commonly employed models including SVM, ANN, LR, DT, K-nearest neighbor (KNN), linear discriminant analysis (LDA), and RF. Their results revealed that FDAF outperformed single feature selection approaches on the Taiwan dataset [29].

## 2.2 Federated Learning in Finance

A growing body of literature across various fields, including healthcare [2, 27], IoT [22], wireless communication networks [31], and energy demand prediction [25], explores the potential impact of FL on enhancing analytical capabilities, while preserving data privacy and security. Authors [15] developed an FL-based prototype to investigate the performance of FL-based credit risk assessments and investigate its potential to reduce disparities in risk calculations between financial institutions. They highlighted the potential of FL in credit risk assessment, allowing financial institutions to collaboratively train models without sharing raw data, thus preserving privacy. They tested the prototype and evaluated the impact of information sharing on risk models for various cases. According to their results, FL can improve performance for smaller financial institutions with limited in-house datasets.

Awosika et al. [4] introduced a novel FL-based fraud detection system that combines FL and eXplainable AI (XAI) techniques to enhance both data privacy and transparency in financial sector. They employed a deep neural network to detect fraudulent transactions across federated databases ensuring high accuracy without compromising data confidentiality. Also, they integrated XAI to provide interpretability for model predictions as it allows financial experts to assess fraud detection results with greater trust and insight.

Li and Wen [20] presented an FL-based approach for credit risk management across multiple banks and addressed data heterogeneity and privacy issues in credit risk modeling. In their experimental framework, they used heterogeneous datasets from a large joint-stock bank and a small city commercial bank. Authors demonstrated that FL can achieve similar or improved performance compared to centralized modeling, while preserving data privacy. They report the potential of FL to enhance credit risk assessment for smaller banks by leveraging collective data.

## 3 System Architecture

FL is a distributed ML approach that allows training models across multiple clients without needing a centralized dataset [1]. Each client trains the model locally using its data; only model updates are shared with a central server. The server aggregates those updates to create a global model and redistributes it to the clients. Therefore, personal data is stored on local devices, preserving privacy and enabling effective learning across the entire network. Communication between the server and clients enhances privacy and optimizes the model's performance across a wide range of datasets.

As illustrated in Fig. 1, FL approach enables various data providers, such as banks and financial institutions, to collaboratively train a shared ML model,

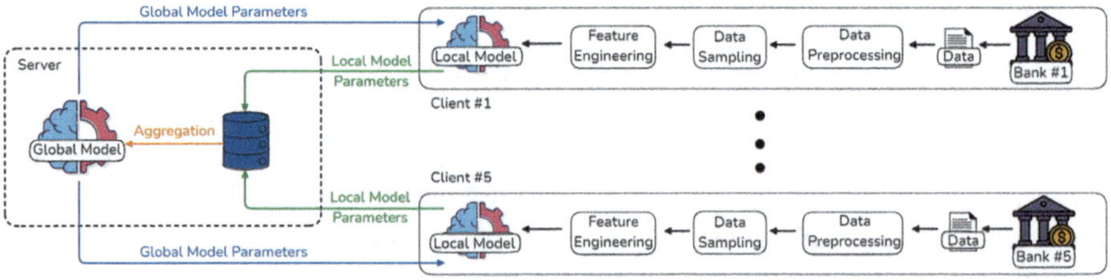


**Fig. 1** Federated learning system architecture. The parameters of the "Local Model" which are trained by each client's local data separately, are sent to the server (green arrow) and aggregated (orange arrow). The resulting "Global Model" is returned to the clients (blue arrow)

while maintaining data privacy. Initially, train and test datasets are split using $DirichletPartitioner$ with each client holding its unique private dataset inaccessible to others. Each client performs preprocessing, sampling, and feature engineering locally. After local training, the clients send model parameters (not raw data) to the central server, aggregating them to generate a global model. The global model is then shared back with the clients, who update their local models to enhance performance in subsequent iterations. In this chapter, we conducted experiments across 10 training rounds, with each iteration resulting in an improved global model that incorporates knowledge from all clients. Moreover, to investigate the impact of FL, we compared two distinct aggregation methods, namely federated averaging (FedAvg) and F1-score weighted aggregation (FedF1).

*Federated averaging* involves updating a global model by averaging the model weights from clients ($K$), with weight for each client, $k$, determined by the amount of data it contributes to the server [21]. Each client trains the model locally and sends the updated weights to the server. The server aggregates those weights, weighted based on the amount of data used by each client during training. The model weights ($w_{t+1}^k$) are aggregated based on the number of data points ($n_k$) from each client, and the total number of data points ($n$) across all clients is divided by the sum to derive the weighted average, namely global model ($w_{t+1}$) (1). The updated global model is redistributed to the clients, after which the process proceeds to the next iteration. FedAvg facilitates global model updates during distributed training, while preserving data locality at the client level.

$$w_{t+1} \leftarrow \sum_{k=1}^{K} \frac{n_k}{n} w_{t+1}^k \tag{1}$$

*F1-score weighted aggregation* has been developed for this study to improve model performance. FedF1 aggregates model weights ($w_{t+1}^k$) from clients ($K$) based on their F1 scores ($f1_k$) (2) over all F1 scores ($F1_t$) and creates the global model ($w_{t+1}$). To enhance the overall model's reliability and stability, it assesses the performance of each client and gives more weight to models with higher F1 scores. FedF1 was utilized due to the significant impact of the imbalanced dataset on model performance. Data sampling methods such as SMOTE and RUS (Sect. 4.4) are used to address this issue, though they often improve one aspect of performance while affecting others. F1 score, as a weighted average of precision and recall (2), balances these metrics. As shown in Algorithm 1 and Algorithm 2, we assign higher weights to clients with more stable and superior F1 scores to improve the overall model.

$$w_{t+1} \leftarrow \sum_{k=1}^{K} \frac{f1_k}{F1_t} w_{t+1}^k \tag{2}$$

**Algorithm 1** FedF1

**Input:**
$K$: Number of clients indexed by $k$,
**Output:**
$w$: Global model weights

Initialize $w_0$
**for** each round $t = 1, 2, \ldots$ **do**
  $m \leftarrow \max(K, 1)$
  $S_t \leftarrow$ (set of $m$ clients)
  **for** each client $k \in S_t$ **in parallel do**
    $w_{t+1}^k, f1_k \leftarrow$ClientUpdate$(k, w_t)$
  **end for**
  $F1_t \leftarrow \sum_{k \in S_t} f1_k$
  $w_{t+1} \leftarrow \sum_{k \in S_t} \frac{f1_k}{F1_t} w_{t+1}^k$
**end for**

**Algorithm 2** ClientUpdate

**Input:**
$k$: Client index,
$w$: Global model weights,
$E$: Number of local epochs
$\eta$: Learning rate,
**Output:**
Updated weights $w$ for the server,
$f1$ score of client

$D_{k_{Train}}, D_{k_{Val}} \leftarrow$ ClientData$(k)$
**for** each local epoch $i$ from 1 to $E$ **do**
  **for** each batch $b \in D_{k_{Train}}$ **do**
    $w \leftarrow w - \eta \nabla \ell(w; b)$
  **end for**
**end for**
$f1 =$ CalcF1Score$(w, D_{k_{Val}})$
**return** $w$, $f1$ to server

# 4 Methodology

In this section, we provide a detailed description of the dataset that we used in this study (Sect. 4.1), critical data splitting strategy (Sect. 4.2), data preprocessing steps (Sect. 4.3), and various ML classifiers (Sect. 4.6) for predicting consumer payment behavior and identifying default risks.

Figure 2 illustrates the training and testing processes for ML models. The dataset was split into training and testing subsets to prevent data leakage and ensure a fair model evaluation. Preprocessing and feature reduction methods were applied to clean the data, reduce complexity, and remove irrelevant features. Five classifiers

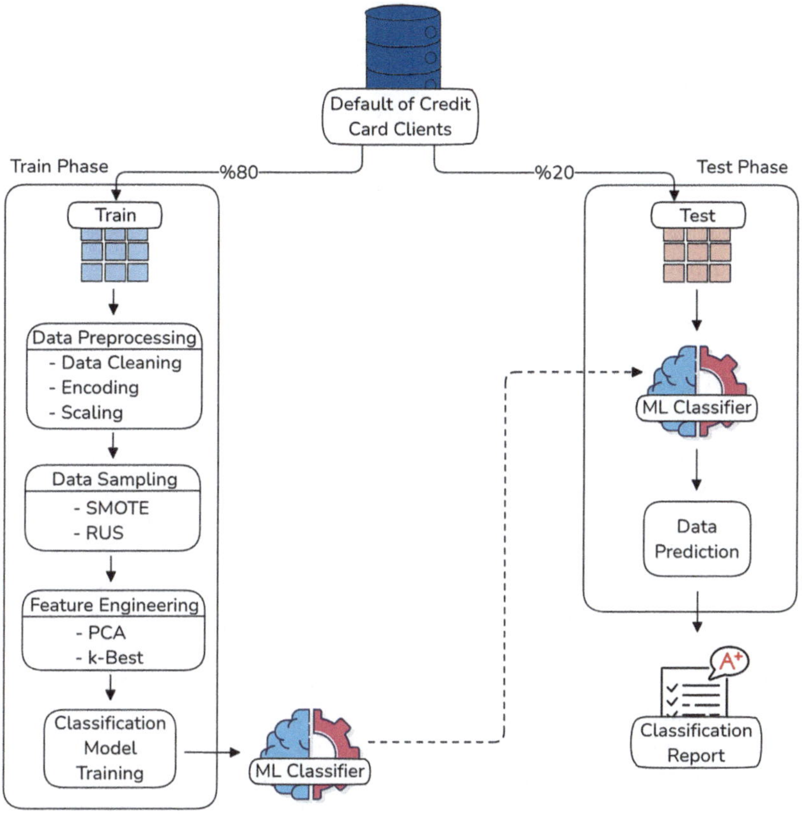


**Fig. 2** Illustration of the proposed experimental setup

were trained with hyperparameter tuning and evaluated 10 times using client-specific test sets with different random seeds to ensure robustness. This study focused on predicting credit card defaults, specifically assessing the performance of Class-1, which represents instances of default.

## *4.1 Dataset*

In this chapter, we utilized the DCCC dataset, which contains customers' default payments data in Taiwan [32], to predict potential default. The dataset is sourced from the UCI ML Repository and dates back to 2005. It includes 30,000 samples, of which 22% (6,636 rows) represents default payments (Table 1). The dataset has no missing values and contains nominal and discrete fields with the binary target column. The dataset's imbalance ratio, as indicated in Table 1, is calculated to be

**Table 1** Value counts for default payment next month

| | No sampling | | SMOTE | | RUS | |
|---|---|---|---|---|---|---|
| Def Pay | 0 | 1 | 0 | 1 | 0 | 1 |
| Count | 23,364 | 6636 | 23,364 | 23,364 | 6636 | 6636 |
| Percentage (%) | 77.88 | 22.12 | 50 | 50 | 50 | 50 |

3:1. Therefore, we applied the SMOTE-RUS methods described in Sect. 4.4 so that the data distribution is equalized.

## 4.2 Train Test Configuration

We argue that test and validation sets should remain completely separate at each step of the study to prevent data leakage and ensure that the evaluation of models remains robust and unbiased. Please note that we did not apply data preprocessing, sampling, and feature reduction steps to the entire dataset at once; instead, we applied them only to the train set and then to the validation and test sets. Similarly, to maintain data privacy in line with FL, we employ a client-specific data preprocessing approach when performing FL-based models. In other words, we allocate data to each client, apply preprocessing exclusively on their local data, and do not let the global model reach or directly train with clients' data. We divided the dataset into three different subsets: train set (70%), validation set (10%), and test set (20%). To manage the randomization and ensure the experiments could be repeated, we employed the $train\_test\_split$ function, using $random\_state$ parameter. Also, we partitioned dataset into clients with unique $random\_states$, ensuring that data partitioning was performed independently for each client. The aim of this strategy is to replicate real-world FL scenarios as closely as possible [34]. For this purpose, we utilized the $DirichletPartitioner$ component of the Flower framework [6] to reflect the real-world data distributions. Each client conducted scaling and encoding techniques exclusively on its own dataset. Additionally, we executed all steps specified in this chapter on a client-by-client basis.

## 4.3 Data Preprocessing

We used preprocessing to increase model performance. To maintain terminology consistency, we eliminated the "ID" column and renamed "PAY_0" to "PAY_1" and "default.payment.next.month" to "def_pay." We scaled the numerical features using Sklearn's [23] $StandardScaler$, excluding the dependent variable (def_pay) and categorical variables such as gender (sex), education level (education), and marital status (marriage). We first fitted the scaler to the training set and then applied it the validation set and test set. To improve model comprehension, we encoded categorical variables using sklearn's $OneHotEncoder$ and made gender, marital status, and education dummy variables. It helps ML models manage categorical data, reducing information loss and enhancing performance.

## 4.4 Data Sampling

Due to the use of an imbalanced dataset in this study, we employed oversampling and undersampling techniques. Table 1 presents the number of samples in the default payment categories within the original dataset.

*Synthetic Minority Oversampling Technique* Class imbalance may harm model training. SMOTE, developed by Chawla et al. [9], addresses class imbalance by generating synthetic samples for the minority class to avoid overfitting from traditional oversampling. SMOTE was used to isolate the target variable (def_pay) from the training data and create synthetic minority class samples for the training data. Please note that we applied SMOTE only to the training set, leaving test and validation sets unchanged to ensure unbiased model evaluation.

*Random Undersampling* RUS is a method employed to address class imbalance by randomly removing samples from the majority class to balance it with the minority class or a predetermined ratio. It is effective for large datasets due to its low computational requirements. We applied RUS to each client's data individually rather than to the entire dataset. After splitting the dataset with $DirichletPartitioner$ algorithm, we used RUS on each client's data to ensure balanced classes.

## 4.5 Feature Reduction

Given the large number of features in the dataset and the suspicion of high correlation among certain features, we employed PCA and k-best methods.

*Principal Component Analysis* is a statistical technique reducing dimensionality by identifying the most significant directions of variation in data. It centers the data, calculates the covariance matrix $C$, and selects the $k$ eigenvectors with the largest eigenvalues to project data into a new subspace. In this chapter, we employed PCA using the train set, and the same transformation was then applied to both validation and test sets. Based on hyperparameter tuning tests, we chose $n\_components = 12$ as the optimum value.

*Select k-best* selects the top $k$ features in a dataset. It uses a statistical test and a user-defined metric to identify feature relevance and the best characteristics. We evaluated categorical data using the $f\_classif$ technique and evaluated numerical data using the $chi2$ method. We set $k$ to 5 and used five most significant features

($PAY\_1$, $PAY\_2$, $PAY\_3$, $PAY\_4$, $PAY\_5$) to conduct model training. This simplified training approach enhanced model performance and forecast accuracy.

## 4.6 Machine Learning Algorithms

*Logistic regression* [12] transforms linear model output to probability using sigmoid function. It is parametric as it utilizes a number of parameters to estimate a function from data, and discriminative as it models the direct relationship between specified features (independent variables) and class labels (dependent variable) using class distinctions [19]. Before applying the algorithm, we conducted hyperparameter tuning utilizing the train and validation sets. During the hyperparameter tuning process, we identified the optimal parameters $C = 0.01$, $max\_iter = 5000$, $penalty = l1$, $class\_weight = None$, and $solver = liblinear$.

*Multilayer perceptron* [17] is a feedforward neural network with fully connected layers, including at least one hidden layer. Through hyperparameter tuning, we set the parameters as follows: Learning_rate is 0.001, hidden_size is 100, and the optimizer used is Adam.

*Support vector machine* [11] is a parametric linear classification algorithm designed to separate two classes through a hyperplane in the data dimension. The kernel trick projects nonlinearly separable data to higher dimensions. FL approach relies on the integration of algorithm parameters based on values provided by clients. Therefore, the number and format of those parameters from the clients must be compatible. For this purpose, we used SVC with a linear kernel. After GridSearch hyperparameter adjustment, we found that $C = 10$, class weight=None, and penalty=l2 were the optimum settings.

*eXtreme Gradient Boosting* [10] bases on gradient boosting principles and sequentially builds several weak learners, usually DT, enhancing and merging them. This approach establishes a robust prediction model through the minimization of errors. In this study, we employed grid search to optimize the parameters $n\_estimators$, $max\_depth$, $learning\_rate$, $objective$, and $tree\_method$ to improve the performance of XGB. The obtained final values were 100, 3, 0.1, logistic, and hist, respectively.

*Random Forest* [18] is an ensemble method combining multiple DTs, with predictions based on majority voting [14]. When employing the RF algorithm, the performance of the model can be significantly impacted by the relevant hyperparameters, such as the number of trees ($n\_estimators$), tree depth ($max\_depth$), and branching criterion ($criterion$). This study used hyperparameter tuning to find the model's optimal parameters. To this end, we employed GridSearchCV and determined the optimal parameters as follows: entropy for criterion, $n\_estimators$ 100, $max\_depth$ 10, $min\_samples\_split$ 5, $min\_samples\_leaf$ 1, $max\_features$ sqrt, and $bootstrap$ true.

# 5 Experimental Results and Discussion

To compare the performance of centralized ML and FL approaches, we evaluated 45 distinct ML models using five ML algorithms. We examined the impact of selecting the best features in k, the reduction of the PCA features, and the data balancing methods (SMOTE and RUS) due to class imbalance. We also tested two distinct FL approaches: FedAvg and F1-weighted aggregation methods. In total, we conducted 135 experiments, each repeated 10 times, and the results were averaged to ensure reproducibility and robustness. Thus, Table 2 summarizes the results of 1350 iterations of model training and testing. We compared the results using widely used evaluation metrics, including accuracy, precision, recall, and F1 score. In this section, we present various perspectives on the interpretation of the results.

**Experiments in View of Different Machine Learning Models** The results in Table 2 indicate that the LR + k-best ML model outperforms others in precision, achieving 73%. With FedAvg, it reaches the highest precision at 75%, while FedF1 records 71%, slightly below centralized learning. When combined RUS method and the FL approaches, LR achieves optimal recall rates of 74% with FedAvg and 75% with FedF1. In centralized learning, MLP + SMOTE achieves the highest recall rate at 74%. Similarly, LR + RUS and LR + K-best + RUS approaches yielded optimal results with FedAvg and FedF1, attaining scores of 74% and 75%, respectively. For F1 score, several models achieve good results at 53%, with RUS or SMOTE. Among centralized models, k-best+RUS performs best models for XGB, and MLP, a trend also observed in FL approaches. Although the majority of models achieve 82% accuracy, the MLP + k-best and RF + k-best ML models yield the highest accuracy for both centralized and FL approaches. The XGB and RF classifiers attained the highest or close accuracy scores.

**Experiments in View of Different Sampling Techniques** ML models using SMOTE or RUS methods exhibit reduced precision and accuracy in both central and FL approaches, whereas improved recall and F1 scores compared to models without sampling. SMOTE enhances the model's ability to identify the minority class by augmenting its samples, but it may increase false positives. Accuracy also declines due to the complexity of classifying balanced data. However, as data stabilize after SMOTE or RUS, the F1 score increases. The highest recall rates are observed for the ML models utilizing SMOTE or RUS, with Central at 74%, FedAvg at 74%, and FedF1 at 75%. The F1 scores were recorded as Central 53%, FedAvg 52%, and FedF1 52%.

**Experiments in View of Feature/Dimension Reduction** In this chapter, we applied PCA with hyperparameter tuning identifying the optimal *n_component* value as 12. As illustrated in Table 2, in CML, PCA generally decreased accuracy for LR and XGB, while it improved for MLP and SVM. The application of PCA typically results in a minor reduction in F1 and accuracy scores overall. In the context of FL, the performance decline may be due to PCA being less informative for certain models or failing to enhance performance through complexity reduction.

**Table 2** Credit card default prediction performance results on DCCC dataset[a]

| Model name | Central ML | | | | FL with FedAvg | | | | FL with FedF1 | | | |
|---|---|---|---|---|---|---|---|---|---|---|---|---|
| | P. | R. | F. | Ac. | P. | R. | F. | Ac. | P. | R. | F. | Ac. |
| LR | 0.71 | 0.23 | 0.35 | 0.81 | 0.71 | 0.16 | 0.26 | 0.80 | 0.69 | 0.23 | 0.34 | 0.81 |
| XGB | 0.66 | 0.36 | 0.46 | 0.82 | 0.68 | 0.35 | 0.46 | 0.82 | 0.66 | 0.30 | 0.40 | 0.81 |
| MLP | 0.44 | 0.19 | 0.23 | 0.76 | 0.63 | 0.37 | 0.46 | 0.81 | 0.32 | 0.39 | 0.34 | 0.68 |
| SVM | 0.38 | 0.62 | 0.46 | 0.68 | 0.37 | 0.64 | 0.46 | 0.68 | 0.37 | 0.63 | 0.46 | 0.68 |
| RF | 0.67 | 0.37 | 0.47 | 0.82 | 0.65 | 0.32 | 0.42 | 0.81 | 0.67 | 0.32 | 0.42 | 0.81 |
| LR + PCA | 0.67 | 0.20 | 0.31 | 0.80 | 0.70 | 0.13 | 0.22 | 0.79 | 0.67 | 0.18 | 0.29 | 0.80 |
| XGB + PCA | 0.61 | 0.33 | 0.43 | 0.81 | 0.64 | 0.28 | 0.39 | 0.81 | 0.63 | 0.28 | 0.38 | 0.81 |
| MLP + PCA | 0.63 | 0.38 | 0.47 | 0.81 | 0.58 | 0.36 | 0.43 | 0.80 | 0.59 | 0.36 | 0.44 | 0.80 |
| SVM + PCA | 0.39 | 0.61 | 0.47 | 0.70 | 0.37 | 0.61 | 0.45 | 0.69 | 0.37 | 0.61 | 0.46 | 0.69 |
| RF + PCA | 0.63 | 0.34 | 0.44 | 0.81 | 0.66 | 0.31 | 0.42 | 0.82 | 0.63 | 0.30 | 0.40 | 0.81 |
| LR + k-best | *0.73* | 0.21 | 0.32 | 0.81 | *0.75* | 0.11 | 0.18 | 0.79 | *0.71* | 0.25 | 0.37 | 0.81 |
| XGB + k-best | 0.69 | 0.36 | 0.48 | *0.82* | 0.68 | 0.29 | 0.39 | 0.81 | 0.69 | 0.30 | 0.40 | 0.81 |
| MLP + k-best | 0.68 | 0.34 | 0.46 | 0.82 | 0.64 | 0.40 | 0.49 | *0.82* | 0.63 | 0.41 | 0.49 | *0.82* |
| SVM + k-best | 0.49 | 0.54 | 0.51 | 0.77 | 0.49 | 0.53 | 0.50 | 0.77 | 0.50 | 0.53 | 0.51 | 0.78 |
| RF + k-best | 0.67 | 0.36 | 0.46 | 0.82 | 0.64 | 0.37 | 0.46 | 0.82 | 0.65 | 0.36 | 0.46 | 0.82 |
| LR + SMOTE | 0.37 | 0.59 | 0.45 | 0.69 | 0.39 | 0.60 | 0.47 | 0.71 | 0.39 | 0.60 | 0.47 | 0.70 |
| XGB + SMOTE | 0.53 | 0.53 | 0.53 | 0.79 | 0.46 | 0.56 | 0.50 | 0.76 | 0.42 | 0.65 | 0.50 | 0.71 |
| MLP + SMOTE | 0.29 | *0.74* | 0.42 | 0.55 | 0.37 | 0.68 | 0.47 | 0.68 | 0.29 | 0.71 | 0.40 | 0.55 |
| SVM + SMOTE | 0.29 | 0.70 | 0.41 | 0.57 | 0.33 | 0.68 | 0.44 | 0.62 | 0.32 | 0.68 | 0.43 | 0.62 |
| RF + SMOTE | 0.49 | 0.56 | 0.52 | 0.77 | 0.53 | 0.51 | *0.52* | 0.79 | 0.52 | 0.53 | 0.52 | 0.79 |
| LR + PCA + SMOTE | 0.37 | 0.64 | 0.47 | 0.68 | 0.30 | 0.56 | 0.39 | 0.62 | 0.32 | 0.56 | 0.40 | 0.63 |
| XGB + PCA + SMOTE | 0.46 | 0.59 | 0.52 | 0.76 | 0.34 | 0.51 | 0.40 | 0.66 | 0.35 | 0.52 | 0.42 | 0.68 |
| MLP + PCA + SMOTE | 0.42 | 0.62 | 0.50 | 0.72 | 0.28 | 0.41 | 0.33 | 0.64 | 0.27 | 0.45 | 0.33 | 0.61 |
| SVM + PCA + SMOTE | 0.38 | 0.62 | 0.47 | 0.69 | 0.31 | 0.54 | 0.39 | 0.64 | 0.33 | 0.55 | 0.41 | 0.64 |
| RF + PCA + SMOTE | 0.50 | 0.55 | 0.53 | 0.78 | 0.48 | 0.53 | 0.50 | 0.77 | 0.49 | 0.54 | 0.51 | 0.78 |
| LR + k-best + SMOTE | 0.48 | 0.54 | 0.51 | 0.77 | 0.47 | 0.56 | 0.50 | 0.76 | 0.46 | 0.58 | 0.50 | 0.74 |
| XGB + k-best + SMOTE | 0.54 | 0.53 | *0.53* | 0.80 | 0.41 | 0.62 | 0.49 | 0.71 | 0.39 | 0.66 | 0.47 | 0.67 |
| MLP + k-best + SMOTE | 0.49 | 0.57 | 0.52 | 0.77 | 0.45 | 0.59 | 0.50 | 0.75 | 0.45 | 0.59 | 0.51 | 0.75 |
| SVM + k-best + SMOTE | 0.49 | 0.55 | 0.52 | 0.77 | 0.50 | 0.52 | 0.50 | 0.78 | 0.49 | 0.55 | 0.51 | 0.77 |
| RF + k-best + SMOTE | 0.54 | 0.52 | 0.53 | 0.80 | 0.55 | 0.49 | 0.51 | 0.80 | 0.54 | 0.50 | *0.52* | 0.80 |
| LR + RUS | 0.37 | 0.67 | 0.47 | 0.67 | 0.28 | *0.74* | 0.40 | 0.53 | 0.29 | *0.75* | 0.41 | 0.53 |
| XGB + RUS | 0.46 | 0.64 | 0.53 | 0.75 | 0.49 | 0.57 | 0.52 | 0.76 | 0.48 | 0.58 | 0.52 | 0.76 |
| MLP + RUS | 0.40 | 0.65 | 0.50 | 0.71 | 0.43 | 0.63 | 0.50 | 0.73 | 0.42 | 0.61 | 0.49 | 0.73 |
| SVM + RUS | 0.39 | 0.64 | 0.48 | 0.69 | 0.38 | 0.65 | 0.47 | 0.69 | 0.38 | 0.65 | 0.47 | 0.69 |
| RF + RUS | 0.47 | 0.62 | 0.53 | 0.76 | 0.45 | 0.64 | 0.52 | 0.75 | 0.45 | 0.63 | 0.52 | 0.75 |
| LR + PCA + RUS | 0.35 | 0.67 | 0.46 | 0.66 | 0.30 | 0.56 | 0.38 | 0.62 | 0.29 | 0.54 | 0.38 | 0.61 |
| XGB + PCA + RUS | 0.45 | 0.62 | 0.52 | 0.75 | 0.33 | 0.53 | 0.41 | 0.65 | 0.37 | 0.55 | 0.44 | 0.68 |
| MLP + PCA + RUS | 0.43 | 0.64 | 0.51 | 0.73 | 0.29 | 0.54 | 0.37 | 0.60 | 0.28 | 0.55 | 0.36 | 0.58 |

(continued)

**Table 2** (continued)

| | Central ML | | | | FL with FedAvg | | | | FL with FedF1 | | | |
|---|---|---|---|---|---|---|---|---|---|---|---|---|
| Model name | P. | R. | F. | Ac. | P. | R. | F. | Ac. | P. | R. | F. | Ac. |
| SVM + PCA + RUS | 0.39 | 0.60 | 0.47 | 0.70 | 0.30 | 0.55 | 0.39 | 0.62 | 0.31 | 0.55 | 0.39 | 0.64 |
| RF + PCA + RUS | 0.47 | 0.59 | 0.53 | 0.77 | 0.44 | 0.60 | 0.50 | 0.75 | 0.46 | 0.60 | 0.52 | 0.76 |
| LR + k-best + RUS | 0.48 | 0.54 | 0.51 | 0.77 | 0.30 | 0.74 | 0.39 | 0.50 | 0.29 | 0.75 | 0.39 | 0.48 |
| XGB + k-best + RUS | 0.50 | 0.57 | 0.53 | 0.76 | 0.47 | 0.58 | 0.51 | 0.76 | 0.48 | 0.59 | 0.52 | 0.76 |
| MLP + k-best + RUS | 0.48 | 0.59 | 0.53 | 0.77 | 0.46 | 0.61 | 0.52 | 0.76 | 0.47 | 0.61 | 0.52 | 0.76 |
| SVM + k-best + RUS | 0.49 | 0.53 | 0.51 | 0.77 | 0.47 | 0.55 | 0.50 | 0.77 | 0.50 | 0.53 | 0.50 | 0.78 |
| RF + k-best + RUS | 0.50 | 0.57 | 0.53 | 0.78 | 0.46 | 0.58 | 0.51 | 0.78 | 0.47 | 0.57 | 0.51 | 0.77 |

[a] The model results that achieve the highest score according to the evaluation metric in each column are shown in *italic*. This means that for each metric, one model is compared with 45 different models column-wise. In cases where there is more than one model with the highest score, the following order is followed: Ac., F., P., R

In comparison to CML, the application of PCA in FL results in greater degradation of model outcomes. This may be related to dividing data into fewer parts (clients) in FL. This may prevent PCA's ability to identify patterns within the dataset. Reducing dataset size, especially in FL, may increase sample variability on components, causing performance variations. ML models trained using centralized and FL methods on the five important features showed considerable accuracy increases. The k-best approach yielded the best results in terms of both accuracy and F1 score, 53% and 52% for CML models and FL models. The LR+k-best model achieved the highest precision score across all models, recording 73%, 75%, and 71% for CML, FedAvg, and FedF1, respectively. Additionally, MLP + k-best and RF + k-best achieved the highest accuracy score of 82% for both approaches.

**Experiments in View of Federated Learning** Figure 3a compares precision scores across three methodologies. Centralized ML models show higher average precision with more consistent performance, as indicated by smaller IQR and Min-Max differences. Nevertheless, the gap between FedF1 models and centralized ML models is minimal. Also, the FedAvg approach achieves the highest precision. The model performance is well-balanced, as the distribution between Q1 (lower quartile), Q2 (median), and Q3 (upper quartile) is nearly equal in the centralized ML. The FedF1 models' overall performance is quite stable and above the median, as proven by the low difference between Q2 and Q3. Figure 3b shows that recall variances of the three methodologies are similar. Centralized ML models and FedF1 approaches exhibit slightly more stability, whereas FedAvg method exhibits a greater variability in values. FedF1 model has a higher Q1 value than the other models, implying that its worst-case performance is generally superior. Overall, FedF1 achieves comparable or superior results to the centralized ML approach, while FedAvg is weaker in terms of recall. Figure 3c shows that the F1-score distribution of the CML model reduces to a narrow range (0.45–0.53), demonstrating a more stable and higher F1 score compared to other approaches.

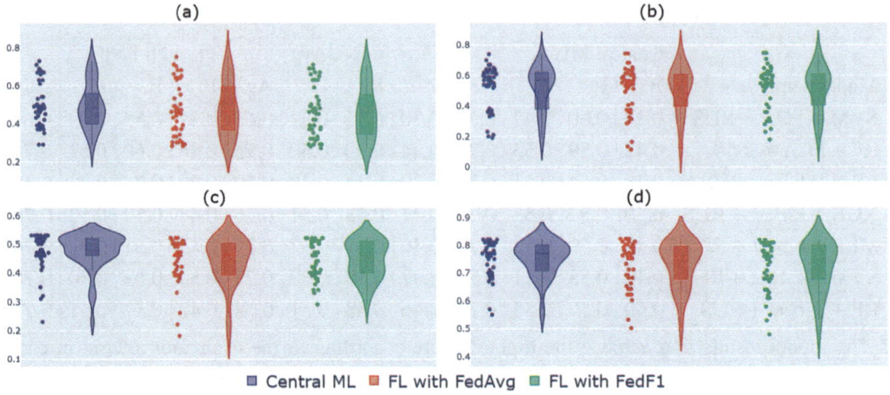


**Fig. 3** "Default Detection" report of precision (**a**), recall (**b**), F1 (**c**), and accuracy (**d**) for CML, FedAvg, and FedF1. From the perspective of both "precision" and "F1" scores, CML and FedF1 achieved similar results, while FedAvg is behind. The "recall" values indicated that all three approaches achieved comparable results. In terms of "accuracy," CML produced little more stable outcomes than FL, although all methods performed well overall

The FedAvg model has a wider F1-score distribution than the others. FedF1 models have an intermediate F1 score compared to other models. The F1 value in the CML approach demonstrates superior scores and greater stability. FedF1 outperformed FedAvg in performance and stability, although not as much as central techniques.

## 6 Conclusion and Future Work

This chapter aims to compare FL and centralized ML methods and highlights the advantages of FL in the financial sector. We investigated the application of FL techniques in credit card default assessment. Our approach demonstrated that FL can effectively protect privacy, while enhancing the performance of predictive models through collaboration among various financial institutions. By utilizing both FedAvg aggregation method and our newly developed FedF1 aggregation method, we achieved significant improvements in model robustness and stability, particularly in the context of imbalanced data. The FedF1 method, designed to encourage clients to achieve superior and more consistent performance, outperformed FedAvg across multiple metrics.

For future work, there are several potential directions to expand upon our current research. First, the scalability of the FL approach should be evaluated in larger and more heterogeneous datasets, representing different types of financial institutions. In addition, client-based performance should be analyzed to examine how well it performs with its limited datasets. Another direction is integrating other data sources, such as real-time financial transaction data, to enhance model performance for detecting anomalies and predicting financial defaults. Moreover, to improve the robustness of this study, future experiments should be conducted using various publicly available datasets.

**Acknowledgments** This work was supported in part by the Scientific and Technological Research Council of Türkiye (TUBITAK) 1515 Frontier R&D Laboratories Support Program for BTS Advanced AI Hub: BTS Autonomous Networks and Data Innovation Lab Project 5239903, and partly by the Scientific Research Projects Coordination Department (BAP), Istanbul Technical University, under the project ITU-BAP MGA-2024 45372.

## References

1. Abreha, H.G., Hayajneh, M., Serhani, M.A.: Federated learning in edge computing: a systematic survey. Sensors **22**(2), 450 (2022). https://doi.org/10.3390/s22020450
2. Antunes, R.S., Da Costa, C.A., Küderle, A., Yari, I.A., Eskofier, B.: Federated learning for healthcare: systematic review and architecture proposal. ACM Trans. Intell. Syst. Technol. **13**(4), 1–23 (2022). https://doi.org/10.1145/3501813
3. Aurna, N., Hossain, M.D., Taenaka, Y., Kadobayashi, Y.: Federated learning-based credit card fraud detection: performance analysis with sampling methods and deep learning algorithms. In: 2023 IEEE International Conference on Cyber Security and Resilience (CSR), pp. 180–186 (2023). https://doi.org/10.1109/CSR57506.2023.10224978
4. Awosika, T., Shukla, R.M., Pranggono, B.: Transparency and privacy: the role of explainable AI and federated learning in financial fraud detection. IEEE Access **12**, 64551–64560 (2024). https://doi.org/10.1109/access.2024.3394528
5. Bakumenko, A., Elragal, A.: Detecting anomalies in financial data using machine learning algorithms. Systems **10**(5), 130 (2022). https://doi.org/10.3390/systems10050130
6. Beutel, D.J., Topal, T., Mathur, A., Qiu, X., Fernandez-Marques, J., Gao, Y., Sani, L., Li, K.H., Parcollet, T., De Gusmão, P.P.B., Lane, N.D.: Flower: A Friendly Federated Learning Research Framework. arXiv (Cornell University) (2020). https://doi.org/10.48550/arxiv.2007.14390
7. Chatterjee, P., Das, D., Rawat, D.B.: Federated learning empowered recommendation model for financial consumer services. IEEE Trans. Consum. Electron. (2023). https://doi.org/10.1109/TCE.2023.3339702
8. Chatterjee, P., Das, D., Rawat, D.B.: Securing financial services with federated learning and blockchain. In: CRC Press eBooks, pp. 178–207 (2024). https://doi.org/10.1201/9781003376712-9
9. Chawla, N.V., Bowyer, K.W., Hall, L.O., Kegelmeyer, W.P.: SMOTE: synthetic minority oversampling technique. JAIR **16**, 321–357 (2002). https://doi.org/10.1613/jair.953

10. Chen, T., Guestrin, C.: XGBoost. In: KDD '16: Proceedings of the 22nd ACM SIGKDD International Conference on Knowledge Discovery and Data Mining, pp. 785–794 (2016). https://doi.org/10.1145/2939672.2939785
11. Cortes, C., Vapnik, V.: Support-vector networks. Mach. Learn. **20**(3), 273–297 (1995). https://doi.org/10.1007/bf00994018
12. Cox, D.R.: The regression analysis of binary sequences. J. R. Stat. Soc. Ser. B (Methodol.) **20**(2), 215–232 (1958). https://doi.org/10.1111/j.2517-6161.1958.tb00292.x
13. Dal Pozzolo, A.: Credit Card Fraud Detection Dataset. Kaggle. https://www.kaggle.com/datasets/mlg-ulb/creditcardfraud (2018)
14. Goldstein, B.A., Hubbard, A.E., Cutler, A., Barcellos, L.F.: An application of Random Forests to a genome-wide association dataset: methodological considerations & new findings. BMC Genomic Data **11**(1) (2010). https://doi.org/10.1186/1471-2156-11-49
15. Goutier, M., Diebel, C., Adam, M., Benlian, A.: Federated learning for credit risk assessment. In: Proceedings of Hawaii International Conference on System Sciences (2024). https://doi.org/10.24251/hicss.2023.048
16. Guo, L., Song, R., Wu, J., Xu, Z., Zhao, F.: Integrating a machine learning-driven fraud detection system based on a risk management framework. In: Proceedings of the 6th International Conference on Computing and Data Science (2024). https://doi.org/10.20944/preprints202406.1756.v1
17. Haykin, S.: Neural Networks: A Comprehensive Foundation, 2nd edn. Prentice Hall, Upper Saddle River, NJ (1998). ISBN: 0132733501
18. Ho, N.T.K.: Random decision forests. In: 3rd International Conference on Document Analysis and Recognition, pp. 278–282 (1995). https://doi.org/10.1109/icdar.1995.598994
19. KayChansiri GitHub - KayChansiri/Demo_logistic_regression_ML. In: GitHub. https://github.com/KayChansiri/Demo_logistic_regression_ML
20. Li, Y., and Wen, G. Research and practice of financial credit risk management based on federated learning. Eng. Lett. **31**(1), 271–278 (2023)
21. McMahan, H.B., Moore, E., Ramage, D., Hampson, S., Arcas, B.A.Y.: Communication-Efficient Learning of Deep Networks from Decentralized Data. arXiv (Cornell University) (2016). https://doi.org/10.48550/arxiv.1602.05629
22. Nguyen, D.C., Ding, M., Pathirana, P.N., Seneviratne, A., Li, J., Poor, H.V.: Federated learning for internet of things: a comprehensive survey. IEEE Commun. Surv. Tutorials **23**(3), 1622–1658 (2021). https://doi.org/10.1109/comst.2021.3075439
23. Pedregosa, F., Varoquaux, G., Gramfort, A., Michel, V., Thirion, B., Grisel, O., Blondel, M., Prettenhofer, P., Weiss, R., Dubourg, V., Vanderplas, J., Passos, A., Cournapeau, D., Brucher, M., Perrot, M., Duchesnay, É.: Scikit-learn: Machine Learning in Python. JMLR (2011). https://doi.org/10.5555/1953048.2078195
24. Ramakrishnan, J., Shaabani, E., Li, C., Sustik, M.A.: Anomaly detection for an ecommerce pricing system. In: KDD '19: Proceedings of the 25th ACM SIGKDD International Conference on Knowledge Discovery & Data Mining (2019). https://doi.org/10.1145/3292500.3330748
25. Saputra, Y.M., Hoang, D.T., Nguyen, D.N., Dutkiewicz, E., Mueck, M.D., Srikanteswara, S.: Energy demand prediction with federated learning for electric vehicle networks. In: 2019 IEEE Conference on Global Communications (GLOBECOM) (2019). https://doi.org/10.1109/globecom38437.2019.9013587
26. Saxena, A., Agarwal, A., Pandey, B.K., Pandey, D.: Examination of the criticality of customer segmentation using unsupervised learning methods. Circ. Econ. Sustain. (2024). https://doi.org/10.1007/s43615-023-00336-4
27. Ülver, B., Yurtoğlu, R.A., Dervişoğlu, H., Halepmollası, R., Haklıdır, M.: Federated learning in predicting heart disease. In: 2023 31st IEEE Signal Processing and Communications Applications (SIU) (2023). https://doi.org/10.1109/siu59756.2023.10223935
28. Varmedja, D., Karanovic, M., Sladojevic, S., Arsenovic, M., Anderla, A.: Credit card fraud detection-machine learning methods. In: 2019 International Symposium on INFOTEH-JAHORINA (INFOTEH), pp. 1–5. IEEE, Piscataway (2019). https://doi.org/10.1109/INFOTEH.2019.8717766

29. Wahab, F., Khan, I., Sabada, S.: Credit card default prediction using ML and DL techniques. IoT CPS **4**, 293–306 (2024). https://doi.org/10.1016/j.iotcps.2024.09.001
30. Wang, J., Bi, S., Xiao, J.: Research on the application of data analysis in predicting financial risk. Finan. Eng. Risk Manage. **7**(5) (2024). https://doi.org/10.23977/ferm.2024.070424
31. Yang, Z., Chen, M., Saad, W., Hong, C.S., Shikh-Bahaei, M.: Energy efficient federated learning over wireless communication networks. IEEE Trans. Wirel. Commun. **20**(3), 1935–1949 (2020). https://doi.org/10.1109/twc.2020.3037554
32. Yeh, I.: Default of Credit Card Clients [Dataset]. UCI Machine Learning Repository (2009). https://doi.org/10.24432/C55S3H
33. Zhou, Y., Chi, G., Liu, J., Xiong, J., Wang, B.: Default discrimination of credit card: feature combination selection based on improved FDAF-score. Exp. Syst. Appl. **206**, 117829 (2022). https://doi.org/10.1016/j.eswa.2022.117829
34. Zhu, H., Xu, J., Liu, S., Jin, Y.: Federated learning on non-IID data: a survey. Neurocomputing **465**, 371–390 (2021). https://doi.org/10.1016/j.neucom.2021.07.098

# Impact of Face Alignment on Face Image Quality

**Eren Onaran, Erdi Sarıtaş, and Hazım Kemal Ekenel**

## 1 Introduction

Face image analysis is one of the most actively studied fields in biometrics [15, 21, 23, 25, 30]. This field branches into multiple subfields, each focused on extracting meaningful features for specific applications, often under real-life conditions. Some of the most commonly studied areas are face recognitions [23, 25], face image quality assessments [21], facial expression recognition [15], and facial attribute classification [30]. However, the performance of these applications is heavily influenced by the conditions under which face images are acquired.

While factors like resolution, noise, and sharpness are often used to assess image quality, they can sometimes be misleading. These general metrics fail to consider face-specific challenges such as pose, occlusion, and illumination. To address these limitations, specialized approaches for assessing the quality of face images have been developed [21].

Face image quality assessment (FIQA) aims to map a face image to a quality score, supporting downstream tasks—most notably face recognition. FIQA plays a critical role in determining whether a face image is suitable for processing [21] and ensures that image quality is effectively accounted for in biometric systems.

Between face detection and feature extraction, an essential step conventionally used in most facial analysis tasks, including face image quality assessment (FIQA),

E. Onaran (✉) · E. Sarıtaş
Department of Computer Engineering, Istanbul Technical University, Istanbul, Turkey
e-mail: onarane21@itu.edu.tr; saritas21@itu.edu.tr

H. K. Ekenel
Department of Computer Engineering, Istanbul Technical University, Istanbul, Turkey

Division of Engineering, NYU Abu Dhabi, Abu Dhabi, UAE
e-mail: ekenel@itu.edu.tr; he2244@nyu.edu

 
B. U. Töreyin et al. (eds.), *8th EAI International Conference on Robotic Sensor Networks*, EAI/Springer Innovations in Communication and Computing,
https://doi.org/10.1007/978-3-031-92143-8_3

is face alignment. Face alignment involves locating predefined landmarks on the face, such as the eyes' centers, the nose's tip, and the mouth's corners. A geometric transformation is computed using these landmarks to ensure that the detected facial features are positioned consistently across all images in the dataset. This standardization results in the uniform relative positioning of facial features [7, 8, 23, 25].

Modern deep-learning-based face detectors further facilitate alignment by integrating facial landmark detection into their pipelines. Models like MTCNN [29], RetinaFace [6], STN [4], and CenterFace [28] have advanced the alignment process, improving both accuracy and robustness in real-world scenarios.

Although face alignment is a common preprocessing step in facial analysis, its broader impact on subsequent tasks, such as quality assessment, remains insufficiently explored. To address this need, our study investigates the effect of alignment on face image quality by comparing the quality scores of cropped and aligned face images. To assess face image quality, we employed several methods, including SER-FIQ [24], FaceQAN [2], DifFIQA [3], and SDD-FIQA [18]. For detection and alignment, we used MTCNN [29] and RetinaFace [6] models across the LFW [13], IJB-B [26], and SCFace [10] benchmark datasets.

To analyze the effect of alignment, we generated two versions of each face image for both detection models: one cropped and one aligned. The cropped images were obtained based on the bounding box coordinates provided by the detection models, while the aligned images were generated using the located facial landmarks. We then calculated the quality scores for both the "Cropped" and "Aligned" images to enable a comparative analysis. Our findings reveal that misalignment consistently reduces face image quality across a variety of conditions. This effect particularly stands out in challenging scenarios, such as surveillance, where images are often of lower quality.

This chapter is organized as follows. Section 2 gives necessary background information and briefly reviews recent advancements. Section 3 explains the methodology and describes the tools used. Section 4 details our experimental setup. Section 5 discusses our findings. Finally, we present our conclusions in Sect. 6.

## 2 Background

This section presents background information and reviews recent advancements in the field to provide the necessary context.

### *2.1 Face Detection and Alignment*

Face detection involves identifying faces within an image and providing their bounding box coordinates, while face alignment focuses on locating predefined landmarks on the detected faces [23, 25]. These landmarks typically include five key points: the corners of the eyes, the tip of the nose, and the corners of the mouth [4, 6, 28, 29].

Recent advancements in deep learning have significantly improved these processes by leveraging the inherent correlation between detection and landmark localization through joint learning methods such as MTCNN [29], RetinaFace [6], STN [4], and CenterFace [28]. These models simultaneously detect faces and predict key facial landmarks, enabling more accurate and efficient alignment. Using these landmarks, face images can be geometrically transformed. In some cases, affine or perspective transformations are applied to correct variations in the angle or tilt of the face, ensuring consistent alignment.

Accurately locating facial landmarks is increasingly difficult under challenging conditions such as poor lighting, extreme poses, occlusions, or low-resolution images. These conditions obscure or distort key facial features, leading to errors in the alignment process, causing misaligned face images and ultimately impacting facial analysis [1, 19].

## 2.2 *Face Image Quality Assessment*

FIQA methods assign quality scores to evaluate how suitable a face image is for processing, leveraging statistical analysis, machine learning, and deep learning models [21]. The quality scores produced by an FIQA method are influenced by factors such as lighting, blur, brightness, contrast, pose, expression, occlusion, resolution, and noise [17]. Due to variations in training methodologies, datasets, and model architectures, different FIQA models respond uniquely to these factors. Consequently, the definition of face image quality is inherently tied to its utility for a specific task and model rather than conforming to a universal definition of image quality.

With quality scores defined based on their utility, FIQA is utilized in various applications, particularly in face recognition systems. In this context, FIQA provides value in several key areas: filtering out low-quality images at the time of capture, optimizing database maintenance by ranking and selecting high-quality images, and calibrating recognition systems to manage variations in image quality [21]. FIQA helps reduce false negatives and enhances overall system performance by identifying potentially poor-quality images before processing.

Recent FIQA methods are predominantly based on deep learning techniques. These methods can be categorized into two main types: model-based methods [2, 3, 14, 18, 22, 24] and regression-based methods [3, 11, 27]. Model-based methods employ a face recognition model to operate directly in the embedding space to calculate quality scores. On the other hand, regression-based methods generate pseudo-labels for quality scores and then train neural networks using these labels.

Model-based FIQA methods leverage embedding-level properties to evaluate the quality of face images by utilizing a face recognition model. Unlike regression-based methods, these approaches assess image quality directly within the feature space of a face recognition model. PFE (probabilistic face embeddings) [22] models face embeddings as probability distributions, producing both a mean vector (embedding) and a variance vector (uncertainty). The quality score is derived from the variance, with lower uncertainty reflecting higher image quality. SER-FIQ [24], an unsupervised method, evaluates face image quality by measuring the consistency of embeddings generated with stochastic dropout. FaceQAN [2] integrates using adversarial perturbations to calculate the pseudo-quality scores. DifFIQA [3] employs denoising diffusion probabilistic models (DDPMs) [12], perturbing facial images through the forward and backward processes of DDPMs to quantify the impact on embeddings. GraFIQs [14] also considers the use of gradients through a face recognition model. To calculate the gradients, they measure the differences in the statistics of batch normalization layers.

Regression-based FIQA methods focus on generating informative pseudo-labels for quality scores. FaceQNet [11] selects an anchor image for each individual using third-party ICAO compliance software. Next, pseudo-quality score labels are calculated as the cosine similarity between the anchor image and the other images of the same individual. PCNet [27] leverages a large number of mated image pairs (distinct images of the same individual), deriving quality labels from the embedding similarities of these pairs. Similarly, SDD-FIQA [18] uses interclass and intraclass distances to compute quality scores as the average Wasserstein distance between mated and non-mated comparison distributions. The DifFIQA [3] can be categorized as both regression-based and model-based. Its main contribution is in the model-based scope, where it also provides a regression model due to the inference time.

## 3 Methodology

We employed MTCNN [29] and RetinaFace [6] for face detection and landmark localization. For assessing the quality of the faces, we used four quality assessment models: SER-FIQ [24] (based on the ArcFace model [5]), FaceQAN [2], DifFIQA [3], and SDD-FIQA [18].

### *3.1 Face Detectors*

- MTCNN [29]: The multitask cascaded convolutional network (MTCNN) is a popular face detection method with three stages that progressively improve face localization. The process starts with the proposal network (P-Net), which scans the image to identify candidate face windows. Due to its broad search, the network often captures many false positives at this stage. Next comes the refine network (R-Net), which takes these initial candidate windows and applies more precise filtering, removing most false positives and keeping only the areas most likely to contain faces. In the final stage, the output network (O-Net) enhances the results further, refining the remaining windows and adding precise bounding boxes with five facial landmarks (two eyes, the nose, and two mouth corners) to support alignment and boost detection accuracy. The cascaded approach helps MTCNN balance efficiency and accuracy, making it well-suited to detect faces in various poses and lighting conditions.

- RetinaFace [6]: RetinaFace is a robust face detection model built on the RetinaNet [16] object detection framework and enhanced with advanced multitask learning techniques for optimized performance. With a feature pyramid network (FPN) backbone, RetinaFace includes context modules at each feature level, allowing it to capture rich contextual details and improve detection accuracy across diverse face sizes and poses. Unlike traditional two-stage face detection models, RetinaFace operates in a single stage, delivering both speed and precision. For each detected face, RetinaFace provides a bounding box, a confidence score, dense 3D face vertices, and five key facial landmarks (eye corners, nose, and mouth corners) to support precise alignment. This unified approach enables RetinaFace to excel at face localization and alignment, even in challenging, real-world scenarios like occlusions, varied lighting, and extreme poses.

### 3.2 *Face Image Quality Assessment*

- SER-FIQ [24]: The stochastic embedding robustness for face image quality (SER-FIQ) method assesses the quality of a face image by examining how consistent the feature embeddings are across multiple forward passes with dropout applied. This approach evaluates image quality based on the stability of embeddings. An image is considered high quality if its embeddings remain consistent across multiple passes through the model. Conversely, when there is a large variance between these embeddings, it indicates higher uncertainty, pointing to a lower-quality image. SER-FIQ measures this by calculating the negative mean Euclidean distance between the embeddings generated in each pass and then applies a sigmoid function to standardize the score.
- FaceQAN [2]: Face quality assessment network (FaceQAN) assesses the quality of an image by examining how well its feature embeddings withstand adversarial noise, giving insight into its resilience in challenging conditions. The process starts by calculating the cosine similarity between the embeddings of the original face image and a noise-injected version. Using the fast gradient sign method (FGSM) [9], the model then applies adversarial noise to the original image over multiple iterations, observing how this affects the embeddings. A quality score is calculated by combining the mean and variance of these cosine similarity scores and the similarity score between the original image and its horizontally flipped version. By focusing on consistency and robustness in this way, FaceQAN provides a quality score where higher values indicate superior stability under adversarial conditions.

- DifFIQA [3]: Denoising diffusion face image quality assessment (DifFIQA) measures face image quality by utilizing both the forward and backward steps of diffusion processes. The method starts by gradually degrading the image through several forward diffusion steps, adding noise in each step. Then, it reconstructs the image using a denoising diffusion probabilistic model (DDPM) [12] during the backward steps. This process is also applied to a horizontally flipped version of the image, resulting in six variations of the original image. These variations are then passed through a face recognition model to extract feature embeddings. Then, cosine similarities are calculated between the embeddings of the original image and each modified version. This process is repeated over multiple iterations, with the average cosine similarity across all iterations as the final quality score. A regression model is trained using these quality scores as pseudo-labels to speed up the assessment process, allowing quick quality evaluation without repeated diffusion steps. In the experiments, a regression model pretrained with pseudo-labels is used.
- SDD-FIQA [18]: Similarity distribution distance face image quality assessment (SDD-FIQA) method evaluates face image quality by analyzing how well an image's embedding can differentiate between samples of the same identity (intraclass) and those of different identities (interclass). Ideally, high-quality images are close to others of the same identity, while clearly separated from different identities. The process begins by creating a positive set (same identity) and a negative set (different identities) for each query image, selecting challenging examples through a hard sampling strategy. Cosine similarity scores between the query image and images in each set are then computed, forming positive and negative distributions. The Wasserstein distance is calculated to measure the separation between these distributions, serving as a pseudo-label that represents image quality. A regression model is then trained on these pseudo-labels, allowing quick quality assessments based on distribution distance—where a greater separation typically indicates higher image quality.

### *3.3 Face Alignment*

Initially, the detection models estimated bounding boxes and five facial landmarks: the left eye, right eye, nose tip, left mouth corner, and right mouth corner. No scaling is applied for the bounding boxes provided by RetinaFace, and the coordinates are used directly. In contrast, for faces detected using MTCNN, the bounding boxes are scaled by a factor of 1.2 to address their tightness, and the faces are cropped accordingly. The datasets generated from these cropped face images, based on MTCNN and RetinaFace, are referred to as "Cropped."

The face images are also processed for alignment using the detected facial landmarks. Each of the five landmarks is mapped to a predetermined position in the output image based on the template used in ArcFace [5]. A similarity transformation matrix is computed to minimize the sum of squared distances between the transformed landmark positions and their corresponding target locations. Notably, this alignment process is applied directly to the original images and operates independently of the cropping operation.

$$T = \arg\min_{T} \sum_{i \in \text{landmarks}} \|T(p_i) - q_i\|^2 \tag{1}$$

where $||.||$ represents the $L_2$ distance, $p_i$ denotes the detected landmark, and $q_i$ is its desired position. The transformation matrix $T$ performs three operations: scaling, rotation, and translation. Mathematically, $T$ is defined as

$$T\left(\begin{bmatrix} x \\ y \end{bmatrix}\right) = \alpha \begin{bmatrix} \cos\theta & -\sin\theta \\ \sin\theta & \cos\theta \end{bmatrix} \begin{bmatrix} x \\ y \end{bmatrix} + \begin{bmatrix} t_x \\ t_y \end{bmatrix} \tag{2}$$

which can also be expressed in homogeneous coordinates as

$$T\left(\begin{bmatrix} x \\ y \\ 1 \end{bmatrix}\right) = \begin{bmatrix} \alpha\cos\theta & -\alpha\sin\theta & t_x \\ \alpha\sin\theta & \alpha\cos\theta & t_y \\ 0 & 0 & 1 \end{bmatrix} \begin{bmatrix} x \\ y \\ 1 \end{bmatrix} \tag{3}$$

Here, $\alpha$ is the scaling factor that ensures the distance between the eyes remains fixed, $\theta$ is the rotation angle that aligns the line connecting the eyes to the horizontal axis, and $t_x$, $t_y$ are the translation parameters used to standardize the face location. The transformation $T$ is applied to the input image to warp it for alignment. Datasets created from these aligned face images are referred to as "Aligned." The overall pipeline is illustrated in Fig. 1.

## 4 Experimental Setup

In this chapter, we utilized three face image datasets: LFW [13], IJB-B [26], and SCFace [10].

### *4.1 Datasets*

- Labeled Faces in the Wild (LFW) [13]: LFW is a widely recognized benchmark dataset for evaluating face recognition and verification models in unconstrained settings. It includes 13,233 images of 5,749 individuals collected from the Internet, with variations in pose, expression, lighting, background, and occlusions. Its primary evaluation procedure consists of 6,000 face image pairs divided into ten folds, with each fold having 300 positive and 300 negative pairs. Then, ten-fold

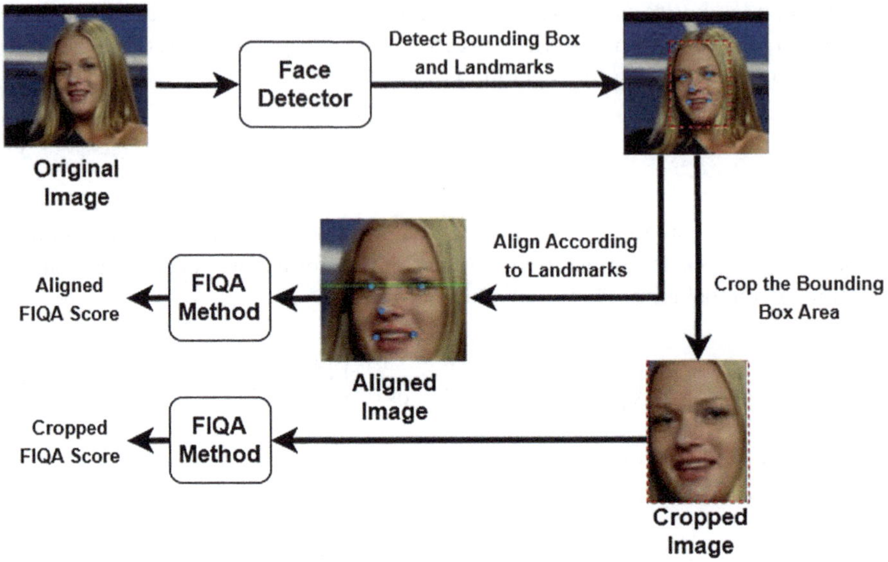


**Fig. 1** The general overview of our pipeline is as follows. A face detector first processes an original image containing a face. The detector outputs a bounding box (red-dotted frame) and facial landmarks (blue dots). The image is then cropped to extract the face region, resulting in what we refer to as the "Cropped Image." In parallel, the input image is aligned based on the detected landmarks, producing the "Aligned Image." This alignment standardizes the face image, such as ensuring the line connecting the eyes (green-dotted line) is parallel to the horizontal axis. Finally, both the cropped and aligned images are evaluated using a FIQA method to estimate their FIQA scores

cross-validation is applied; for each fold, the nine other folds are used to find an optimal threshold, and the current fold is used to calculate the test performance.
- The IARPA Janus Benchmark-B (IJB-B) [26]: IJB-B is a comprehensive benchmark dataset designed for unconstrained face recognition. It consists of 11,754 images and 7,011 videos featuring 1,845 subjects. The dataset covers a range of variations, including pose and quality, with detailed annotations on occlusions, skin tone, and eye/nose location in the bounding boxes. In the evaluation procedure of IJB-B, the 1-1 verification protocol is used generally, where similarity scores are calculated between predetermined pairs. Then, these similarity scores are used to calculate TAR@FAR (true accept rate at fixed false accept rates) or ROC (receiver operating characteristic) curves.
- Surveillance Cameras Face Database (SCFace) [10]: SCFace is a specialized benchmark dataset created for surveillance applications. It features images of 130 subjects captured at three distances (4.2 m, 2.6 m, and 1.0 m) using five different video surveillance cameras of varying quality, resulting in 15 images per subject. Each subject also has a standard mugshot image. For each combination of distance and camera, captured face images are compared with mugshot images,

**Fig. 2** Sample original images from LFW [13], IJB-B [26], and SCFace [10]

and the identity of the closest match is assigned. For each distance, the accuracy across all camera settings is averaged to evaluate performance.

Sample original images from each dataset are shown in Fig. 2. For each original image, the bounding box and landmark coordinates are determined using MTCNN and RetinaFace. Then, "Cropped" and "Aligned" versions of these images are generated. Quality scores for each version are then computed using the specified FIQA methods. Sample outputs are displayed in Fig. 3, where the original images are presented in the first column. The corresponding detector and whether the image is cropped or aligned are indicated at the top of each column, while the FIQA scores for each version are provided below.

## 5 Results

We began our analysis by first investigating the input image variations and individual FIQA results shown in Fig. 3. Examining the quality scores, we find that the cropped versions consistently score lower across all examples, reflecting the negative impact of misalignment. Moreover, this effect is more apparent when the original images are more misaligned.

Quality score distributions for the LFW and IJB-B datasets are presented in Figs. 4 and 5, respectively. Each figure includes four subplots, corresponding to the outputs of different FIQA methods: SER-FIQ [24] (top-left), FaceQAN [2] (top-right), DifFIQA [3] (bottom-left), and SDD-FIQA [18] (bottom-right). In each

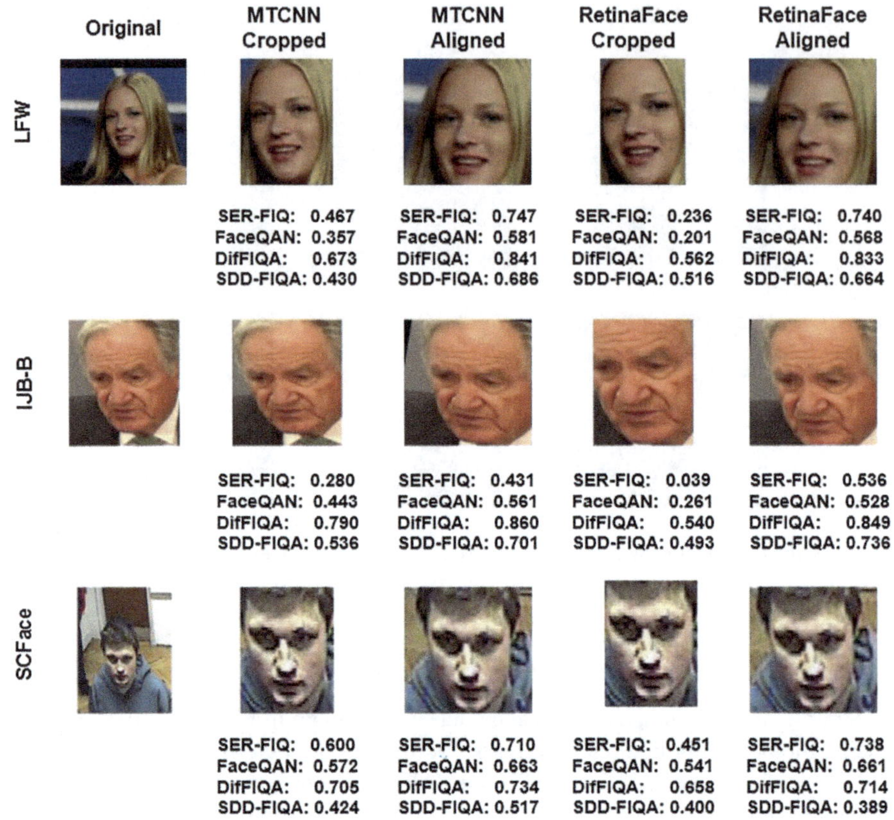


**Fig. 3** Sample original images from LFW [13], IJB-B [26], and SCFace [10], their cropped and aligned versions. The results of each FIQA method are written below in each cropped and aligned image

subplot, the y-axis represents density, while the x-axis shows the quality scores. The plots feature four results: MTCNN [29] cropped (red-dotted line), MTCNN aligned (solid red line), RetinaFace [6] cropped (blue-dotted line), and RetinaFace aligned (solid blue line).

For the LFW results in Fig. 4, the quality score distributions of aligned images have higher averages and lower variances than the ones of cropped images. The quality score distributions of aligned images are very similar, whether MTCNN or RetinaFace is used for landmark localization, while cropped images show slight variations in their distributions. In the SER-FIQ results (Fig. 4a), the quality scores have a high density near zero, followed by a nearly uniform distribution up to 0.6. In contrast, the quality scores for aligned images follow a Gaussian distribution with averages close to 0.8. For the remaining plots—Fig. 4b–d—the quality score distributions for both cropped and aligned images exhibit approximate Gaussian

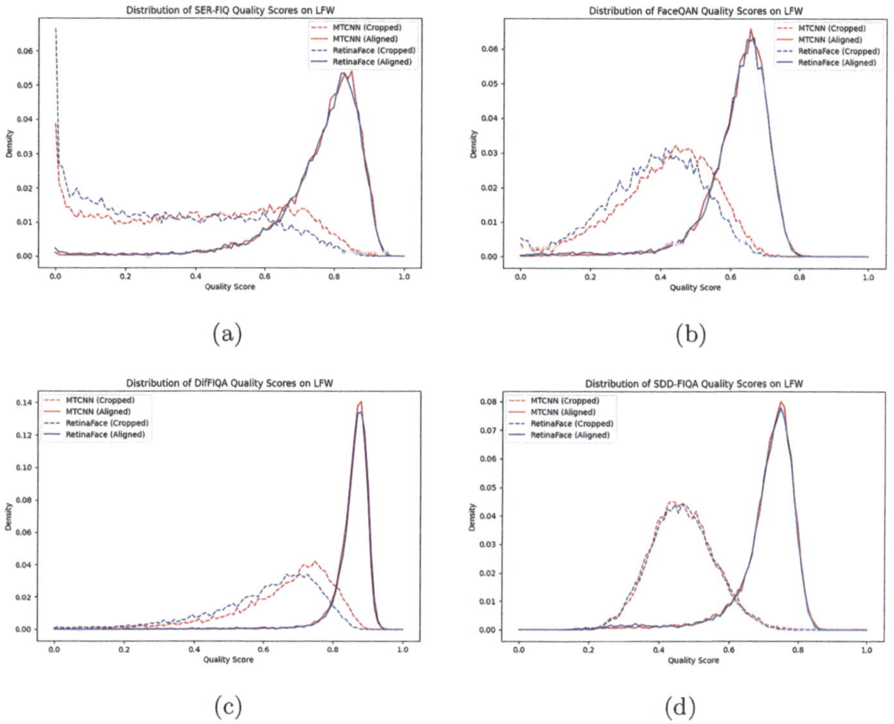


**Fig. 4** Quality score distributions for LFW [13] dataset. (**a**) SER-FIQ. (**b**) FaceQAN. (**c**) DifFIQA. (**d**) SDD-FIQA

shapes. The results can be summarized based on the means of the quality score distributions. FaceQAN's mean quality score is approximately 0.45 for cropped images and 0.7 for aligned images. For DifFIQA, the mean quality score is around 0.7 for cropped images and 0.85 for aligned images. For SDD-FIQA, the mean score is approximately 0.45 for cropped images and 0.75 for aligned images.

In the IJB-B results shown in Fig. 5, the quality score distributions for aligned images are nearly identical, while the distributions for cropped images show more notable distinctions. For the SER-FIQ results in Fig. 5a, the density of quality scores near zero is much higher compared to the one in Fig. 4a, and the scores beyond this range exhibit a more flat, low-density distribution. Moreover, for the aligned images, we can see Gaussian distributions with the means close to 0.75. For FaceQAN (Fig. 5b), the quality scores for cropped images are notably lower and look like a monotonically decreasing function. In contrast, the quality score distributions of aligned images form a curved shape with means close to 0.55. For the DifFIQA results (Fig. 5c), the quality score distribution for cropped images has a mean near 0.5 and high variance, while the quality score distribution for aligned images has a mean near 0.8 and low variance. For SDD-FIQA (Fig. 5d), the distributions for cropped images show lower variances than those for aligned images. The average

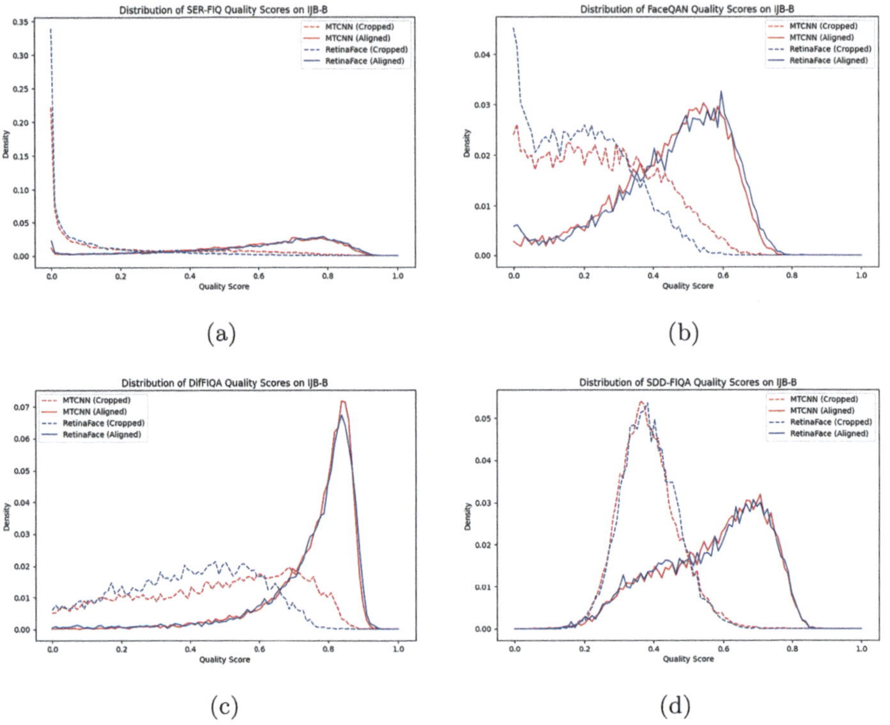


**Fig. 5** Quality score distributions for IJB-B [26] dataset. (**a**) SER-FIQ. (**b**) FaceQAN. (**c**) DifFIQA. (**d**) SDD-FIQA

quality scores are approximately 0.4 for cropped images and around 0.7 for aligned images. These results from two datasets—Figs. 4 and 5—clearly show that cropped images consistently achieve lower mean quality scores compared to aligned images.

The SCFace results are shown in Table 1. The table presents the average quality scores for each FIQA method at specific distances across different scenarios, including the detection method and whether the images are aligned or cropped. Consistent with previous findings, aligned face images exhibit higher mean quality scores than cropped face images. Additionally, the quality scores decrease as the distance at which the image is captured increases. The contribution to face image quality is most pronounced in the SER-FIQ results, followed by FaceQAN and DifFIQA. Among the FIQA methods, SDD-FIQA shows the least sensitivity to alignment by producing close quality scores between cropped and aligned images. The results indicate that alignment is a crucial step for FIQA in surveillance.

**Table 1** SCFace [10] dataset results

| FIQA method | Distance | MTCNN Cropped | MTCNN Aligned | RetinaFace Cropped | RetinaFace Aligned |
|---|---|---|---|---|---|
| SER-FIQ | 1.0 m | 0.600 | **0.789** | 0.448 | **0.792** |
| | 2.6 m | 0.483 | **0.705** | 0.285 | **0.711** |
| | 4.2 m | 0.187 | **0.474** | 0.070 | **0.426** |
| FaceQAN | 1.0 m | 0.548 | **0.665** | 0.475 | **0.666** |
| | 2.6 m | 0.497 | **0.586** | 0.396 | **0.588** |
| | 4.2 m | 0.331 | **0.451** | 0.198 | **0.407** |
| DifFIQA | 1.0 m | 0.712 | **0.800** | 0.633 | **0.802** |
| | 2.6 m | 0.630 | **0.715** | 0.531 | **0.710** |
| | 4.2 m | 0.461 | **0.578** | 0.316 | **0.543** |
| SDD-FIQA | 1.0 m | 0.419 | **0.584** | 0.413 | **0.579** |
| | 2.6 m | 0.371 | **0.435** | 0.360 | **0.422** |
| | 4.2 m | 0.331 | **0.342** | 0.325 | **0.331** |

The bold values indicate higher quality score values, demonstrating that face alignment consistently increases the quality scores measured by all face image quality assessment methods.

### *5.1 Discussion*

Faces are often captured upright, making them almost naturally aligned after cropping with a well-predicted bounding box. However, our results clearly show that using only cropped images produces inferior quality scores. Similar to face recognition [1, 19], this behavior may result from the FIQA methods being trained with aligned images. However, this much quality drop when using the cropped images still remains unexpected. The results demonstrate that the FIQA methods are sensitive to alignment.

The images in the selected datasets were collected under real-world conditions. Therefore, they contain various factors that degrade image quality, such as distance, lighting, or occlusions. Moreover, when these factors are combined, their impact amplifies each other [20], resulting in the loss of vital facial details. These degradation factors can lead to faulty localization of landmarks, which can result in misalignment. We can draw an indirect relation between alignment and face image quality; as the quality of images becomes lower, it is harder to estimate the correct positions of the landmarks, resulting in misalignment and, in turn, lower quality scores.

## 6 Conclusion

Face alignment is crucial for preparing face images for feature extraction and significantly influences face recognition performance [7, 8]. However, how the alignment affects the face image quality has not been explored in detail. In this chapter, we assessed the impact of face alignment by utilizing quality scores from SER-FIQ, FaceQAN, DifFIQA, and SDD-FIQA. We conducted our analysis on the benchmark datasets, including LFW, IJB-B, and SCFace. For bounding box and facial landmark localization, we used the MTCNN and RetinaFace models. To analyze the impact of the alignment on image quality, we compared the quality scores of cropped images with the aligned ones.

We analyzed quality score distributions on the LFW and IJB-B datasets and reviewed average quality scores at different distances in the SCFace dataset. Misalignment had a major negative effect on quality scores in all the datasets. Even though this quality score reduction is significant for relatively normal real-world condition datasets like LFW and IJB-B, the effect is more apparent in challenging scenarios like surveillance, as evidenced by the results of the SCFace dataset, especially at greater capturing distances. Thus, the alignment process could be more crucial for surveillance scenarios where face images are likely low quality. Our findings indicate that misaligned face images—only cropped versions—produce lower quality scores than properly aligned ones.

It is essential to remember that the FIQA methods were trained using aligned face images. Even though image quality may not directly relate to misalignment, the results show that FIQA methods are sensitive to misaligned face images. For some facial analysis applications, it might be desirable to minimize the quality score variations of these models when dealing with faces that are not ideally aligned.

**Acknowledgments** This research is supported by the TÜBİTAK 2247-C Intern Researcher Scholarship Programme (STAR) and the Meetween Project, funded by the European Union's Horizon Europe Research and Innovation Programme under Grant Agreement No. 101135798.

## References

1. An, Z., Deng, W., Hu, J., Zhong, Y., Zhao, Y.: APA: adaptive pose alignment for pose-invariant face recognition. IEEE Access **7**, 14653–14670 (2019)
2. Babnik, Ž., Peer, P., Štruc, V.: FaceQAN: face image quality assessment through adversarial noise exploration. In: 2022 26th International Conference on Pattern Recognition (ICPR), pp. 748–754. IEEE, Piscataway (2022)
3. Babnik, Ž., Peer, P., Štruc, V.: DifFIQA: face image quality assessment using denoising diffusion probabilistic models. In: 2023 IEEE International Joint Conference on Biometrics (IJCB), pp. 1–10. IEEE, Piscataway (2023)
4. Chen, D., Hua, G., Wen, F., Sun, J.: Supervised transformer network for efficient face detection. In: Computer Vision–ECCV 2016: 14th European Conference, Amsterdam, October 11–14, 2016, Proceedings, Part V 14, pp. 122–138. Springer, Berlin (2016)
5. Deng, J., Guo, J., Xue, N., Zafeiriou, S.: ArcFace: additive angular margin loss for deep face recognition. In: Proceedings of the IEEE/CVF Conference on Computer Vision and Pattern Recognition, pp. 4690–4699 (2019)
6. Deng, J., Guo, J., Ververas, E., Kotsia, I., Zafeiriou, S.: RetinaFace: single-shot multi-level face localisation in the wild. In: Proceedings of the IEEE/CVF Conference on Computer Vision and Pattern Recognition, pp. 5203–5212 (2020)

7. Ekenel, H.K., Stiefelhagen, R.: Face alignment by minimizing the closest classification distance. In: 2009 IEEE 3rd International Conference on Biometrics: Theory, Applications, and Systems, pp. 1–6. IEEE, Piscataway (2009)
8. Ekenel, H.K., Stiefelhagen, R.: Why is facial occlusion a challenging problem? In: Advances in Biometrics: Third International Conference, ICB 2009, Alghero, June 2–5, 2009. Proceedings 3, pp. 299–308. Springer, Berlin (2009)
9. Goodfellow, I.J., Shlens, J., Szegedy, C.: Explaining and harnessing adversarial examples. In: International Conference on Learning Representations (2015)
10. Grgic, M., Delac, K., Grgic, S.: SCFace–surveillance cameras face database. Multimed. Tools Appl. **51**, 863–879 (2011)
11. Hernandez-Ortega, J., Galbally, J., Fierrez, J., Haraksim, R., Beslay, L.: FaceQnet: quality assessment for face recognition based on deep learning. In: 2019 International Conference on Biometrics (ICB), pp. 1–8. IEEE, Piscataway (2019)
12. Ho, J., Jain, A., Abbeel, P.: Denoising diffusion probabilistic models. Adv. Neural Inf. Proces. Syst. **33**, 6840–6851 (2020)
13. Huang, G.B., Mattar, M., Berg, T., Learned-Miller, E.: Labeled faces in the wild: a database for studying face recognition in unconstrained environments. In: Workshop on faces in 'Real-Life' Images: Detection, Alignment, and Recognition (2008)
14. Kolf, J.N., Damer, N., Boutros, F.: GraFIQs: face image quality assessment using gradient magnitudes. In: Proceedings of the IEEE/CVF Conference on Computer Vision and Pattern Recognition, pp. 1490–1499 (2024)
15. Li, S., Deng, W.: Deep facial expression recognition: a survey. IEEE Trans. Affect. Comput. **13**(3), 1195–1215 (2020)
16. Lin, T.Y., Goyal, P., Girshick, R., He, K., Dollár, P.: Focal loss for dense object detection. In: Proceedings of the IEEE International Conference on Computer Vision (ICCV), Venice, 22–29 October 2017, pp. 2980–2988. https://doi.org/10.1109/ICCV.2017.324
17. Lopez Paya, L., Cordoba, P., Sanchez Perez, A., Barrachina, J., Benavent-Lledo, M., Mulero-Pérez, D., Garcia-Rodriguez, J.: Face recognition bias assessment through quality estimation models. Electronics **12**(22), 4649 (2023)
18. Ou, F.Z., Chen, X., Zhang, R., Huang, Y., Li, S., Li, J., Li, Y., Cao, L., Wang, Y.G.: SDD-FIQA: unsupervised face image quality assessment with similarity distribution distance. In: Proceedings of the IEEE/CVF Conference on Computer Vision and Pattern Recognition, pp. 7670–7679 (2021)
19. Saadabadi, M.S.E., Malakshan, S.R., Dabouei, A., Nasrabadi, N.M.: ARoFace: alignment robustness to improve low-quality face recognition. In: European Conference on Computer Vision, pp. 308–327 (2025)
20. Sarıtaş, E., Ekenel, H.K.: Analyzing the effect of combined degradations on face recognition. In: 2024 IEEE 18th International Conference on Automatic Face and Gesture Recognition (FG), pp. 1–5. IEEE, Piscataway (2024)
21. Schlett, T., Rathgeb, C., Henniger, O., Galbally, J., Fierrez, J., Busch, C.: Face image quality assessment: a literature survey. ACM Comput. Surv. **54**(10s), 1–49 (2022)
22. Shi, Y., Jain, A.K.: Probabilistic face embeddings. In: Proceedings of the IEEE/CVF International Conference on Computer Vision, pp. 6902–6911 (2019)
23. Taskiran, M., Kahraman, N., Erdem, C.E.: Face recognition: past, present and future (a review). Digit. Sig. Proces. **106**, 102809 (2020)
24. Terhorst, P., Kolf, J.N., Damer, N., Kirchbuchner, F., Kuijper, A.: SER-FIQ: unsupervised estimation of face image quality based on stochastic embedding robustness. In: Proceedings of the IEEE/CVF Conference on Computer Vision and Pattern Recognition, pp. 5651–5660 (2020)
25. Wang, X., Peng, J., Zhang, S., Chen, B., Wang, Y., Guo, Y.: A survey of face recognition. Preprint. arXiv:2212.13038 (2022)
26. Whitelam, C., Taborsky, E., Blanton, A., Maze, B., Adams, J., Miller, T., Kalka, N., Jain, A.K., Duncan, J.A., Allen, K., et al.: IARPA Janus Benchmark-B face dataset. In: Proceedings of the IEEE Conference on Computer Vision and Pattern Recognition Workshops, pp. 90–98 (2017)

27. Xie, W., Byrne, J., Zisserman, A.: Inducing predictive uncertainty estimation for face verification. In: BMVC (2020)
28. Xu, Y., Yan, W., Yang, G., Luo, J., Li, T., He, J.: CenterFace: joint face detection and alignment using face as point. Sci. Program. **2020**(1), 7845384 (2020)
29. Zhang, K., Zhang, Z., Li, Z., Qiao, Y.: Joint face detection and alignment using multitask cascaded convolutional networks. IEEE Sig. Proces. Lett. **23**(10), 1499–1503 (2016)
30. Zheng, X., Guo, Y., Huang, H., Li, Y., He, R.: A survey of deep facial attribute analysis. Int. J. Comput. Vis. **128**, 2002–2034 (2020)

# Pituitary Gland Segmentation from Pre-processed Brain MRI Slice with VGG-UNet

**Seifedine Kadry, Sahar Yassine, Hong Lin, Venkatesan Rajinikanth, and Ömer Melih Gül**

## 1 Introduction

Artificial Intelligence (AI)-facilitated data analysis denotes the application of certain AI methodologies to scrutinize, analyze, and extract insights from extensive datasets. This technology enhances traditional data management methods, rendering them more efficient and effective. The recent AI techniques like the Machine Learning (ML) and Deep Learning (DL) methods are commonly considered in the data evaluation task, due to its efficiency. Compared to the ML-methods, the DL-techniques are widely adopted in recent years to handle a variety of data like numerical, signals, images, and videos. Further, the DL-models provide a better result on the chosen database compared to the ML-techniques [1, 2].

Recently, the application of the DL-models in healthcare domain is gradually rising due to its clinical significance. Further, the DL-models works well on both

S. Kadry (✉) · S. Yassine
Department of Applied Data Science, Noroff University College, Kristiansand, Norway
e-mail: Seifedin.kadry@noroff.no; Sahar.yassine@noroff.no

H. Lin
Department of Computer Science and Engineering Technology, University of Houston-Downtown, Houston, TX, USA
e-mail: linh@uhd.edu

V. Rajinikanth
Department of Computer Science and Engineering, Division of Research and Innovation, Saveetha School of Engineering, Saveetha Institute of Medical and Technical Sciences (SIMATS), Chennai, TN, India

Ö. M. Gül
Informatics Institute, Istanbul Technical University (ITU), Istanbul, Turkey
e-mail: omgul@itu.edu.tr



B. U. Töreyin et al. (eds.), *8th EAI International Conference on Robotic Sensor Networks*, EAI/Springer Innovations in Communication and Computing,
https://doi.org/10.1007/978-3-031-92143-8_4

the gray- and RGB scale medical images of a chosen pixel dimension. Further, a number of pre-trained architectures are available to implement for the chosen task. When the pre-trained models are trained and validated on the chosen database, it will help to provide an agreeable result during the testing, and this model will work well when the similar database is considered for further examination. A number of DL-based medical image examination methods available in the literature confirm its merit in analyzing the chosen medical image of a particular modality for effective disease diagnosis and treatment planning process. The existing DL-models provide a better result on the classification as well as the segmentation process compared to the available traditional approaches [3–5].

Application of the DL-models in the healthcare domain is gradually rising since these models will help to provide necessary result during the disease diagnosis. This study seeks to employ a DL-model to analyze 2D brain MRI slices. The chosen DL-segmentation method is applied to extract and assess the Region of Interest (RoI) from brain MRI slices. This research considered the Pituitary Gland (PG) segmentation task from the chosen brain MRI slice.

In humans, the PG plays a crucial role in regulating hormones that affect various biological functions. Issues with the PG may present as hormonal imbalances, neoplasm, and anatomical irregularities. The PG-problems commonly begins mostly with age group 30 to 40 years and appropriate monitoring and treatment will help to maintain appropriate healthcare. The earlier works on the PG examination using the brain MRI can be accessed from [6–8]. During this examination task, the PG is extracted and evaluated using a chosen image segmentation scheme [9, 10].

This research proposed the VGG-UNet-supported segmentation of the PG from the sagittal-plane brain MRI slice. The various phases in the proposed methodology includes the following: (1) image and mask collection, (2) ITK-snap supported 3D to 2D conversion [11, 12] and resizing these 2D slices to $224 \times 224 \times 1$ pixels, (3) training and performance confirmation of VGG-UNet model, (4) comparing the RoI and mask to compute necessary metrics, and (5) confirming the advantage of DL-segmentation technique.

This work considered the VGG19-supported encoder-decoder to construct the VGG-UNet model as presented in [13]. The proposed DL-model is initially tested on the unprocessed images and the obtained metrics are verified. Later, the chosen images are pre-processed using the Kapur's Entropy and Butterfly Algorithm (KE + BA)-based tri-level thresholding and computed metrics with these images are also documented for the evaluation. The experimental outcome confirms that the result of the VGG-UNet is better on the pre-processed images compared to the unprocessed images. Further, a relative analysis among the VGG-UNet on other DL-models like UNet, UNet+, and UNet++ also confirms that the VGG-UNet provides a better outcome on the chosen image database [14, 15].

The major part of this study includes:

1. Developing a DL-segmentation proposal to segment the PG from brain MRI slice.
2. Pre-processing the brain MRI slice using tri-level thresholding with KE + BA to enhance the RoI from the chosen MRI slice.

3. Comparing the performance of VGG-UNet with, UNet, UNet+, and UNet++.

The other part of this work is arranged as follows: Section 2 presents the context, Section 3 demonstrates the methodology, Section 4 and 5 shows the result and conclusion.

## 2 Context

In human, the PG plays a vital role and its abnormality will lead to various health issues. One of the common issues which causes the PG is the Pituitary Adenoma (PA). Hence, necessary medication is necessary when an individual is the Pituitary Adenoma (PA), a benign tumor arising from the PG. The PA can extensively impact hormone production and lead to a range of health issues based on the harshness.

Due to its significance, a number of PG segmentation and assessment procedures are discussed in the literature. The earlier works implemented the chosen segmentation process to extract PG from the brain MRI slice and the extracted RoI is then examined by the doctor to plan the necessary treatment to reduce the harshness of the PA.

In the literature, a number of segmentation methods are proposed to extract the PG from the MRI slice for further examination. The earlier works implemented the chosen segmentation technique to effectively extract the PG [16, 17] and a detailed review on the existing methods can be found in [8]. The proposed work aims to implement a DL-model to effectively extract the PG with better accuracy to confirm its clinical significance.

## 3 Methodology

From the image processing literature, it can be observed that the choice of a particular methodology depends on the image to be examined and the complexity of the task. Compared to the methods employed for the brain tumor segmentation from the brain MRI, the extraction of the PG is quite complex, since the visibility of the RoI is poor. Hence, this research work proposed a DL-tool and to further improve the segmentation accuracy, the chosen MRI slices are pre-processed using a tri-level thresholding based on KE + BA.

Figure 1 depicts the proposed technique to successfully segment the RoI from MRI slice. The proposed scheme consists of two phases: phase 1 depicts the techniques executed to get the necessary 2D MRI slice for the examination. The necessary image and the mask is collected from the data repository [18] and then the ITK-snap is then considered for 3D to 2D conversion and the sagittal-plane MRI slice is extracted and considered for the examination. Before the analysis, every image and the mask are resized to $224 \times 224 \times 1$ pixels. Further, the chosen images

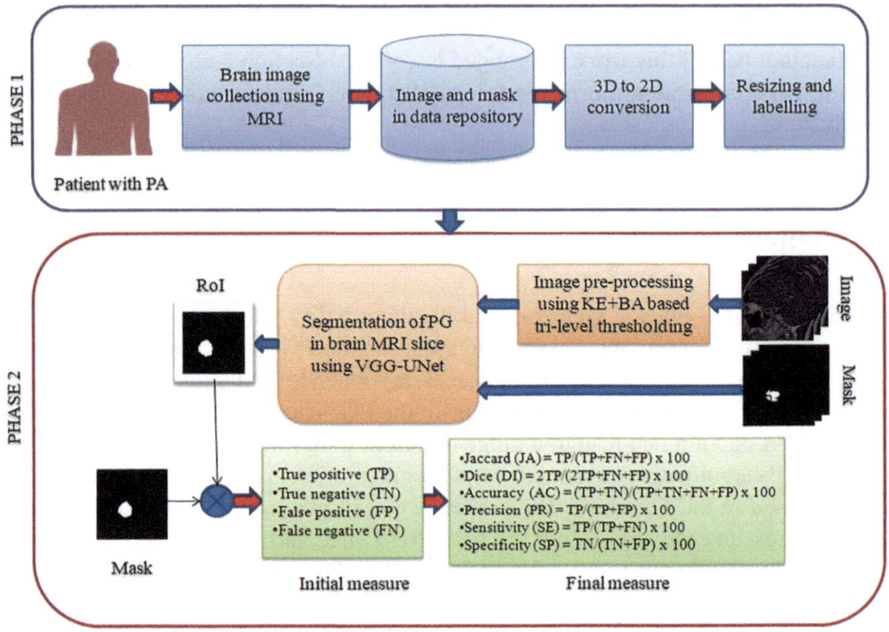


**Fig. 1** Structure of the proposed PG extraction and evaluation procedure using VGG-UNet model

are pre-processed with KE + BA to improve the visibility of the PG. Finally, these images are considered to train and verify the merit of VGG-UNet.

The second phase of this tool implements the DL-segmentation procedure to extract the RoI with a better accuracy. Comparison of RoI with mask provides necessary image. The pixel-wise comparison implemented between the RoI and mask presents the necessary measures like TP, TN, FP, and FN as shown in figure. From these values, the necessary image metrics are then computed. This procedure is then implemented for all the test images (100 slices), and the overall performance of the executed methodology is verified using the mean ± standard deviation (M ± SD) of the final evaluation metrics considered in this research.

## *3.1 Brain MRI Database*

The clinical significance of the automated brain MRI examination task depends mainly on the choice of the image database. This work considered the clinical grade MRI image available in [18] for the investigation. This database consists of the 3D MRI data of 55 volunteers along with the associated binary mask.

After collecting the image and the mask, 2D slices are extracted using the ITK-snap and extracted image is then evaluated. This MRI dataset consists of

**Table 1** The brain MRI data considered in the PG examination task

| | Number of images | | | |
|---|---|---|---|---|
| Data (224 × 224 × 1) | Total | Training | Validation | Testing |
| Image | 600 | 400 | 100 | 100 |
| Mask | 600 | 400 | 100 | 100 |

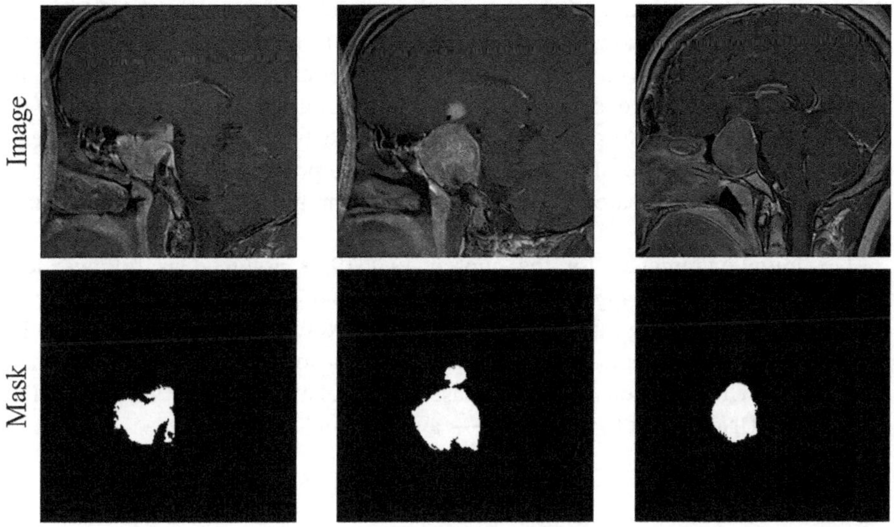


**Fig. 2** The sample image and the mask of the chosen brain MRI database

a complete sagittal-plane and the cropped axial-and coronal-plane. Hence, this research considered only the sagittal-plane 2D MRI slice for the examination. These images are then resized to 224 × 224 × 1 pixels before implementing the proposed technique to examine the PG.

The total number of images considered in this research is depicted in Table 1. The sample image and the mask of this database are shown in Fig. 2. From the image, it can be noted that the visibility of the PG is less and its pixel intensity is similar to the bone region present in the MRI slice. Hence, identification and extraction of the PG from the chosen image needs some appropriate technique and hence, this work implemented the segmentation after pre-processing the MRI slices with KE + BA-based approach.

## *3.2 VGG-UNet Model*

Extraction and evaluation of the RoI is an essential practice in the chosen image examination task. In the literature, a number of traditional and modern methods are discussed for accurate RoI segmentation [19, 20]. Most of the traditional methods are semi-automatic procedures and needs a manual operator to initiate the process [21]. Hence, in recent days, the automatic segmentation approach based on the DL-technique is widely considered, and these methods work well on a number of gray and RGB scale images [22].

This research considered the VGG-UNet model discussed in [13] to extract the PG from the chosen brain MRI slice. In this model, the pre-trained VGG19 acts as the backbone for encoder-decoder, and it helps to extract the PG with a better accuracy. The complete structure of the VGG-UNet can be found in [13] and the encoder section helps to convert the image pixel into features and the decoder converts the features into pixel which supports in extracting the PG with a better accuracy. The outcome of this segmentation helps to provide a binary section of the RoI, which is then compared with its mask to confirm the merit of the implemented scheme.

The initial parameters for the chosen VGG19 is assigned as follows: total epochs = 100, monitoring metrics = accuracy and loss, pooling = Max, optimizer = Adam, learning rate = 1e-4, activation = ReLu, and batch size = 8. Other essential information about the VGG19 model can also be found in [23]. The merit of the VGG-UNet is then confirmed based on a relative analysis with other DL-models like UNet, UNet+, and UNet++ available in [24, 25].

### *3.3 Image Pre-processing*

Image enhancement is one of the common practices considered to enhance the information in the chosen test images. Image thresholding is one of the enhancement technique considered to improve the pixel information in the chosen image and the multi-level thresholding is a common process to identify and enhance the image based on the pixel-grouping concept [26]. Due to its merit, a number of image examination tasks considered the thresholding procedure to improve the pixel information in gray/RGB scale images [27]. Entropy-supported image examination is a common practice and hence, this research considered the KE's tri-level thresholding to improve the PG region.

This work implemented the BA-based tri-level thresholding with the KE and the necessary information about the implemented BA can be found in [28]. Proposed study employed the Brownian-walk BA (BBA) discussed by Rajinikanth et al. (2024) to identify the optimal threshold which maximizes the KE for the chosen image. The initial algorithm parameters for this BA are assigned as in [29], and this work helps to enhance the PG from the chosen image and the attained result is represented in Fig. 3. Figure 3a, b shows the chosen image and the mask. Figure 3c shows the pre-processed image with the KE + BA and this confirms that the visibility of the PG is better in this image compared to Fig. 3a. This region is then extracted with the VGG-UNet model and considered for evaluation.

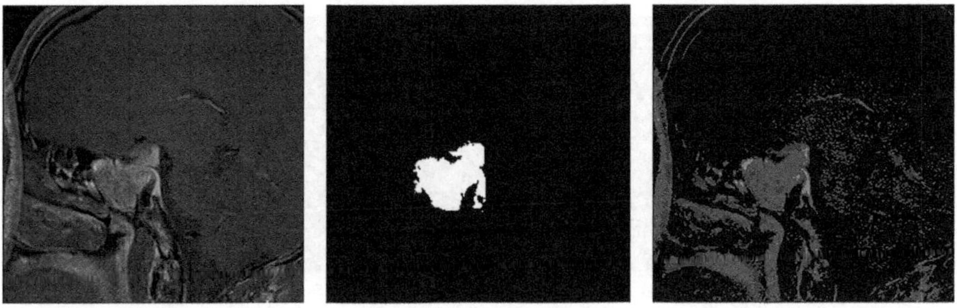


**Fig. 3** Implementing the KE + BA-based image thresholding to enhance the chosen brain MRI slice. (**a**) Unprocessed. (**b**) Mask. (**c**) Pre-processed

### *3.4 Performance Evaluation*

Performance of developed system is verified using the achieved performance metrics. To compute these metrics, a pixel-wise comparison among the RoI and the mask is executed to get the initial metrics like TP, TN, FP, and FN. From these values, the final metrics like JA, DI, AC, PR, SE, and SP are then computed and based on these values; the merit of the developed scheme on the chosen database is confirmed [30].

## 4 Results and Discussion

This section of the research presents the experimental investigation and its outcome achieved on the chosen brain MRI database.

Initially, the VGG-UNet model's segmentation performance is tested using the unprocessed MRI slices and the achieved results are recorded, and this process is then repeated using other chosen DL-segmentation methods chosen in this study. The results of this experiment confirms that the VGG-UNet provided a lesser values of JA (>85%), DI (>93%), and AC (>93%) during the PG segmentation and evaluation task.

To achieve a better result, the proposed work is repeated using the pre-processed images and the achieved results confirms an improvement in the chosen performance metrics. Figure 4 presents the training and testing result achieved with VGG-UNet for a chosen epoch value of 100. Figure 4a shows the combination of the image and mask during training process. Figure 4b, c shows the accuracy Vs epoch and loss Vs epoch achieved with this process, and Fig. 4d presents the testing result for a chosen MRI slice. These confirm that the obtained result with the VGG-UNet is fine, and the similar process is repeated with all the test images (100 numbers) considered in this study. Similar process is repeated for other chosen DL-models

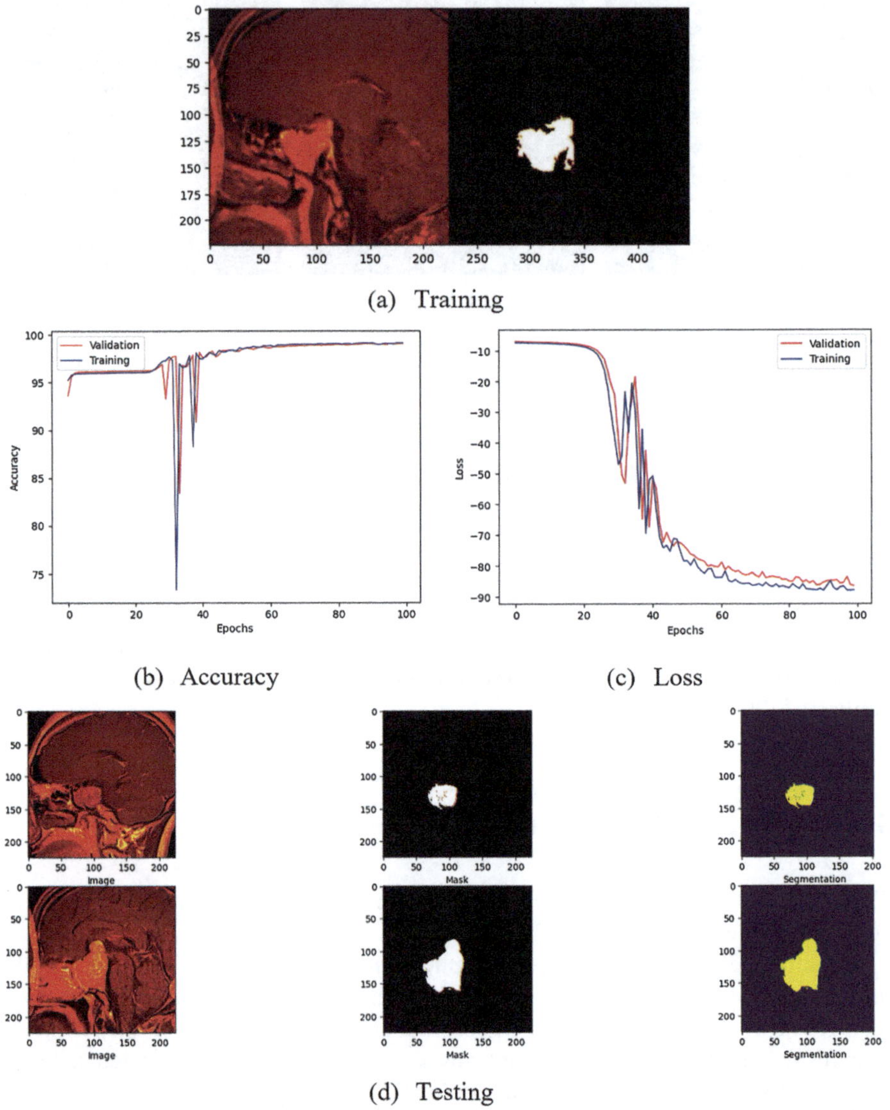


**Fig. 4** Experimental outcome achieved with VGG-UNet for a chosen epoch of 100. (**a**) Training. (**b**) Accuracy. (**c**) Loss. (**d**) Testing

and the achieved results are illustrated in Fig. 5. Figure 5a, b shows the chosen sample image and mask. Figure 5c–f shows the segmentation result with the chosen DL-models.

The overall performance metrics achieved with the testing data is recorded as in Table 2, and these results confirm that the performance of the VGG-UNet

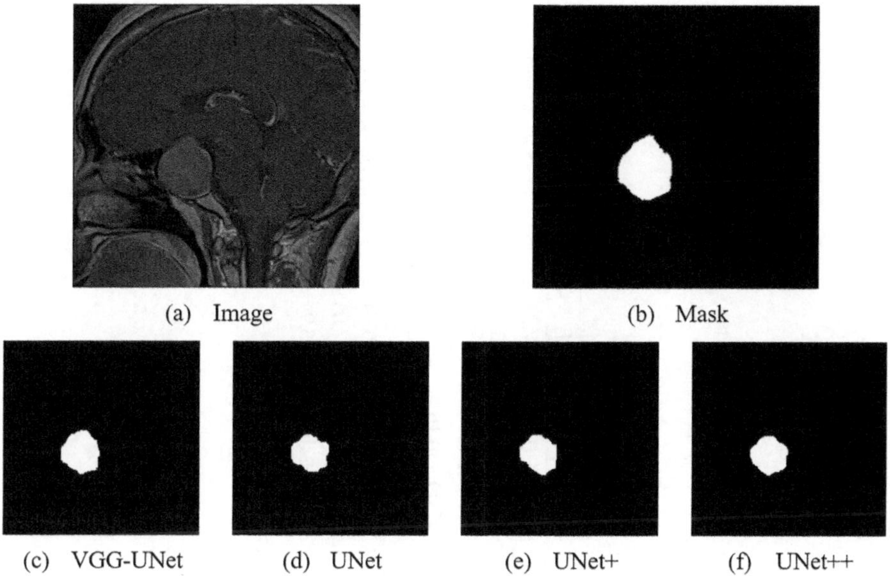


**Fig. 5** Various DL-model results achieved for a chosen test MRI slice. (**a**) Image. (**b**) Mask. (**c**) VGG-UNet. (**d**) UNet. (**e**) UNet+. (**f**) UNet++

model is better compared to other DL-models employed in this study. The overall outcome achieved with the pre-processed image is better compared to unprocessed data-based result, and this confirms that the thresholding followed by the DL-segmentation helps to obtain a better outcome on the chosen brain MRI examination task.

To assess the overall validity of the obtained result, a Glyph-plot is generated utilizing the data from Table 2, and the resultant output is illustrated in Fig. 6. Figure 6a displays the results for unprocessed data, whereas Fig. 6b illustrates the results for pre-processed data. The VGG-UNet facilitated superior results for these two datasets.

This research implemented a DL-segmentation method on the clinical grade MRI slice, and this technique helped to achieve a better result during the PG segmentation task. This technique helped to achieve an accuracy of >97% on the chosen model. In the future, this value can be improved by considering other DL-segmentation methods in the literature. Further, the pre-processing based on the KE can also be replaced with other entropy-based methods found in the literature.

**Table 2** The overall performance metrics computed with testing results

| Image | Model | JA | DI | AC | PR | SE | SP |
|---|---|---|---|---|---|---|---|
| Unprocessed | VGG-UNet | 85.13 ± 1.92 | 93.18 ± 0.13 | 93.85 ± 1.05 | 93.46 ± 0.16 | 92.74 ± 0.25 | 93.07 ± 1.15 |
| | UNet | 84.75 ± 1.16 | 92.76 ± 1.72 | 92.74 ± 0.82 | 92.88 ± 1.03 | 92.41 ± 0.61 | 92.65 ± 0.22 |
| | UNet+ | 85.11 ± 1.15 | 92.84 ± 0.41 | 93.17 ± 0.31 | 92.90 ± 0.27 | 92.07 ± 0.19 | 92.49 ± 0.22 |
| | UNet++ | 84.89 ± 0.17 | 92.91 ± 0.22 | 93.28 ± 0.35 | 92.79 ± 0.16 | 92.18 ± 0.14 | 92.83 ± 0.19 |
| Pre-processed | VGG-UNet | 91.37 ± 0.14 | 96.83 ± 0.04 | 97.08 ± 0.02 | 96.88 ± 0.13 | 97.21 ± 0.07 | 96.95 ± 0.03 |
| | UNet | 90.74 ± 0.18 | 95.73 ± 0.13 | 96.59 ± 0.13 | 96.15 ± 0.14 | 96.74 ± 0.15 | 96.42 ± 0.13 |
| | UNet+ | 91.03 ± 0.16 | 95.98 ± 0.08 | 96.42 ± 0.08 | 96.42 ± 0.21 | 97.05 ± 0.08 | 96.38 ± 0.06 |
| | UNet++ | 91.16 ± 0.11 | 96.33 ± 0.12 | 96.81 ± 0.05 | 96.38 ± 0.18 | 96.94 ± 0.11 | 96.61 ± 0.04 |

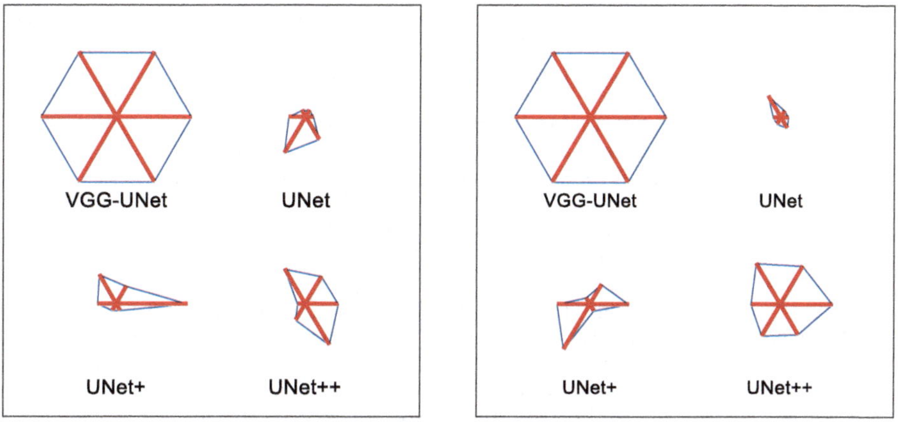


(a) Unprocessed data (b) Pre-processed data

**Fig. 6** Glyph-plot to verify the overall performance. (**a**) Unprocessed data. (**b**) Pre-processed data

# 5 Conclusion

Recently, several AI algorithms have been developed to analyze disease-related information inside selected medical images, and then evaluated by physicians for treatment planning and implementation. This study presented a way for extracting and assessing the PG from the selected brain MRI slice. This study utilized the VGG-UNet, constructed on the pre-trained VGG19 backbone, to attain improved segmentation outcomes. Upon acquiring the requisite 2D MRI slices in the sagittal-plane, each picture is shrunk to $224 \times 224 \times 1$ pixels, and these images are subsequently utilized to train and evaluate the efficacy of the selected deep learning models. This work also considered the image pre-processing based on the tri-level thresholding implemented with KE + BA to enhance the RoI from the chosen MRI slice. The enhanced image helped to provide the better segmentation outcome for the chosen DL-models of this research. The performance of the developed scheme is tested on both the unprocessed and pre-processed images, and the overall result by the pre-processed image is better and it helped to achieve a segmentation accuracy of >97% on the chosen MRI database. In the future, the performance of the proposed scheme can be improved by replacing the KE-based thresholding with other entropy-based methods found in the literature.

# References

1. Castiglioni, I., Rundo, L., Codari, M., Di Leo, G., Salvatore, C., Interlenghi, M., Sardanelli, F.: AI applications to medical images: from machine learning to deep learning. Phys. Med. **83**, 9–24 (2021)

2. Gheisari, M., Ebrahimzadeh, F., Rahimi, M., Moazzamigodarzi, M., Liu, Y., Dutta Pramanik, P.K., Kosari, S.: Deep learning: applications, architectures, models, tools, and frameworks: a comprehensive survey. CAAI Trans. Intell. Technol. **8**(3), 581–606 (2023)
3. Atmakuru, A., Chakraborty, S., Faust, O., Salvi, M., Barua, P.D., Molinari, F., Homaira, N.: Deep learning in radiology for lung cancer diagnostics: a systematic review of classification, segmentation, and predictive modeling techniques. Expert Syst. Appl. **255**, 124665 (2024)
4. Mahmood, T., Rehman, A., Saba, T., Nadeem, L., Bahaj, S.A.O.: Recent advancements and future prospects in active deep learning for medical image segmentation and classification. IEEE Access. **11**, 113623 (2023). https://doi.org/10.1109/ACCESS.2023.3313977
5. Archana, R., Jeevaraj, P.E.: Deep learning models for digital image processing: a review. Artif. Intell. Rev. **57**(1), 11 (2024)
6. Kim, M., Kim, H.S., Kim, H.J., Park, J.E., Park, S.Y., Kim, Y.H., Lebel, M.R.: Thin-slice pituitary MRI with deep learning–based reconstruction: diagnostic performance in a postoperative setting. Radiology. **298**(1), 114–122 (2021)
7. Staartjes, V.E., Serra, C., Muscas, G., Maldaner, N., Akeret, K., van Niftrik, C.H., Regli, L.: Utility of deep neural networks in predicting gross-total resection after transsphenoidal surgery for pituitary adenoma: a pilot study. Neurosurg. Focus. **45**(5), E12 (2018)
8. Saha, A., Tso, S., Rabski, J., Sadeghian, A., Cusimano, M.D.: Machine learning applications in imaging analysis for patients with pituitary tumors: a review of the current literature and future directions. Pituitary. **23**, 273–293 (2020)
9. Choi, U.S., Sung, Y.W., Ogawa, S.: deepPGSegNet: MRI-based pituitary gland segmentation using deep learning. Front. Endocrinol. **15**, 1338743 (2024)
10. Wang, H., Zhang, W., Li, S., Fan, Y., Feng, M., Wang, R.: Development and evaluation of deep learning-based automated segmentation of pituitary adenoma in clinical task. J. Clin. Endocrinol. Metabol. **106**(9), 2535–2546 (2021)
11. http://www.itksnap.org/pmwiki/pmwiki.php
12. Yushkevich, P.A., Piven, J., Hazlett, H.C., Smith, R.G., Ho, S., Gee, J.C., Gerig, G.: User-guided 3D active contour segmentation of anatomical structures: significantly improved efficiency and reliability. NeuroImage. **31**(3), 1116–1128 (2006)
13. Khan, M.A., Rajinikanth, V., Satapathy, S.C., Taniar, D., Mohanty, J.R., Tariq, U., Damaševičius, R.: VGG19 network assisted joint segmentation and classification of lung nodules in CT images. Diagnostics. **11**(12), 2208 (2021)
14. Mohan, R., Prabha, S., Rajinikanth, V.: Mitochondrial network segmentation from fluorescence microscopy images using VGG-UNet. In: 2024 Tenth International Conference on Bio Signals, Images, and Instrumentation (ICBSII), pp. 1–4. IEEE (2024, March)
15. Mohan, R., Arunmozhi, R., Rajinikanth, V.: Deep-learning segmentation and recognition of tooth in Thresholded panoramic X-ray. In: 2023 Winter Summit on Smart Computing and Networks (WiSSCoN), pp. 1–5 (2023)
16. Shu, X., Zhou, Y., Li, F., Zhou, T., Meng, X., Wang, F., Xu, B.: Three-dimensional semantic segmentation of pituitary adenomas based on the deep learning framework-nnU-net: a clinical perspective. Micromachines. **12**(12), 1473 (2021)
17. Dai, C., Sun, B., Wang, R., Kang, J.: The application of artificial intelligence and machine learning in pituitary adenomas. Front. Oncol. **11**, 784819 (2021)
18. Wang, M., Wang, H.: Pituitary Adenoma MRI Segmentation Dataset. IEEE Dataport (2023). https://doi.org/10.21227/66ks-t035
19. Ramadasan, S., Vijayakumar, K., Prabha, S., Karthickeien, E.: Forest region extraction and evaluation from satellite images using CNN segmentation. In: 2024 International Conference on Advances in Computing, Communication and Applied Informatics (ACCAI), pp. 1–5. IEEE (2024, May)
20. Mohan, R., Ramadasan, S., Vijayakumar, K.: Development of a novel UDL-model for dental abnormality detection from digital photographs. In: 2024 Ninth International Conference on Science Technology Engineering and Mathematics (ICONSTEM), pp. 1–6. IEEE (2024, April)

21. González-Villà, S., Oliver, A., Valverde, S., Wang, L., Zwiggelaar, R., Lladó, X.: A review on brain structures segmentation in magnetic resonance imaging. Artif. Intell. Med. **73**, 45–69 (2016)
22. Minaee, S., Boykov, Y., Porikli, F., Plaza, A., Kehtarnavaz, N., Terzopoulos, D.: Image segmentation using deep learning: a survey. IEEE Trans. Pattern Anal. Mach. Intell. **44**(7), 3523–3542 (2021)
23. https://keras.io/api/applications/vgg/
24. Bousias Alexakis, E., Armenakis, C.: Evaluation of UNet and UNet++ architectures in high resolution image change detection applications. Int. Arch. Photogramm. Remote. Sens. Spat. Inf. Sci. **43**, 1507–1514 (2020)
25. Mohan, R., Abouhawwash, M., Arunmozhi, R., Rajinikanth, V.: Automatic segmentation of underwater images with Shannon's thresholding and UNet variants. In: 2023 Winter Summit on Smart Computing and Networks (WiSSCoN), pp. 1–6. IEEE (2023)
26. Yazid, H., Basah, S.N., Rahim, S.A., Safar, M.J.A., Basaruddin, K.S.: Performance analysis of entropy thresholding for successful image segmentation. Multimed. Tools Appl. **81**(5), 6433–6450 (2022)
27. Jardim, S., António, J., Mora, C.: Image thresholding approaches for medical image segmentation-short literature review. Proc. Comput. Sci. **219**, 1485–1492 (2023)
28. Arora, S., Singh, S.: Butterfly optimization algorithm: a novel approach for global optimization. Soft. Comput. **23**, 715–734 (2019)
29. Rajinikanth, V., Biju, R., Mittal, N., Mittal, V., Askar, S.S., Abouhawwash, M.: COVID-19 detection in lung CT slices using Brownian-butterfly-algorithm optimized lightweight deep features. Heliyon. **10**(5), e27509 (2024)
30. Agrawal, R., Sharma, M.: Review of segmentation methods for brain tissue with magnetic resonance images. Int. J. Comput. Network Inform. Secur. **6**(4), 55 (2014)

# Effects of Communication Channels on Explainable Food Recommendation Systems

**Berkecan Koçyiğit, Berk Buzcu, Davide Calvaresi, Zehra Kesemen, Joris Hulstijn, Igor Tchappi, and Reyhan Aydoğan**

## 1 Introduction

With the increasing integration of artificial intelligence (AI) technologies into everyday life, there is a growing demand for systems that not only provide precise recommendations but can also deliver transparent, understandable explanations [4]. Explainable AI (XAI) has been employed to address this need by making complex algorithmic processes more accessible to the end user, thus promoting user trust and understanding [2]. However, explainability alone does not compose the effectiveness of the content, but also how the explanations are communicated plays a crucial role in how users perceive, engage with, and ultimately accept these recommendations [7]. Previous research has shown that physical presence of a robot

B. Koçyiğit (✉) · Z. Kesemen
Özyeğin University, Nişantepe, İstanbul, Turkey
e-mail: berkecan.kocyigit@ozu.edu.tr; zehra.kesemen@ozu.edu.tr

B. Buzcu · D. Calvaresi
HES-SO Valais-Wallis, Sierre, Switzerland
e-mail: berk.buzcu@ozu.edu.tr; davide.calvaresi@hevs.ch

J. Hulstijn
Utrecht University, Utrecht, Netherlands
e-mail: j.hulstijn@uu.nl

I. Tchappi
University of Luxembourg, Esch-sur-Alzette, Luxembourg
e-mail: igor.tchappi@uni.lu

R. Aydoğan
Özyeğin University, Nişantepe, İstanbul, Turkey

Delft University of Technology (TU Delft), Delft, Netherlands
e-mail: reyhan.aydogan@ozyegin.edu.tr

 
B. U. Töreyin et al. (eds.), *8th EAI International Conference on Robotic Sensor Networks*, EAI/Springer Innovations in Communication and Computing,
https://doi.org/10.1007/978-3-031-92143-8_5

can increase user engagement by providing a richer, more interactive experience that fosters a sense of connectedness [13]. Moreover, there is a notable effect of different communication channels (e.g., physical embodiment versus purely digital) for users' within personalized meal recommendation systems [15]. However, there is still a gap in understanding how embodiment translates to practical applications, particularly in scenarios where complex, personalized information needs to be communicated efficiently and effectively. To investigate this, we have developed two different interface configurations: (i) one based on a standard web interface accessible with a tablet and (ii) another that complements this tablet interface with a humanoid robot capable of providing food recommendations along with explanations through speech and gestures. The robot introduces physical embodiment and aims to create a more human-like interaction by utilizing multimodal communication. By examining user reactions and interactions with both systems, we aim to determine whether physical embodiment influences user engagement and affects perceptions of the recommendations, including acceptance and satisfaction with the system.

The remainder of this chapter is organized as follows. Section 2 presents related work. Section 3 describes the setup of the study, detailing the components and configurations used. Section 4 provides an evaluation and discussion of the experimental results. Section 5 concludes the paper and suggests future research.

## 2 Related Work

The increasing adoption of artificial intelligence in day-to-day applications has developed an interest in understanding the most effective ways to deliver personalized and explainable recommendations. In particular, research has focused on the communication mediums: the means of presenting the recommendations and corresponding explanations to the user. Various communication mediums, such as web-based systems [4, 9, 11] and mobile interfaces [3, 14, 16], have been studied for their effects on recommender systems. Moving beyond these conventional mediums, recent research has focused on robots as a promising alternative for recommendation delivery, particularly in scenarios where physical presence may improve effectiveness with enhanced user engagement.

Kamei et al. [9] experiment with robots recommending items based on customers' purchasing behaviors tracked by networked sensors in a shop. Using multiple robots, they show that participants lingered longer near shelves when robots interacted with them, often mirroring previous purchasing behaviors. Herse et al. [6] investigate embodiment's role in social robots' persuasiveness within a service setting. They conduct an experiment comparing a human, robot, and kiosk for restaurant recommendations. The results reveal that human-like embodiment enhances persuasiveness, though only with specific recommendation phrasing. They suggest human-like traits can boost recommendation impact, with a dependence on the choice of language. Sakai et al. [15] conduct an experiment with robots in both virtual and embodied setting with visitors in conversations about food preferences before recommending dishes. While behavioral differences were minimal, the study found that physical robots notably improved satisfaction and agreement with recommendations, suggesting that embodiment has a positive impact on user engagement. Interestingly, in our setting, user engagement is higher when participants interact solely with a conventional tablet interface compared to when they interact with the robot and tablet together.

Embodiment research also takes place in education and negotiation. Çakan et al. [5] study the difference in negotiation styles, when interacting with virtual robots and physically embodied robots. Human participants took the negotiation more seriously against physically embodied robots and made more collaborative moves in the virtual robot setting. Survey responses indicate that participants perceived the robot as more human-like when it is physically embodied.

Köse et al. [10] investigate embodiment and gesture effects in child-robot interactions with a child-sized humanoid robot in an interactive drumming game. They studied three forms of communication: (i) direct interaction with an embodied system, (ii) a sound only, and (iii) a real-time virtual avatar. Through mixed-design measures, data from the experiments reveal that physical embodiment, especially with gestures, significantly enhances interaction quality, performance, and user enjoyment. Ultimately, a physically embodied robot with gesture capabilities enhances the perceived intelligence of the agent and improves engagement in human-computer interactions. In a similar setup, [12] compared various forms of embodied negotiation opponents, to test the hypothesis that the appearance of the robots would change participants' impressions and attitudes. Ultimately, it did not alter the final session results.

In contrast, our study focuses on recommender systems. We examine how explainable recommender systems are experienced through different interfaces, namely a conventional tablet and a robot combined with a tablet, and how these interfaces affect user engagement, satisfaction, and overall system performance.

## 3 Research Setup and Hypotheses

In this chapter, we utilize the personalized explainable recommendation framework developed by Buzcu et al. [4], designed to support users in making food choices that meet both health requirements and personal preferences. The interaction between users and system is governed by the protocol illustrated in Fig. 1. The interaction is initiated by a user request consisting of several constraints, and the system provides an explanation along with the given recommendation. The user can provide feedback about recommended recipes and their corresponding explanations or end the interaction by either accepting or terminating the session without any agreement. With the given feedback, the recommendation strategy revises its recommendation, thus engaging in a subtle negotiation. This process lasts until they reach a termination condition.

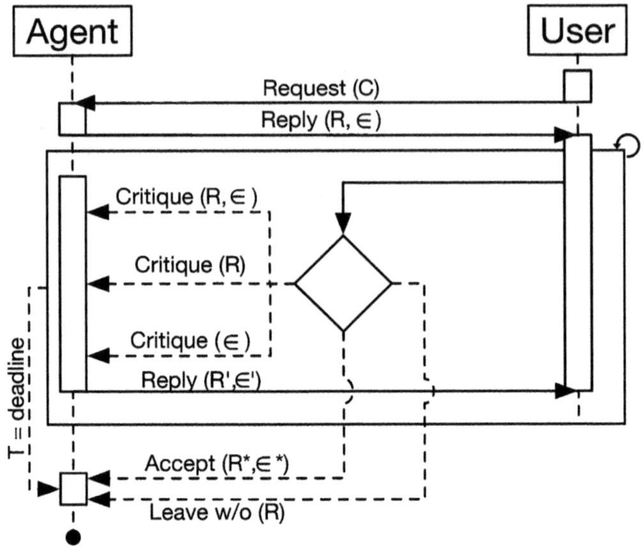


**Fig. 1** FIPA description of the negotiation protocol where C corresponds to the user constraints, R is a recipe recommended by the agent, and $\epsilon$ is an explanation that comes with the recipe

The recommendation generation process begins with filtering. The system processes each user's dietary restrictions, allergies, and ingredient preferences and filters out items that do not align directly. For instance, vegan users will only receive plant-based recipes. This filtering is enabled by an RDF ontology-based database, which defines complex relationships between food entities and user dietary restrictions to refine the available choices to meet each group of users' requirements. The system evaluates the remaining items using a multicriteria utility function, which rates each option based on (i) *nutritional value*, (ii) *active metabolic rate (AMR)*, and (iii) *user satisfaction scores*. Ultimately, these scores are combined via their respective weights to calculate an overall additive utility score for each item, which determines a heuristic ranking among the filtered options. The highest-ranked item that has not been recently recommended is then selected to be recommended. The system generates post hoc explanations to clarify the reasoning behind the choice from the selected recommendation. This step includes two types of explanations:

- **Item and User-Based Explanations**: Using historical data, a decision tree is constructed to identify the key features that influenced the recommendation in a post hoc manner.
- **Contrastive Explanations**: The system compares the chosen item with a similar alternative that did not meet the decision criteria or was filtered otherwise. This contrast is used to show why the recommended item is a more preferable option by highlighting features.

So far, [4] has implemented the recommendation framework with a web interface. Figure 2 shows the web interface as in the recommendation state. The web interface provides a medium for users to interact with the recommender system with the means of allowed actions in the protocol (see Fig. 1). The interface allows users to see recommended recipes in detail (e.g., name of the recipe, recipe ingredients, nutritional information) along with the explanation of the recommendation and give feedback about both recommendation and explanation. The recipe feedback section consists of options and the recipe ingredients. Users can pick the appropriate option (e.g., "I don't like...," "I ate the following recently...") with preferred ingredients. The explanation feedback can also be given by selecting the preferred option in the explanation feedback section. Users have three options to respond to the given recipe: (i) *accepting the recommendation and ending the session*, (ii) *submitting the feedback and requesting new recommendation*, and (iii) *terminating the session*.

Apart from the content of the explanations, how they are delivered also plays a crucial role in their effectiveness. Therefore, it is essential to investigate which communication medium/modality would establish more effective interaction with the user when building personalized explainable systems. This chapter mainly focuses on personalized explainable food recommendation systems and examines the effect of communication medium/modality (i.e., the effect of physical embodiment and textual/speech-based communication). Consequently, this chapter aims to investigate the following research hypothesis through user studies. For sake of readability, hereafter, tablet (touch-screen-based devices) only is referred to as a conventional interface:

- **Hypothesis 1 (H1)**: Incorporating a physically embodied robot into an explainable food recommendation system affects user engagement. *(Metric: Amount of user feedback)*
- **Hypothesis 2 (H2)**: Incorporating a physically embodied robot into a conventional interface shifts the original recommendation perception and acceptance. *(Metric: Recommendation acceptance rate)*
- **Hypothesis 3 (H3)**: Interacting with a physically embodied robot with a conventional interface could result in different satisfaction levels with the personalization of recommendations compared to conventional interfaces. *(Metric: Qualitative user feedback)*

Accordingly, we compare two settings in which the user interacts with the system via only a tablet (only visual and textual data is delivered) in one setting, whereas they interact with a QT robot via speech in addition to the use of a tablet. Note that we utilize the same strategies for recommendation and explanation generation in both settings. Ultimately, we aim to study user engagement, acceptability of the explanations, and communication effectiveness by comparing the interactions with the user when the explanations are delivered by a tablet-based medium or a physical humanoid robot with a tablet interface.

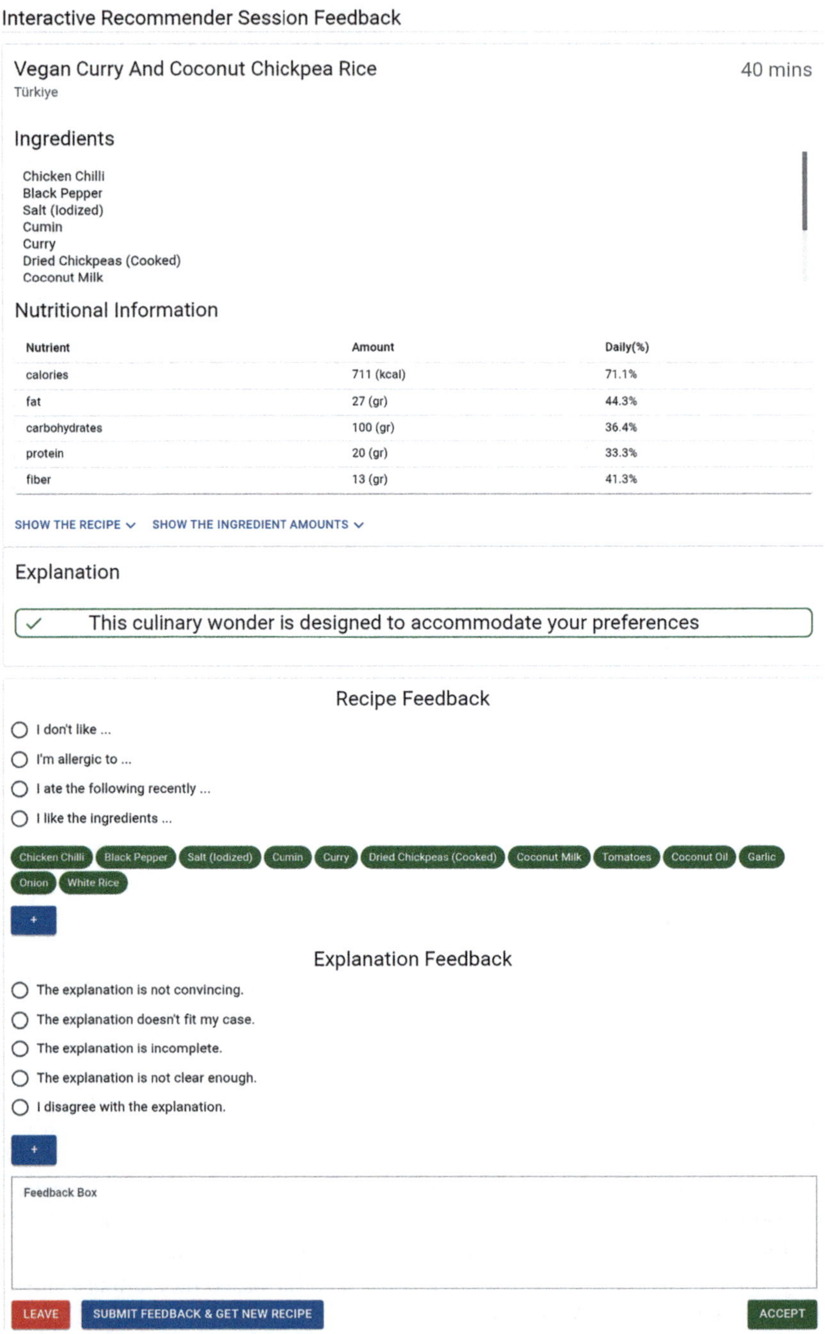


**Fig. 2** Tablet web interface

# 4 Experimental Evaluation

In this section, we first introduce our methodology (Sect. 4.1) and discuss our results with the experimental settings (Sects. 4.2 and 4.3).

## *4.1 Methodology*

In order to test our hypotheses, we utilize a QT robot[1] supporting various interaction services such as text to speech, speech to text, emotions display, and gestures. Human-robot interaction requires speech communication, emotions, and gestures to capture the users' attention and create a more human-like interaction. On the other hand, grasping the provided material within the dialogue and reasoning them may require some visual deliverables. That is, if we provide the same context in the form of visual and textual, it might be easier for a human user to evaluate. While determining what content should be delivered in textual format via tablet and what content to be delivered by the robot via speech, we conducted a pilot user study where the users give feedback about the effectiveness of communication. We observed that users had difficulty in understanding the details of the given recipe or keeping in mind the related details affecting their decisions. We found out that utilizing speech-based dialogue for short explanations and structured user feedback creates a rapport between system and users. Therefore, we employ the speech-based interaction for the following tasks:

- Greeting the user (e.g., "Hello! My name is QT. I am a nutritionist! I am here to help you about your food selection! Can you hear me?" )
- Providing the name of the recipe and its corresponding explanations verbally (e.g., "Great! I can hear you too! Now I recommend you Meatless Potato Meal. Because I think this culinary marvel brings delight as an affordable option. Did you like this recipe?")
- Receiving structured feedback (e.g., "I am curious to hear your thoughts! Would you like to give me a feedback before we continue?" )

On the other hand, the content of the full recipe is displayed on the tablet screen as shown in Fig. 3b so that the users can assess the recommendation. Moreover, the robot also uses gestures and expresses feelings, emotions via facial expressions alongside the recommendations to add human likeliness to the system for a more fluent experience. Furthermore, the user can track the interaction states from the tablet interface by displaying the speech recognition outputs simultaneously and options for structured explanations. It is worth noting that both settings (only tablet versus robot and tablet) provide the same content to the users. The only difference

[1] https://luxai.com/humanoid-social-robot-for-research-and-teaching/

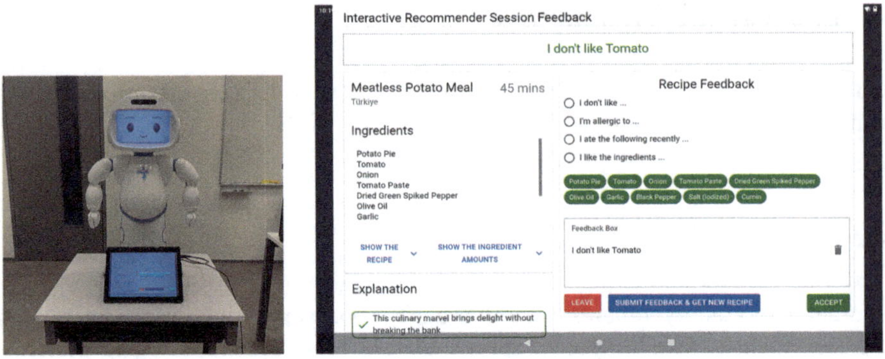


(a) (b)

**Fig. 3** Robot with tablet session setup and tablet interface

is that some of the visible content changes with the interaction state in the robot setting, while all information is displayed on the tablet in the only tablet setting.

## *4.2 Experiment Setup*

The primary objective of our experiment is to analyze the effects of different communication channels on the efficiency of the recommender system. As described above, we have two communication settings of our food recommender: *QT robot with tablet* and *only tablet*. Therefore, our study comprises two subject groups: One only experienced the tablet, whereas the other experienced the robot with the tablet in order to get a food recipe recommendation along with its explanations. We employ between-subject design for our experimental setup to reduce (i) learning effect, (ii) cognitive overload, and (iii) comparison bias so that each participant interacted with only one setting. During our experiment, each participant goes through the steps depicted by Fig. 4:

1. **Preexperiment Survey**: Participants fill out a survey to provide demographic information and their initial perceptions of robots and technology. This survey includes questions about age, gender, education level, and familiarity with technology.
2. **Experiment Sessions**: Participants (i) are informed about the underlying setting through a brief explanatory video, and (ii) practice with a short demo version to familiarize how to interact with the system effectively. Then they start the experiment choose a food preference profile from a pool of predefined profiles (e.g., vegetarian, fast-food lover, sports). According to the chosen profile, the system makes recommendations along with their explanations. Note that the content of the recommendations is generated in the same way for the same

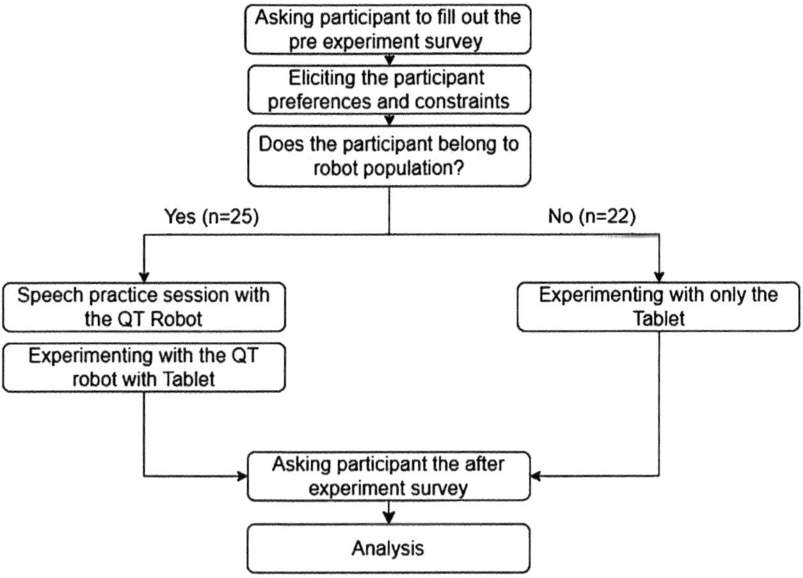


**Fig. 4** Experiment procedure

profile. Only the communication channels are different for different subject groups.

3. **Post-experiment Survey**: After their interaction with the system, they fill out post-experiment survey regarding the quality of the recommendations and explanations as well as the likability of the interaction.
4. **Informal Interviews**: At the end of the experiment, participants are interviewed informally to gather qualitative data on their experiences.

In the experiments, we used a 5-point Likert scale to measure participants' responses. After collecting the data, we assessed the distribution to determine the appropriate statistical tests for analysis. Using the Kolmogorov-Smirnov test of normality, we found that our data did not follow a normal distribution. This lack of normality precludes the use of parametric tests, which generally assume a normal distribution of data. The previous lead to select Mann-Whitney U Test, a nonparametric alternative suitable for comparing differences between two independent groups without assuming normal distribution.

The study was conducted at Özyeğin University and involved 59 participants, including students and employees. Participation was voluntary, with optional credits for social sciences students. In order to validate participants' attention, we included a question inside the preexperiment survey that asks, "If you are paying attention, please select 2." Nine participants failed this test. Moreover, three participants had to end the experiment due to personal circumstances. Thus, 47 participants were evaluated for the experiment. Participants were divided into two groups. The robot

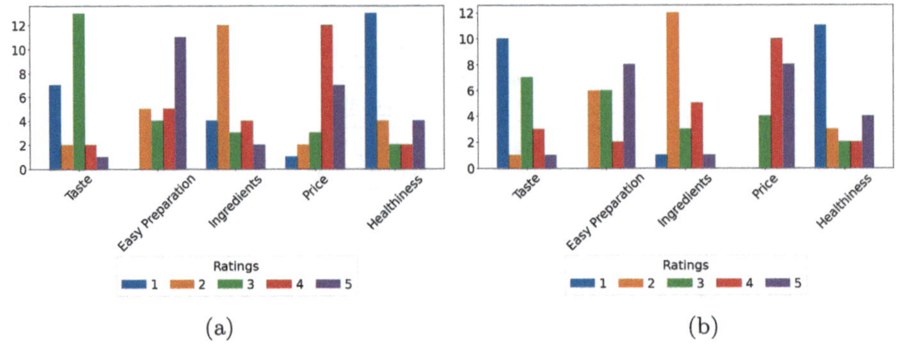


**Fig. 5** Histogram analysis of the preexperiment survey for each group. (**a**) Robot with tablet. (**b**) Tablet only

with tablet group consisted of 25 participants: 15 male, 10 female; 20 bachelor's, 4 master's, and 1 PhD student; 9 aged 18–21, 12 aged 22–25, 3 aged 26–30, and 1 aged 36–40; 13 from engineering, 2 from mathematics, and 10 from social sciences. The only tablet group included 22 participants: 13 male, 9 female; 16 bachelor's, 5 master's, and 1 PhD student; 7 aged 18–21, 13 aged 22–25, and 2 aged 26–30; 13 from engineering, 1 from mathematics, and 8 from social sciences. The university ethics committee approved the study.[2]

Finally, Fig. 5 presents a histogram analysis of the questionnaire results, where participants rated factors on a scale from 1 to 5, with one indicating the highest importance. The analysis reveals that, in both groups, the healthiness of a recipe is considered to be the most vital factor in choosing it (52% for robot with tablet group and 50% for tablet-only group). In contrast, 44% of participants within the robot with tablet group rated ease of preparation as the least important factor, and 36% of the participants voted equally price and easy preparation to be the least important factor. The users' priorities on the decision criteria for choosing different recipes are similar for both groups.

## 4.3 *Results and Discussions*

Within the analysis of self-explanatory systems, success is commonly measured through two categories of metrics: objective and subjective [8]. On one hand, objective metrics are derived from participants' actions during their interactions with the system, including measures like success rate (e.g., percentage of accepted

[2] Participant data is anonymized and securely stored. They can withdraw anytime without consequences. The study follows ethical guidelines to ensure no physical or psychological harm to participants.

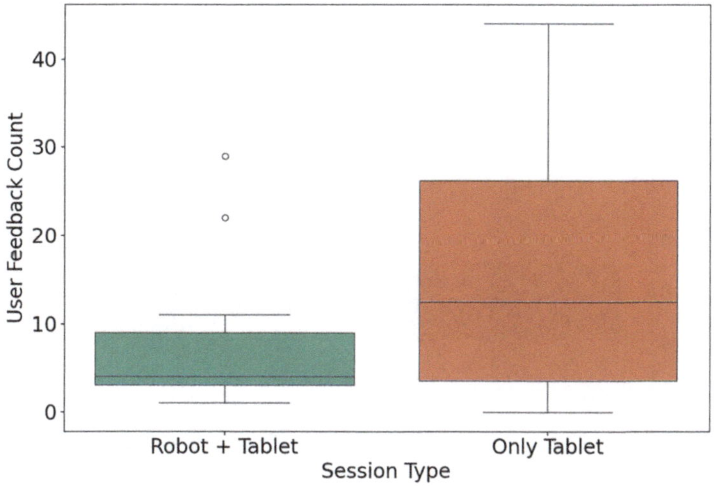


**Fig. 6** Amount of feedback per type of session

recipes), the number of interaction rounds per session, or feedback during sessions. Subjective metrics, on the other hand, are gathered from qualitative data such as the post-experiment surveys (see Fig. 8), where participants rate aspects such as perceived effectiveness, satisfaction, and ease of use. These subjective ratings assess whether the incorporating of the robot affects the user satisfaction with the recommendations, explanations, and the general working of the system.

First and foremost, we consider the users' feedback to the system's recommendation. Recall that the feedback is comprised of any verbal (robot with tablet session) or nonverbal (only tablet session) user response to a given recommendation, which we take as an indicator of engagement levels across groups. Meanwhile, it is difficult to compare the quality of feedback, and we can still consider the amount of feedback where the higher number means the user is engaged more with the system since they actively partake. This gives us a quantifiable measure of user engagement while testing *H1*. Figure 6 shows the boxplot of the feedback counts per group. We note that there is a significant statistical difference between the groups ($p = 0.04 < 0.05$) where the tablet group provided more feedback to the system. We observe that the users engaged more in the tablet-only group in contrast to the robot with tablet group, thus satisfying the *H1* (on average, 14.72 vs. 6.52, respectively).

On the other hand, higher acceptance counts could indicate that users find the recommendations more convincing, so they might be more willing to adopt them. Acceptance counts are defined as the number of times users explicitly agree with a food recipe. Ultimately, the comparative study between the groups would allow us to assess whether physical embodiment affects users' acceptance of the system's recommendations. Our results indicate that the users accepted 68% of the recommendations, whereas tablet acceptance was 95%. Consequently, we note that this supports the *H2* as we found a significant difference among the two groups.

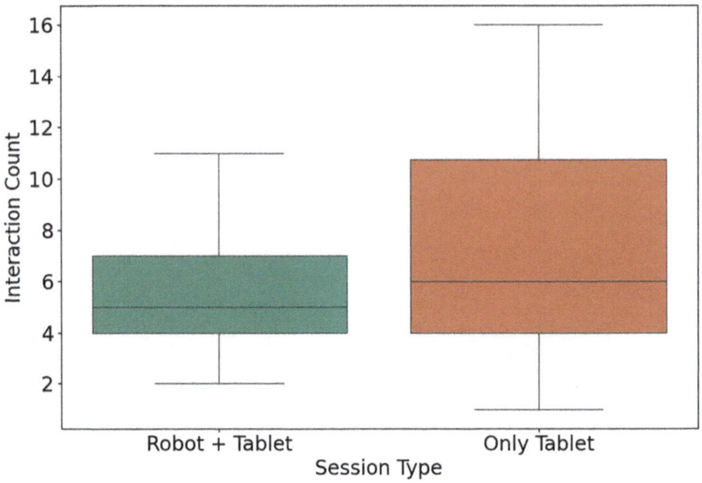


**Fig. 7** Amount of interaction per type of session

The acceptance rate and feedback counts could be related to the fact that robots are scrutinized by humans since embodiment brings additional expectations from the system [1]. This could also signal a cognitive load (i.e., may get tired from interacting with the robot)and more tendency to terminate the interaction.

Finally, we consider the number of turns between the groups, as shown in Fig. 7. Here, we define turns as each time a user responds to a recommendation. As a side observation, we noticed a minor difference between the groups ($p = 0.72$, 6.76 vs. 5.94 interactions on average). This may suggest different engagement levels, potentially influenced by user expectations and the mentioned cognitive load during interactions.

Figure 8 illustrates the results of our post-experiment survey questions with regards to our subjective metrics. We link some of the questions to *H3* (Q1, Q11, Q12, Q13, Q14, and Q16) that assess various aspects of recommendation satisfaction and personalization between the groups. These questions provide insights into participants' perceived relevance and accuracy of the recommendations, which reflects the user satisfaction with the system's communication method when compared between the groups. Q1 ($p = 0.69$, on average, 3.73 vs. 3.80) gauges initial liking for the recommended items, setting a foundation for perceived quality, and Q11 ($p = 0.26$, 3.36 for the tablet only vs. 3.76 for the robot with tablet) assesses alignment with the scenario, highlighting satisfaction with situational appropriateness. Q12 ($p = 0.54$, 3.5 vs. 3.68) and Q13 ($p = 0.20$, 3.54 vs. 3.92) measure participants' perceptions of how accurately and personally the system recognized and incorporated their individual preferences. Finally, Q14 ($p = 0.24$, 3.46 vs. 3.76) and Q16 ($p = 0.03$, 3.41 vs. 4.08) focus on the perceived effectiveness of the personalization process, specifically regarding dietary

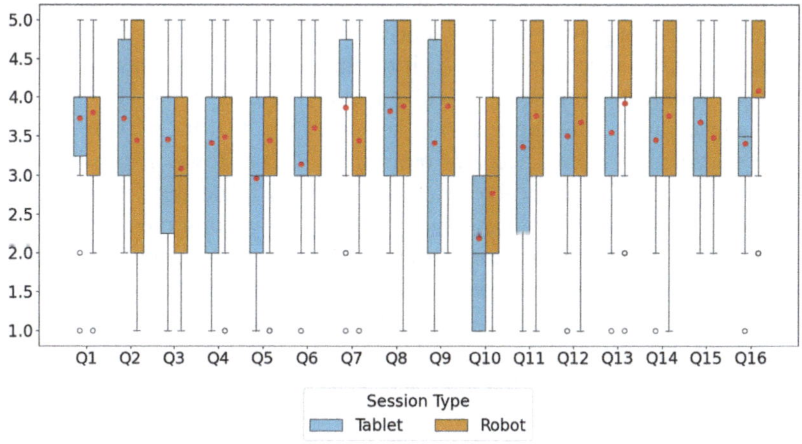


**Q1** I liked the recommended recipes.
**Q2** I enjoyed using the recommender system.
**Q3** Given interface was comfortable to navigate.
**Q4** I liked the explanations that were given.
**Q5** I find the given explanations convincing.
**Q6** I find the given explanations intuitive.
**Q7** The interaction with the recommender was fluent.
**Q8** I enjoyed the interaction with the recommender.
**Q9** I would like to engage in such an interaction in the future.
**Q10** The interaction with the recommender was frustrating for me.
**Q11** The recommendations were right for the given scenario.
**Q12** Rate the food recommendation system's ability to recognize your food preferences accurately.
**Q13** Recommendations were personalized according to my preferences.
**Q14** Rate the effectiveness of the food recommendation system in personalizing food recommendations based on your profile.
**Q15** What is your opinion about the duration of the experiment?
**Q16** How well did the food recommendation system address your specific dietary preferences and constraints?

**Fig. 8** Post-experiment survey results for only tablet and robot with tablet

preferences and constraints. Ultimately, *H3* is only partially supported since we found a significant difference for Q16.

Besides the mentioned questions, we observe no significant difference among the questions. However, we would like to note that there was a higher average for the robot with tablet group among Q4 ($p = 0.92$, 3.41 vs. 3.48), Q5 ($p = 0.16$, 2.95 vs. 3.44), Q6 ($p = 0.09$, 3.13 vs. 3.60). These results could signal that the explanations had a slightly higher impact among the participants when they are delivered in an embodied manner. On the other hand, users reported higher scores in Q2 ($p = 0.66$, 3.73 vs. 3.44) and Q3 ($p = 0.27$, 3.45 vs. 3.08). The frustration with the interaction measured by Q10 ($p = 0.12$, 2.18 vs. 2.76) and higher average for the robot supports our finding with the objective metrics. This indicates that either system was not perceived particularly more interesting than the other.

During our unstructured interviews, it became clear that the design of the state machine communication in the robot with tablet configuration may cause impatience for some participants. This aligned with the survey results for Q10 (Fig. 8) where 7 out of 25 participants gave a score of more than 3. The slower pace of interaction combined with occasional speech recognition issues made the session seem less fluent compared to the tablet-only configuration. It was observed that the simple nature of the tablet-only setup made it easy for users to provide feedback, which explains why engagement remained high. In addition, some participants struggled to engage with the system due to their varying levels of English, which may have impacted their ability to fully engage. These challenges highlight the need for more user-friendly communication strategies and better support for non-native speakers.

## 5 Conclusion and Future Work

This chapter investigated the effects of different communication mediums, a tablet-only interface, and a robot-enhanced setup on the user experience within a personalized food recommendation system. Differences were observed in the frequency of user engagement between the two setups, indicating that the choice of communication medium can significantly impact how actively users participate in interactions. Acceptance levels also varied across the setups, suggesting that communication medium influences users' willingness to adopt the system's recommendations. On the other hand, we looked at how users perceived the system's alignment with their preferences and needs, focusing on their satisfaction with the interaction. The robot-enhanced setup affected user satisfaction minimally. Satisfaction remained largely consistent across both setups, suggesting that physical embodiment alone did not significantly enhance users' contentment or sense of alignment with the system's functionality. These findings emphasize the importance of selecting suitable communication mediums to optimize user experience and perceived effectiveness in recommendation systems. Future designs may benefit from simplifying communication flows in embodied setups to minimize user impatience, enhancing response fluency, and implementing strategies that address language barriers. These improvements could further increase accessibility and engagement across diverse user backgrounds, supporting the development of more intuitive and responsive interactive systems.

**Acknowledgments** This work has been supported by the grant CHIST-ERA-19-XAI-005, and by the Swiss National Science Foundation (G.A. 20CH21_195530), the Luxembourg National Research Fund (G.A. INTER/CHIST/19/14589586), and the Scientific and Research Council of Turkey (TÜBİTAK, G.A. 120N680).

## References


1. Ahmad, M.I., Bernotat, J., Lohan, K.S., Eyssel, F.: Trust and cognitive load during human-robot interaction. ArXiv, abs/1909.05160 (2019). https://api.semanticscholar.org/CorpusID:202558882
2. Arrieta, A.B., Díaz-Rodríguez, N., Del Ser, J., Bennetot, A., Tabik, S., Barbado, A., Garcia, S., Gil-Lopez, S., Molina, D., Benjamins, R., Chatila, R., Herrera, F.: Explainable artificial intelligence (XAI): concepts, taxonomies, opportunities and challenges toward responsible AI. Inf. Fusion **58**, 82–115 (2020). ISSN 1566-2535. https://doi.org/10.1016/j.inffus.2019.12.012
3. Buzcu, B., Pannatier, Y., Aydoğan, R., Schumacher, M.I., Calbimonte, J.P., Calvaresi, D.: A framework for explainable multi-purpose virtual assistants: a nutrition-focused case study. In: Calvaresi, D., Najjar, A., Omicini, A., Carli, R., Ciatto, G., Aydogan, R., Hulstijn, J., Främling, K. (eds.) Explainable and Transparent AI and Multi-Agent Systems - 6th International Workshop, EXTRAAMAS 2024, Revised Selected Papers, Lecture Notes in Computer Science (including subseries Lecture Notes in Artificial Intelligence and Lecture Notes in Bioinformatics), pp. 58–78. Springer, Berlin (2024)
4. Buzcu, B., Tessa, M., Tchappi, I., Najjar, A., Hulstijn, J., Calvaresi, D., Aydogan, R.: Towards interactive explanation-based nutrition virtual coaching systems. Auton. Agent. Multi-Agent Syst. **38**(5), (2024). https://doi.org/10.1007/s10458-023-09634-5
5. Çakan, U., Keskin, M.O., Aydoğan, R.: Effects of agent's embodiment in human-agent negotiations. In: Proceedings of the 23rd ACM International Conference on Intelligent Virtual Agents, IVA '23, New York, NY. Association for Computing Machinery, New York (2023). ISBN 9781450399944. https://doi.org/10.1145/3570945.3607362
6. Herse, S., Vitale, J., Ebrahimian, D., Tonkin, M., Ojha, S., Sidra, S., Johnston, B., Phillips, S., Gudi, S.L.K.C., Clark, J., Judge, W., Williams, M.-A.: Bon appetit! robot persuasion for food recommendation. In: Companion of the 2018 ACM/IEEE International Conference on Human-Robot Interaction, HRI '18, New York, NY, pp. 125–126. Association for Computing Machinery, New York (2018). ISBN 9781450356152. https://doi.org/10.1145/3173386.3177028
7. Herse, S. Vitale, J., Tonkin, M., Ebrahimian, D., Ojha, S., Johnston, B., Judge, W., Williams, M.-A.: Do you trust me, blindly? Factors influencing trust towards a robot recommender system. In: 2018 27th IEEE International Symposium on Robot and Human Interactive Communication (RO-MAN), pp. 7–14 (2018). https://doi.org/10.1109/ROMAN.2018.8525581
8. Hulstijn, J., Tchappi, I., Najjar, A., Aydoğan, R.: Metrics for evaluating explainable recommender systems. In: AAMAS, EXTRAAMAS 2023, London, May 29, 2023. Springer, Berlin (2023)
9. Kamei, K., Shinozawa, K., Ikeda, T., Utsumi, A., Miyashita, T., Hagita, N.: Recommendation from robots in a real-world retail shop. In: International Conference on Multimodal Interfaces and the Workshop on Machine Learning for Multimodal Interaction, ICMI-MLMI '10, New York, NY, 2010. Association for Computing Machinery, New York (2010). ISBN 9781450304146. https://doi.org/10.1145/1891903.1891929
10. Köse, H., Ferrari, E., Dautenhahn, K., Syrdal, D.S., Nehaniv, C.: Effects of embodiment and gestures on social interaction in drumming games with a humanoid robot. Adv. Robot. **23**, 1951–1996 (2009). https://doi.org/10.1163/016918609X12518783330360
11. Lawo, D., Neifer, T., Esau, M., Stevens, G.: Buying the "right" thing: designing food recommender systems with critical consumers. In: Proceedings of the 2021 CHI Conference on Human Factors in Computing Systems, CHI '21, New York, NY. Association for Computing Machinery, New York (2021)
12. Onur Keskin, M., Akay, S., Dogan, A., Koçyigit, B., Kanero, J., Aydogan, R.: You look nice, but I am here to negotiate: the influence of robot appearance on negotiation dynamics. In: Companion of the 2024 ACM/IEEE International Conference on Human-Robot Interaction, HRI '24, New York, NY, pp. 598–602. Association for Computing Machinery, New York (2024). ISBN 9798400703232. https://doi.org/10.1145/3610978.3640759

13. Rossi, A., Rossi, S.: Engaged by a bartender robot: recommendation and personalisation in human-robot interaction. In: Adjunct Proceedings of the 29th ACM Conference on User Modeling, Adaptation and Personalization, UMAP '21, New York, NY, pp. 115–119. Association for Computing Machinery, New York (2021). ISBN 9781450383677. https://doi.org/10.1145/3450614.3463423
14. Rzepka, C., Berger, B.: User interaction with AI-enabled systems: a systematic review of IS research (2018). ICIS 2018 Proceedings. 7. https://aisel.aisnet.org/icis2018/general/Presentations/7
15. Sakai, K., Nakamura, Y., Yoshikawa, Y., Ishiguro, H.: Effect of robot embodiment on satisfaction with recommendations in shopping malls. IEEE Robot. Autom. Lett. **7**(1), 366–372 (2022). https://doi.org/10.1109/LRA.2021.3128233
16. Yang, W.-S., Cheng, H.-C., Dia, J.-B.: A location-aware recommender system for mobile shopping environments. Exp. Syst. Appl. **34**, 437–445 (2008). https://doi.org/10.1016/j.eswa.2006.09.033

# Trajectory Planning for eVTOL Aircraft with High-Fidelity HV Battery Modeling and SOC Estimation

**Mostafa Abdelkhalek and Ismail Bayezit**

## 1 Introduction

Urban air mobility (UAM) seeks to transform urban transportation by alleviating congestion through electric vertical takeoff and landing (eVTOL) vehicles. These multi-rotor aircraft, renowned for their vertical takeoff, landing capabilities, and maneuverability, offer significant potential to reduce traffic, noise, and emissions [9, 11]. Despite these advantages, challenges persist in optimizing the efficiency and range of eVTOL systems [1, 3, 12, 15]. With their ability to support high-payload, long-endurance operations, eVTOLs are well-suited for roles as sky taxis or personal air vehicles (PAVs) for trips exceeding 32 km [17]. Additionally, these vehicles hold critical value in disaster response, facilitating search and rescue, damage assessment, and emergency supply delivery [5–7].

To realize these applications, eVTOL aircraft require energy-dense battery systems capable of sustaining substantial power demands during flight [17]. Current battery technologies, however, face limitations in energy density, power delivery, and lifespan, particularly under the rigorous conditions of takeoff and hovering [17]. Lithium-ion (Li-ion) batteries, specifically Nickel Manganese Cobalt (NMC) chemistries like NMC-700-112, are considered optimal for eVTOL applications due to their high energy density, power output, and cycle life [13]. These low-voltage cells, connected in series and parallel configurations, form high-voltage

M. Abdelkhalek
Faculty of Aeronautics and Astronautics, Istanbul Technical University, Istanbul, Turkiye
e-mail: abdelkhalek22@itu.edu.tr

I. Bayezit (✉)
Faculty of Aeronautics and Astronautics, Istanbul Technical University, Istanbul, Turkiye
e-mail: bayezit@itu.edu.tr



B. U. Töreyin et al. (eds.), *8th EAI International Conference on Robotic Sensor Networks*, EAI/Springer Innovations in Communication and Computing,
https://doi.org/10.1007/978-3-031-92143-8_6

**Table 1** NMC-700-112 battery cell specification

| | |
|---|---|
| Battery cell type | 3.7 V 58 Ah NMC battery cell |
| Model number | L148N58A |
| Nominal voltage | 3.7 V |
| Nominal capacity | 58 Ah |
| Cycle life | ≥2000 cycles |
| Working voltage | 2.75–4.35 V |
| Internal resistance | $0.6 \sim 0.8\,\text{m}\Omega$ |
| End of discharge voltage | 2.75 V |
| Dimensions (mm) | $L148.24\,\text{mm} * W26.66\,\text{mm} * H105.9\,\text{mm}$ |
| Weight | $926 \pm 20\,\text{g}$ |
| Shell material | Aluminum |

(HV) battery packs tailored to meet eVTOL energy and operational requirements. The NMC-700-112 battery cell specifications are outlined in Table 1. Despite their advantages, developing robust battery systems that deliver adequate payload capacities while maintaining efficiency and safety remains a significant hurdle. This paper addresses this challenge by modeling and evaluating the performance of HV Li-ion battery packs specifically designed for eVTOL sky taxis.

Effective battery pack design requires precise modeling of individual cells and their configurations to optimize performance and safety [8, 10, 16]. The widespread adoption of battery systems in electric vehicles has led to the development of battery management systems (BMS), which are crucial for monitoring battery health, optimizing energy usage, and providing critical state-of-charge (SOC) information [2]. These systems enable real-time decision-making for mission trajectory planning and fault management [2]. This chapter presents an innovative HV battery pack model based on NMC-700-112 cells, specifically designed for eVTOL applications. The model integrates key parameters such as energy density, capacity, and payload to optimize operational range, weight, and power delivery. Thrust requirements for various flight phases—climb, cruise, and descent—are derived from dynamic vehicle behavior, with a proportional-integral-derivative (PID) controller ensuring precise current regulation to meet trajectory demands. The proposed approach, illustrated in Fig. 1, minimizes energy consumption and provides a scalable, energy-efficient solution for urban air transport.

## 2 Problem Definition

Battery performance, particularly in terms of energy density and power output, is a key determinant of the operational viability of eVTOL aircraft. Lithium-ion batteries are widely regarded as the leading energy storage technology for eVTOL applications due to their high energy density, scalability in high-voltage configurations, and long cycle life [3, 15]. Specifically, Nickel Manganese Cobalt

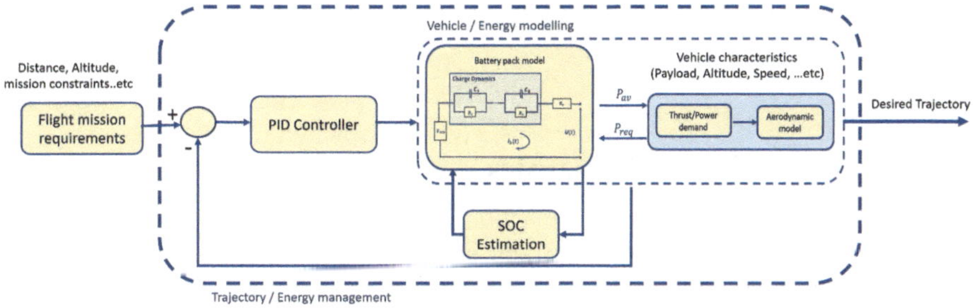


**Fig. 1** Integrated trajectory/energy management system model

(NMC) chemistry has proven ideal for electric aviation systems, offering a superior balance of energy density, power delivery, and thermal stability [13].

The unique operational profile of eVTOL vehicles, characterized by rapid power demands during takeoff, hovering, and landing, imposes significant stress on battery systems. These demands exacerbate challenges related to thermal management, capacity degradation, and maintaining consistent performance under variable loads [11, 14]. Consequently, developing battery systems that balance energy density, power output, and durability is critical for sustainable eVTOL operations.

Modeling lithium-ion batteries at cell and pack levels has advanced significantly, providing system-wide insights into performance under diverse conditions. The resistor-capacitor ($R - C$) framework is particularly effective for representing short- and long-term battery behavior, capturing key parameters such as internal resistance, open-circuit voltage (OCV), and state of charge (SOC) [2, 8]. Accurate SOC estimation is essential for safe and efficient energy management, especially in high-stakes applications like eVTOLs. While traditional methods such as Coulomb counting offer simplicity, advanced techniques like the extended Kalman filter (EKF) have emerged as the standard for dynamic SOC estimation, owing to their ability to handle nonlinearities and measurement noise [11, 14].

Previous studies have explored electric vehicles' energy demands under nominal and non-nominal conditions. For example, [4] analyzed energy requirements for hovering and urban missions but did not integrate real-time battery current profiles or dynamic vehicle behaviors into a unified model. Additionally, existing trajectory optimization frameworks often overlook the interplay between battery limitations and mission-specific demands, leaving a gap in ensuring energy efficiency and reliability in eVTOL applications.

In this chapter, a high-voltage lithium-ion battery pack model, based on NMC-700-112 cells, has been developed to address these challenges. This model integrates SOC estimation using EKF, dynamic vehicle behavior modeling, and specific flight mission requirements into a cohesive framework. Furthermore, a proportional-integral-derivative (PID) controller regulates the battery's current profile to align with precise thrust demands during various flight phases. The proposed approach, illustrated in Fig. 1, ensures optimal energy management, enhances battery utilization, and supports efficient trajectory planning tailored to the operational needs of eVTOL systems.

## 3 Flying Car Vehicle Dynamics

The dynamic model comprises two key components: the thrust analysis for the flying car and its aerodynamic modeling. The thrust analysis involves converting operational parameters into thrust output, while the aerodynamic model establishes the correlation between the applied forces and the required power. Integrating the aerodynamic and thrust models is crucial for simulating the overall flight dynamics of the flying car. The combined model must solve the equations of motion, which include the forces and moments acting on the vehicle, to predict its trajectory and stability under various flight conditions.

### *3.1 Thrust Model*

Thrust analysis is a pivotal element in flying car design, as it determines the force necessary to propel the vehicle through the air while counterbalancing the aerodynamic forces [4, 12]. Thrust is precisely chosen to offset gravitational force (weight, $W$), resist aerodynamic drag, $D$, interact with lift $L$, and facilitate the desired acceleration ($\ddot{x}$) for different maneuvers such as hovering, vertical ascent, forward flight, and descent. Particularly, $L$ is the force that acts perpendicular to the direction of motion, which keeps the vehicle airborne. On the other hand, $D$ is the force acting parallel and opposite to the direction of motion, which resists the vehicle's forward movement. They are represented mathematically as follows:

$$
\begin{aligned}
L &= \frac{1}{2}\rho v^2 S C_L \\
D &= \frac{1}{2}\rho v^2 S C_D
\end{aligned}
\tag{1}
$$

where $\rho$ is the air density ($kg/m^3$), $S$ is the wing area ($m^2$), and $v$ is the velocity of the vehicle relative to the surrounding air (m/s). Also, note that $C_L$ is the lift coefficient which depends on the shape of the airfoil and the angle of attack, while $C_D$ is the drag coefficient, influenced by the vehicle's shape, surface roughness, and other factors.

The dynamic behavior of a flying car under these conditions can be modeled using Newton's second law, which provides a framework for analyzing the motion of the vehicle's centroid [4, 12]. The vector equation governing the forces during various flight regimes is fundamental to understanding and predicting the vehicle's

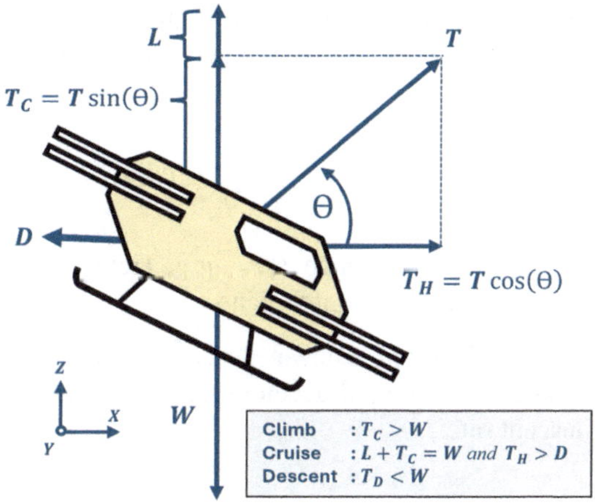


**Fig. 2** Flying car dynamics model during straightforward flight

performance. The generalized form of Newton's second law, applied to the forces acting on a flying car, is expressed as

$$\begin{aligned}\sum F &= m\vec{\ddot{x}} \\ &= \vec{W} - \vec{D} + \vec{L} + \vec{T}.\end{aligned} \tag{2}$$

This equation can be decomposed to reflect the different components and conditions of flight as shown in Fig. 2:

**Hovering** In hovering, the vehicle remains stationary in the air. The thrust $T$ must exactly balance the vehicle's weight $W = mg$, while drag and lift have minimal effect since there is no significant forward or vertical motion:

$$\begin{aligned}T_h &= W \\ &= mg\end{aligned} \tag{3}$$

where $m$ is the mass of the vehicle and $g$ is the gravitational constant.

**Vertical Climb** During the vertical climb, thrust $T_c$ must not only balance the weight but also overcome drag and provide the vertical acceleration $\ddot{x}_c$. Lift in this scenario may assist, depending on the vehicle's design:

$$T_c = mg + D + m\ddot{x}_c - L \tag{4}$$

This equation shows that the required thrust increases with vertical acceleration and drag, reduced by any lift contribution.

**Forward Flight** In forward flight, thrust $T_F$ must overcome drag $D$ and ensure adequate forward horizontal acceleration $\ddot{x}_F$. Simultaneously, lift $L$ must balance the weight:

$$T_F = \frac{D + m\ddot{x}_F}{\cos(\theta)} \tag{5}$$

For vertical balance $L = mg$. Here, $\theta$ represents the angle of the flight path relative to the horizontal plane.

**Vertical Descent** During descent, thrust $T_D$ is reduced to allow gravity to contribute to downward acceleration, while drag and lift play roles in controlling the descent rate:

$$T_D = mg - D - m\ddot{x}_D + L \tag{6}$$

$\ddot{x}_D$ is the acceleration during descent.

## 3.2 *Aerodynamic Model*

Aerodynamic modeling of a flying car highlights the complex relationship between thrust and power consumption, essential for evaluating the vehicle's efficiency and performance. This model applies momentum theory to assess various flight phases—hovering, vertical ascent, forward motion, and vertical descent—each posing unique challenges due to differing flow field states that affect aerodynamic forces and energy requirements [4, 12]. By examining the momentum exchange between the vehicle and surrounding air, this approach predicts the power necessary to achieve the required thrust across different flight segments.

**Hovering** The power required for hovering $P_h$ can be expressed using the momentum theory as

$$P_h = \frac{T_h^{3/2}}{\sqrt{2\rho A}} \tag{7}$$

**Vertical Climb** During vertical ascent, the vehicle needs to overcome its weight and additional drag. The power required $P_c$ in vertical climb is

$$P_c = \frac{(T_c + D)^{3/2}}{\sqrt{2\rho A}} + T_c v_c \tag{8}$$

Similarly, for forward flight, the power required $P_F$ is affected by both forward horizontal thrust $T_F$ and drag forces components and forward flight velocity $v_F$.

**Vertical Descent** For vertical descent, the power required $P_D$ is reduced since gravity aids the descent. The equation can be expressed as

$$P_D = \frac{(T_D - D)^{3/2}}{\sqrt{2\rho A}} - T_D v_D \quad (9)$$

# 4 Battery Pack Model

## 4.1 Battery Cell Specification

We employed the NMC-700-112 battery pack in our model due to its high energy density, which enhances flight range and endurance, especially under heavy payloads. Its strong power output supports critical phases like takeoff and hover, while thermal stability ensures safe operation under high loads. With a long cycle life and scalable high-voltage configurations, this battery is well-suited for eVTOL applications. Detailed performance, technical specifications, and safety considerations are provided in the $L148N58A$ datasheet, summarized in Table 1.

## 4.2 Battery Cell Model

Various lithium cell models, including electrochemical models and equivalent circuit models (ECM), have been developed to simulate cell behavior [8]. Electrochemical models provide detailed internal dynamics but are computationally intensive, making them impractical for system-level applications. ECM, however, correlates electrochemical phenomena with circuit elements, offering reduced computational demand while capturing nonlinear behaviors, as shown in Fig. 3. This makes ECM ideal for powertrain electrification and system-level modeling.

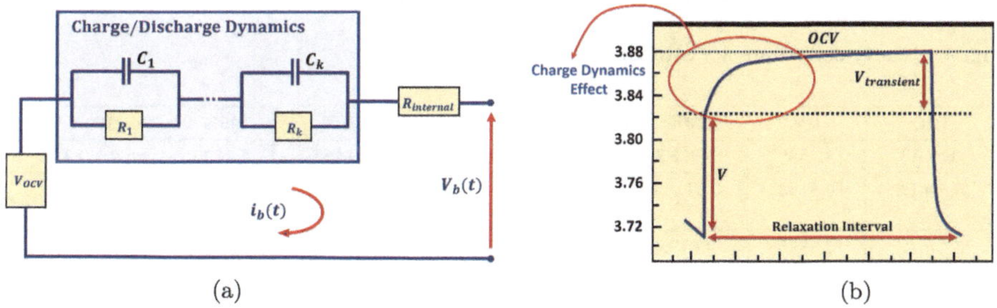


**Fig. 3** Generic equivalent circuit model for the battery cell. (**a**) (R-C)$_k$ ECM battery cell model. (**b**) Battery dynamics

### 4.3 Battery Cell Parameters Estimation

To get an accurate cell model, we need to carefully estimate each element of the equivalent circuit of Fig. 3 as a function of SOC and temperature. For simplicity, the charge/discharge dynamics are modeled using a single R-C circuit, as incorporating additional R-C circuits would unnecessarily complicate the model [8]. Our analysis comprises two stages: parameter estimation/validation and simulation. In the parameter estimation stage, discharge profiles were simulated iteratively and compared with experimental data. Figure 4 shows the algorithm used for estimating the cell model parameters.

The simulation was carried out using Simscape language and blocks. The system model depicts a single RC block, and each circuit element was implemented as a subsystem consisting of standard and custom electrical blocks for calculating the circuit element properties.

We used that model to enable the estimation process since a simple charging circuit model was connected to the Simscape ECM using an ideal current source and a voltage sensor. The parameter estimation tool in Simulink design optimization (SDO) was utilized to determine the cell parameters mentioned earlier. This tool enabled a more efficient estimation of the system parameters, which is critical for accurate modeling and simulation of complex electrical circuits. The estimated cell parameters are shown in Fig. 5.

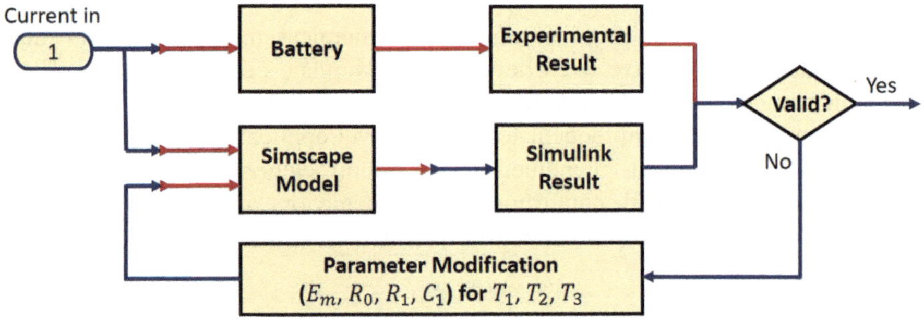


**Fig. 4** Battery cell parameter estimation algorithm

| SOC | 0% | 10% | 25% | 50% | 75% | 90% | 100% |
|---|---|---|---|---|---|---|---|
| OCV (v) | 3.4545 | 3.5785 | 3.6384 | 3.7143 | 3.8751 | 4.0696 | 4.2782 |
| $R_0$ (mΩ) | 0.3153 | 0.41282 | 3.8519 | 3.5318 | 3.9364 | 4.1389 | 5.2128 |
| $R_1$ (mΩ) | 0.3336 | 3.0199 | 2.9055 | 2.5832 | 2.5649 | 2.8376 | 3.5908 |
| $Tau_1$ (s) | 13.5414 | 79.0571 | 62.0276 | 32.9159 | 49.3802 | 37.8523 | 41.8525 |

**Fig. 5** Battery cell parameter estimated values

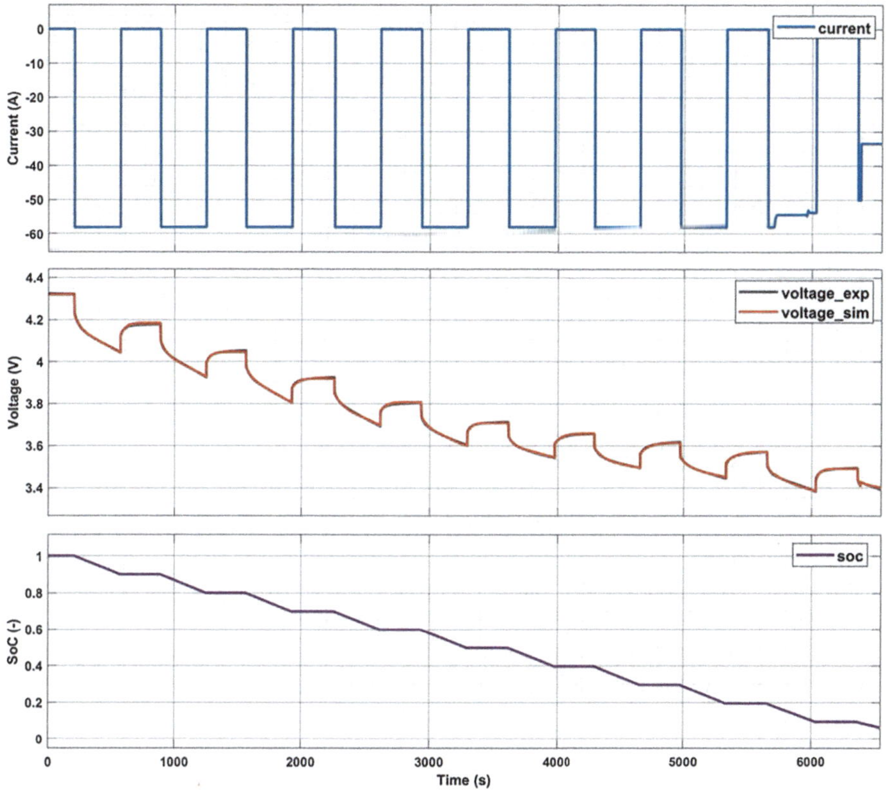


**Fig. 6** Simulated vs. measured cell voltage after one R-C model parameters estimation

Figure 6 shows the simulation accurately replicating the discharge current profile and tracking the OCV decline. The inset demonstrates the optimizer's ability to capture system dynamics, with a single R-C circuit effectively modeling the battery's behavior. This close alignment between simulated and measured voltage profiles underscores the model's utility for control algorithm and system-level development.

## 5 Battery Pack Model

This section describes the construction of a Simscape system model for an NMC-700-112 prismatic lithium-ion battery pack with cell balancing circuits, using the Simscape battery toolbox. Designed for high-voltage packs exceeding 700V, the model includes parallel assemblies of cells in series. To build this comprehensive system, cell, parallel assembly, module, and module assembly objects are config-

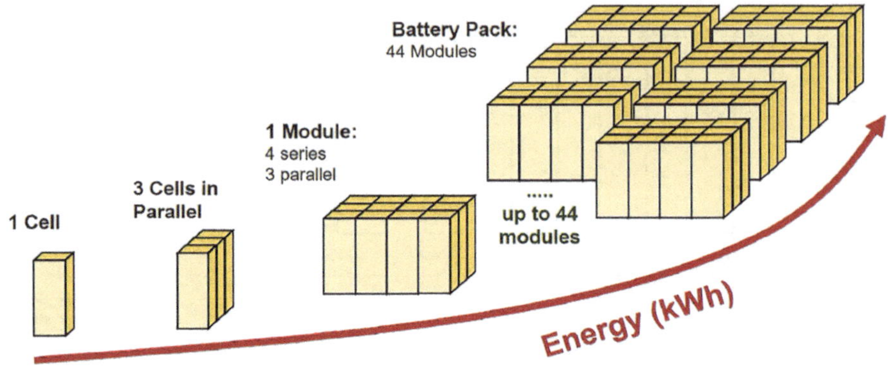


**Fig. 7** Process to build the battery pack model

ured, enabling the use of a built-in battery function to automate Simscape model generation for these battery objects. Figure 7 illustrates the steps involved in this modeling process.

After we have created the battery objects, we need to convert them into a Simscape block library (.lib) to use them in a model-based design. The built-in battery function is used to create a library that contains the Simscape battery model object. The battery pack model block is shown in Fig. 8.

## 6 State-of-Charge Estimation

Effective battery management is essential for electric vehicle performance, requiring continuous SOC monitoring to assess reliability across diverse conditions. Accurate SOC estimation prevents deep discharge and prolongs battery life. Here, we introduce the most commonly used SOC estimation method.

### *6.1 Extended Kalman Filter*

The extended Kalman filter (EKF) is widely used for SOC estimation in BMS, especially for nonlinear systems. It integrates battery voltage, current, and temperature data, predicting SOC with improved accuracy over Coulomb counting (CC) by accounting for nonlinear effects and aging. Though EKF requires more computational resources, it is often combined with CC to enhance precision.

The equations representing the equivalent circuit cell model dynamics, incorporating the additional state of the terminal resistance ($R_0$), are as follows:

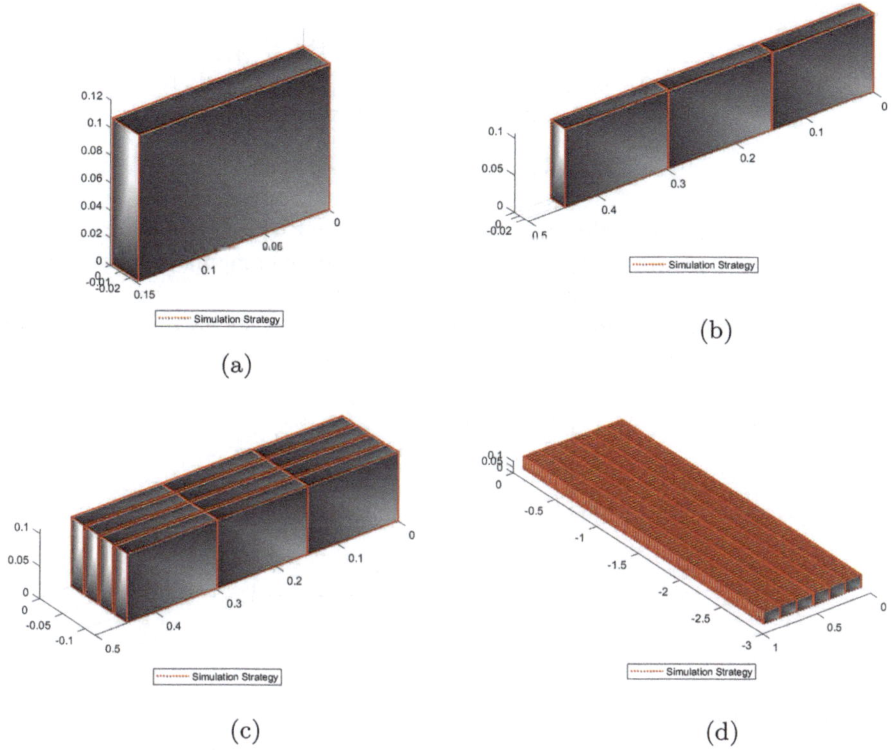


**Fig. 8** Battery pack implementation using Simulink Simscape battery toolbox. (**a**) One cell. (**b**) Parallel assembly (three cells in parallel). (**c**) One module (4 parallel assemblies in series). (**d**) Battery pack (44 modules in series)

$$
\begin{aligned}
&\frac{\mathrm{d}}{\mathrm{d}t} SOC = -\frac{i}{3600 * AH} \\
&\frac{\mathrm{d}}{\mathrm{d}t} V_1 = \frac{i}{C_1(SOC)} - \frac{V_1}{R_1(SOC) * C_1(SOC)} \\
&\frac{\mathrm{d}}{\mathrm{d}t} R_0 = 0 \\
&V_t = V_{OCV}(SOC) - i * R_0 - V_1
\end{aligned} \tag{10}
$$

where SOC denotes the state of charge, $i$ represents the electric current, $V_{OCV}$ signifies the open-circuit voltage, $V_t$ denotes the voltage across the terminal, AH represents the ampere-hour rating, $R_1$ represents the primary polarization resistance, $C_1$ represents the parallel resistance-capacitance ($R - C$), and $AH$ represents the rating ampere-hour. The states and the observation function are expressed as follows:

$$x = \begin{bmatrix} SOC & V_1 & R_0 \end{bmatrix}$$
$$f(x,i) = \begin{bmatrix} \frac{-i}{3600*AH} \\ \frac{i}{C_1(SOC)} - \frac{V_1}{R_1(SOC)C_1(SOC)} \end{bmatrix}$$
$$h(x,i) = V_{OCV}(SOC) - i * R_0 - V_1 \quad (11)$$

In order to approximate the nonlinear system, the EKF algorithm necessitates linearization at each time step. This involves the online computation of Jacobians by the algorithm to achieve the desired linearization.

Upon discretization with $T_s$ as the sample time, the Jacobians of the estimation of SOC for the battery are computed as follows:

$$F = \begin{bmatrix} 1 & 0 & 0 \\ 0 & e^{\frac{-T_s}{R_1C_1}} & 0 \\ 0 & 0 & 1 \end{bmatrix}$$
$$H = \begin{bmatrix} \frac{-i}{3600*AH} \\ \frac{i}{C_1(SOC)} - \frac{V_1}{R_1(SOC)C_1(SOC)} \end{bmatrix} \quad (12)$$

## 7 Numerical Analysis

The numerical analysis aims to evaluate the energy efficiency and trajectory control performance of the eVTOL system under varying flight mission profiles. The analysis integrates battery dynamics, PID-controlled current regulation, and real-time SOC estimation.

### *7.1 Flight Mission*

The flight mission profile in this study explores altitude options of 100, 200, and 500 m, with climb times ranging from 40 to 45 seconds and descent times between 30 and 35 seconds. For an altitude of 100 m and a cruising speed of approximately 40 m/s, the energy demand is calculated using our algorithm, which optimizes the trajectory through a PID controller based on mission requirements. The PID controller gains were tuned with the following parameters: $K_p = 0.85$, $K_i = 0.1$, and $K_d = 0.05$. These values were selected to ensure stable flight paths under varying aerodynamic loads, optimize power consumption during climb and descent, and enable rapid corrections for trajectory deviations. The energy demands for the

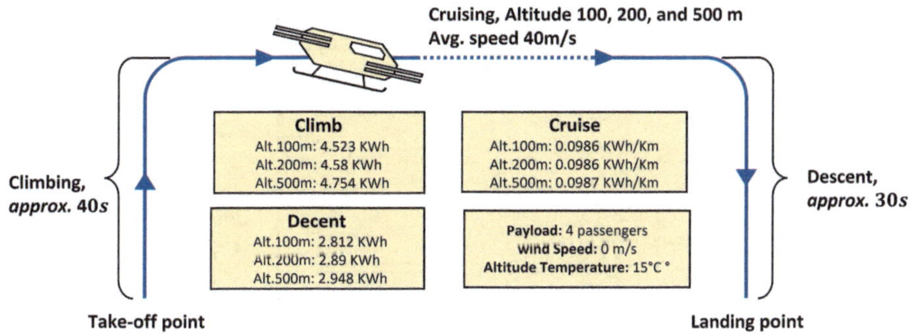


**Fig. 9** Flight mission

**Table 2** Simulation parameters

| Parameters | Value | Unit |
|---|---|---|
| Battery pack weight (528 cells) | 490 | $Kg$ |
| Payload (four passengers) | 350 | $Kg$ |
| Total weight | 2110 | $Kg$ |
| Number of blades in one rotor | 2 | Blades |
| Propeller blade radius | 160 | $mm$ |
| Motor winding resistance | 0.1 | $\Omega$ |
| Motor speed constant (EMF) | 0.00246 | $V/RPM$ |
| Propeller blade chord length | 190 | $mm$ |
| Projected frontal area | 0.0304 | $m^2$ |
| Acceleration of gravity | 9.8 | $m/s^2$ |
| Air density | 1.225 | $Kg/m^3$ |
| Horizontal drag coefficient | 0.0334 | – |
| Cruise height (altitude) | 100, 200, 500 | $m$ |
| Rotor blade figure of merit | 0.7 | – |
| Rotors efficiency | 0.95 | – |
| Battery capacity | 174 | $Ah$ |

climb and descent are 4.523 kWh and 2.812 kWh, respectively. During the cruise, the energy consumption is 0.0986 kWh per kilometer. Simulation results for flight missions at altitudes of 100 m, 200 m, and 500 m indicate optimal energy usage through PID-controlled trajectory adjustments. This framework ensures accurate modeling of the interplay between aerodynamic forces and battery performance. This profile, illustrated in Fig. 9, provides a foundation for assessing the energy needs of eVTOL vehicles under typical operational scenarios. Additional simulation parameters are detailed in Table 2.

In Fig. 10, the results illustrate optimized current profiles across the climb, cruise, and descent phases at altitudes of 100, 200, and 500 m, showing the PID controller's role in energy efficiency. The current ramps up rapidly for the climb phase to meet

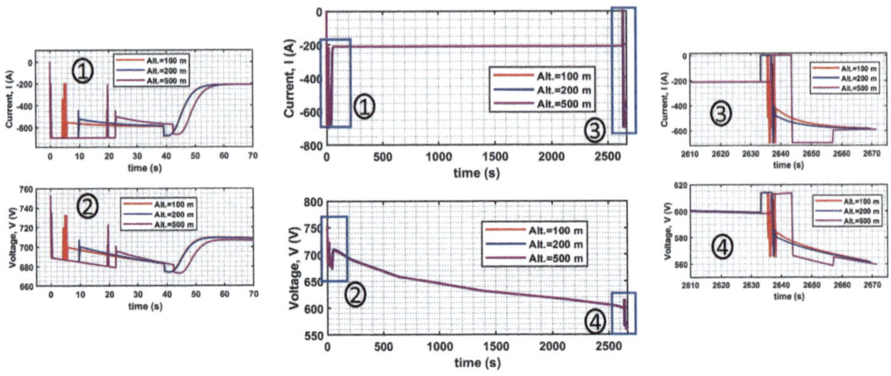


**Fig. 10** Current profile of the flight missions 100, 200, and 500 m altitude

ascent demands and stabilize. The current shows a slight lead at higher altitudes, allowing smooth altitude reach, while lower altitudes exhibit a minor lag for faster stabilization. During the cruise, the current is kept steady to efficiently support the 40 m/s speed. In descent, the controller moderates the profile to leverage gravity, with slight leads or lags based on the altitude to ensure smooth descent. This demonstrates the controller's adaptive modulation, optimizing battery usage across flight phases.

In Fig. 11, the SOC estimation figure for the 100-m altitude trajectory demonstrates the battery's SOC degradation due to the discharge current profile during the mission. This figure highlights how the current profile impacts SOC, with a gradual decline reflecting energy consumption across flight phases. Additionally, the EKF estimation error is presented in a separate figure, showing minimal error bounded within $+/-$ sigma limits, indicating the EKF's high accuracy in tracking SOC throughout the trajectory. This accurate SOC estimation, within constrained error bounds, underscores the reliability of the EKF in managing battery performance for eVTOL applications.

Figure 12 shows optimized flight paths at 100, 200, and 500 m, demonstrating smooth climbs, steady cruising at 40 m/s, and controlled descents, highlighting the PID controller's efficiency and the eVTOL's adaptability for urban air mobility.

## 8 Conclusion

This study developed a high-voltage battery pack model using the R-C framework for NMC-700-112 cells, with EKF-based SOC estimation and PID-controlled current profiles ensuring energy-efficient flight. The model demonstrated minimal

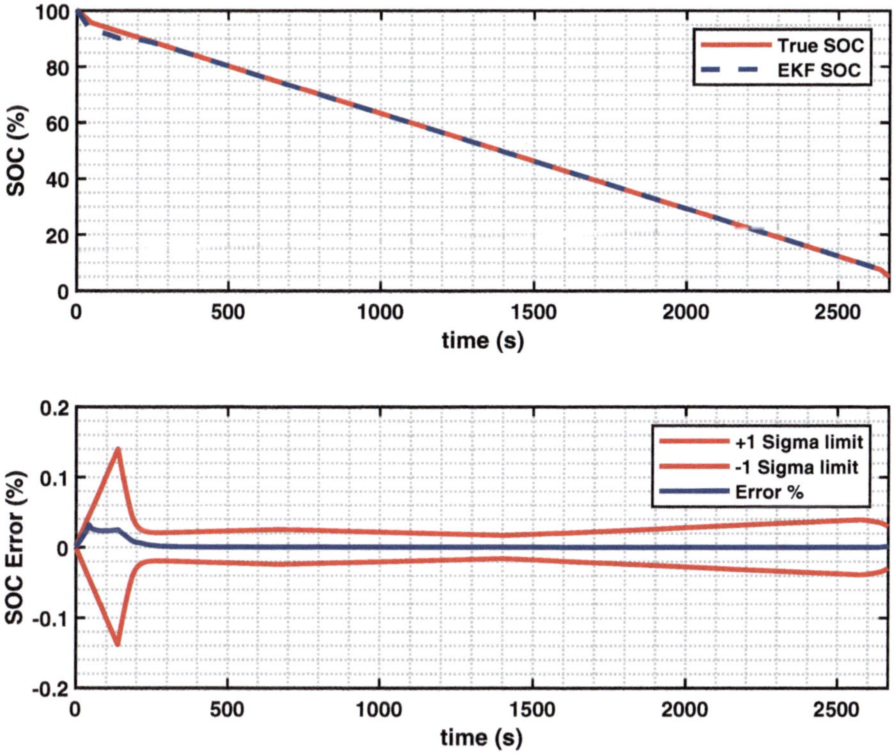


**Fig. 11** SOC estimation using EKF algorithm for trajectory altitude 100 m

EKF errors and adaptable control, confirming its potential for eVTOL use in urban air mobility. Future work will address thermal dynamics and environmental factors to enhance real-world applicability.

**Acknowledgments** This research study is conducted at the Model-Based Design and Control Systems Laboratory under the auspices of Istanbul Technical University. This project is financially supported by the Scientific Research Projects Coordination Unit of Istanbul Technical University (ITU) HIZDEP Research Fund with Project ID: 46503.

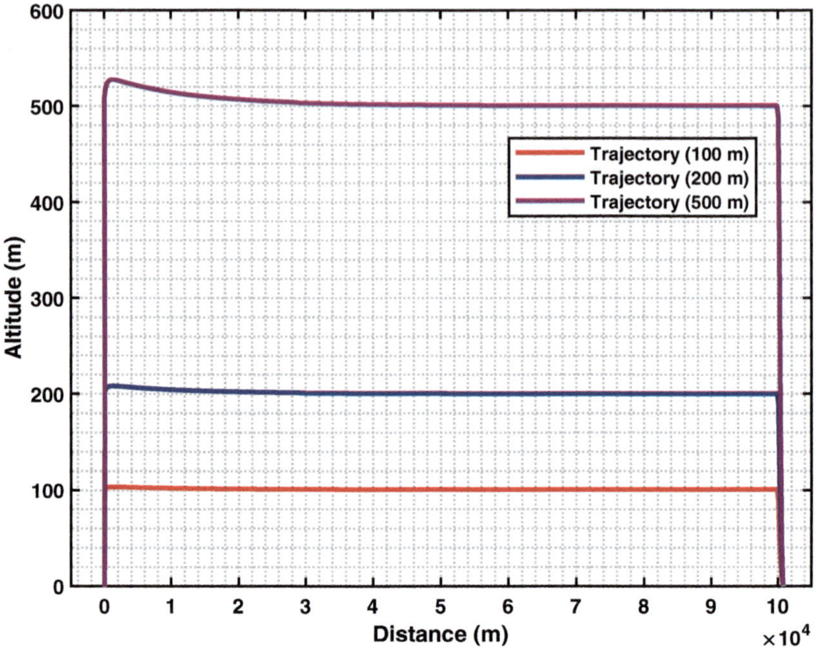


**Fig. 12** Desired trajectory generated for 100, 200, 500 m altitudes

## References


1. Abdelkhalek, M., Bayezit, A.B., Bayezit, I., Bircan, Y., Cakir, A., Kurtoglu, F., Mandaci, D., Erunlu, G., Fidan, B.: E-VCU software toolbox for ARM Cortex-R4 processor based electric vehicle control. In: 2023 XXIX International Conference on Information, Communication and Automation Technologies (ICAT), pp. 1–6. IEEE, Piscataway (2023)
2. Dini, P., Colicelli, A., Saponara, S.: Review on modeling and SOC/SOH estimation of batteries for automotive applications. Batteries **10**(1), 34 (2024)
3. Garrow, L.A., German, B.J., Leonard, C.E.: Urban air mobility: a comprehensive review and comparative analysis with autonomous and electric ground transportation for informing future research. Transp. Res. Part C Emerging Technol. **132**, 103377 (2021)
4. Hagag, N., Gasche, S., Jäger, F., Kallies, C.: Energy demand analysis for eVTOLs in cluttered and dynamic environments based on adaptive trajectory prediction. In: 2024 Integrated Communications, Navigation and Surveillance Conference (ICNS), pp. 1–15. IEEE, Piscataway (2024)
5. Helmy, M., Arslan, H.: Utilization of aerial heterogeneous cellular networks: signal-to-interference ratio analysis. J. Commun. Networks **20**(5), 484–495 (2018)
6. Helmy, M., Baykaş, T., Arslan, H.: Optimization of aerial base station location in lap for disaster situations. In: 2015 IEEE Conference on Standards for Communications and Networking (CSCN), pp. 240–244. IEEE, Piscataway (2015)
7. Helmy, M., Ankaralı, Z.E., Siala, M., Baykaş, T., Arslan, H.: Dynamic utilization of low-altitude platforms in aerial heterogeneous cellular networks. In: 2017 IEEE 18th Wireless and Microwave Technology Conference (WAMICON), pp. 1–6. IEEE, Piscataway (2017)

8. Huria, T., Ceraolo, M., Gazzarri, J., Jackey, R.: High fidelity electrical model with thermal dependence for characterization and simulation of high power lithium battery cells. In: 2012 IEEE International Electric Vehicle Conference, pp. 1–8. IEEE, Piscataway (2012)
9. Kasliwal, A., Furbush, N.J., Gawron, J.H., McBride, J.R., Wallington, T.J., De Kleine, R.D., Kim, H.C., Keoleian, G.A.: Role of flying cars in sustainable mobility. Nat. Commun. **10**(1), 1555 (2019)
10. Kroeze, R.C., Krein, P.T.: Electrical battery model for use in dynamic electric vehicle simulations. In: 2008 IEEE Power Electronics Specialists Conference, pp. 1336–1342. IEEE, Piscataway (2008)
11. Liu, M., Hao, H., Sun, X., Qu, X., Wang, K., Qian, Y., Hao, X., Xun, D., Geng, J., Dou, H., et al.: Exploring the key technologies needed for the commercialization of electric flying cars: a levelized cost and profitability analysis. Energy **303**, 131991 (2024). ISSN 0360-5442. https://doi.org/10.1016/j.energy.2024.131991
12. Long, Q., Ma, J., Jiang, F., Webster, C.J.: Demand analysis in urban air mobility: a literature review. J. Air Trans. Manage. **112**, 102436 (2023)
13. Roberts, C., Petrovich, S., Ebrahimi, K.: Battery modeling for emulators in vehicle test cell. Batteries **10**(6), 199 (2024)
14. Sierra, G., Orchard, M., Goebel, K., Kulkarni, C.: Battery health management for small-size rotary-wing electric unmanned aerial vehicles: an efficient approach for constrained computing platforms. Reliab. Eng. Syst. Saf. **182**, 166–178 (2019)
15. Straubinger, A., Rothfeld, R., Shamiyeh, M., Büchter, K.D., Kaiser, J., Plötner, K.O.: An overview of current research and developments in urban air mobility–setting the scene for UAM introduction. J. Air Trans. Manage. **87**, 101852 (2020)
16. Tahir, M.U., Moaz, T., Khan, H.A., Zaffar, N.A., Khan, I.: Accurate modeling of Li-ion cells applied to LiFePO 4 and NMC chemistries. In: 2020 IEEE Texas Power and Energy Conference (TPEC), pp. 1–6. IEEE, Piscataway (2020)
17. Ullman, D.G., Homer, V., Ouellette, R.: Comparing electric sky taxi visions. 1–31 (2017). https://doi.org/10.13140/RG.2.2.14819.50729

# Deep Reinforcement Learning Based Uplink Resource Allocation in Open RAN Systems

**Ali Yıldırım, Hasan Anıl Akyıldız, İbrahim Hökelek, and Hakan Ali Çırpan**

## 1 Introduction

Critical features of 5G, such as high data rates, reduced latency, and the ability to connect a large number of users simultaneously, have introduced significant new use cases. 6G is anticipated to further enhance these capabilities, making mobile networks more immersive and integrated. In the context of robotics, 6G will enable collaboration with intelligent robots, known as "cobots." To support this, efficiently scheduling limited resources for users—including robots and cobots—becomes increasingly important.

Resource scheduling is a highly challenging task in mobile wireless networks, where the number of users along with their channel conditions can rapidly change due to the mobility, and the peculiarities of wireless medium including path loss, fading, and shadowing. The emerging application and services in the next-generation networks have stringent requirements in terms of throughput, reliabil-

A. Yıldırım (✉)
Communication and Information Technologies Business Sector (HBT), ASELSAN Inc., Ankara, Turkey

Electronics and Communication Department, Istanbul Technical University, İstanbul, Turkey
e-mail: aliyildirim@aselsan.com.tr; yildirimal18@itu.edu.tr

H. A. Akyıldız · H. A. Çırpan
Electronics and Communication Department, Istanbul Technical University, İstanbul, Turkey
e-mail: akyildizh@itu.edu.tr; cirpanh@itu.edu.tr

İ. Hökelek
Electronics and Communication Department, Istanbul Technical University, İstanbul, Turkey

Research Center for Advanced Technologies on Informatics and Information Security, TÜBİTAK BİLGEM, İzmit, Turkey
e-mail: ibrahim.hokelek@tubitak.gov.tr; hokelek@itu.edu.tr



B. U. Töreyin et al. (eds.), *8th EAI International Conference on Robotic Sensor Networks*, EAI/Springer Innovations in Communication and Computing,
https://doi.org/10.1007/978-3-031-92143-8_7

ity, and latency. Traditional resource allocation approaches are not sufficient to meet these highly diversified expectations, and hence, new resource scheduling approaches, which are more flexible, robust, and efficient, are needed.

Recently, some researchers have focused on using AI/ML, specifically reinforcement learning methods, for addressing the challenges in mobile communication networks such as beamforming, energy efficiency, and network slicing. There has been an extensive research on using reinforcement learning for coarse- and fine-grained resource allocation [1]. Coarse approaches generally utilize deep reinforcement learning (DRL) to allocate resources to network slices with differing quality-of-service considerations. Reference [2] proposes a hierarchical structure for allocating resources to ultra-reliable low-latency communication (URLLC) and enhanced mobile broadband (eMBB) slices at a higher level of design. A deep Q-network (DQN) is used as the RL agent in this study, unlike [3], where the authors employ the advantage actor critic (A2C) algorithm for upper-level resource allocation. The open interfaces defined by the open radio access network (O-RAN) architecture provide seamless integration of AI/ML capabilities into mobile wireless communication networks. For example, in [4], the researchers develop an ML-based xApp to find the optimal resource allocation policy for network slices. In addition to coarse approaches, fine-grained approaches such as deep reinforcement learning for downlink scheduling, in particular, have been the focus of many studies. In [5], the authors propose a double deep Q-network (DDQN) agent for resource allocation and demonstrated that the scheduler's performance is comparable to traditional scheduling algorithms. The resource allocation problem in the downlink is solved by a DQN scheduler in [6], which takes total throughput and fairness into account.

There are relatively fewer studies for uplink resource allocation compared to downlink scheduling in the literature. For example, a deep deterministic policy gradient (DDPG) approach using buffer state information is implemented for resource allocation in non-orthogonal multiple access (NOMA) in [7]. Reference [8] extends uplink resource allocation in the NOMA problem to meet the stringent requirements of URLLC, proposing a solution that enhances the state-of-the-art proximal policy optimization (PPO) algorithm. In [9], the researchers address uplink scheduling in cellular networks under URLLC constraints, proposing a deep Q-learning trained agent for grant-free (GF) access to uplink resources. Additionally, it is important to note that DRL-aided resource allocation within the O-RAN architecture has recently gained the attention of researchers [4, 10, 11]. Consequently, there remains a research gap in DRL-based uplink scheduling for the next-generation mobile networks supporting the O-RAN architecture. The key contributions of this chapter can be outlined as follows:

- DRL-based uplink resource allocation problem in 5G and beyond networks is formulated.
- An AI/ML-aided scheduling algorithm, which can be directly utilized in the O-RAN architecture by leveraging the rApp and xApp structure, is designed.

- An integrated simulation platform is developed using MATLAB's 5G and reinforcement learning toolboxes, where the proposed DRL-based resource allocation agents are implemented, and their performance results are reported.

This chapter is organized as follows: Sect. 2 presents the system model; Sect. 3 details the RL-based resource allocation method; Sect. 4 covers training and comparisons with traditional approaches; Sect. 5 discusses configuration impacts and scalability; and Sect. 6 concludes and suggests future research.

## 2 System Model

This chapter addresses the resource allocation problem using the DRL agents, offering a flexible framework for operators to implement their high-level network policies. The DRL agents act as a wireless medium access layer scheduler, assigning uplink resource blocks to users. One of the DRL agents is designed to maximize the system throughput, while the second one aims to jointly optimize both the total data rate achieved by the users and the fairness of the resource allocation process. An appropriate xApp can be employed in the system to achieve the objectives of high-level network policies. This xApp employment can also be dynamic, where the operator may switch from one policy to another and hence from one xApp to another. As a result, the performance objectives based on the requirements derived from these time-varying policies can be achieved by switching from one trained DRL agent to another one. Note that the trained DRL agent can also be run within dApp, which is suggested by reference [12] for real-time applications, if the fine-grained resource allocation is needed.

The system model primarily consists of the following components:

- *rApp*: An application running on the non-real-time RAN intelligent controller (Non-RT RIC) to enhance network efficiency. In this context, it is used to receive policies from the operator and convey them to the near-real-time RAN intelligent controller (near-RT RIC).
- *xApp*: An application running on the near-RT RIC to further improve network efficiency. In this chapter, the xApp is used to receive policies from the rApp, execute the chosen DRL agent, and interact with the scheduler in the open distributed unit (O-DU) to influence resource allocation.
- *DRL agents*: Reinforcement learning agents that take environmental observations from the O-DU and make scheduling decisions.
- *Resource scheduler*: A specific software component of the O-DU that allocates RBs to users.
- *User Equipment (UE)*: Devices with uplink data stored in their buffers, awaiting transmission.

As depicted in Fig. 1, rApp is an application running on Non-RT RIC, through which the network operator inputs its policies. These policies may vary, ranging

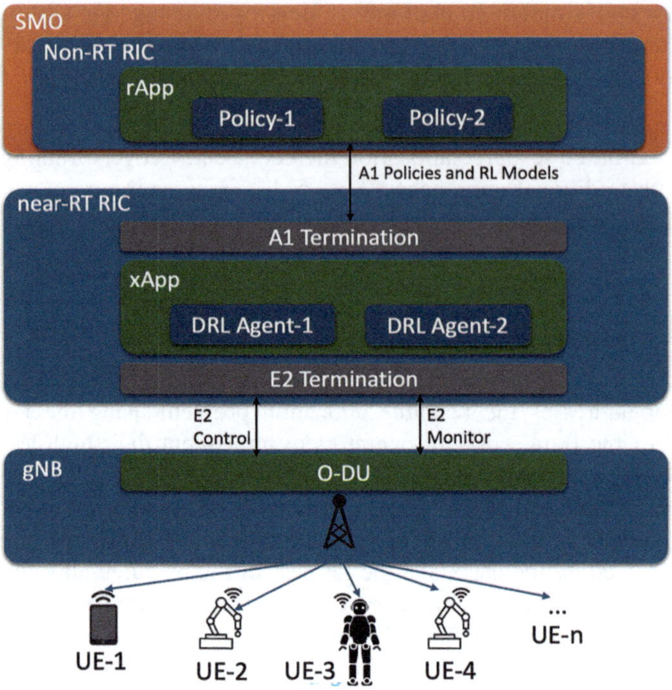


**Fig. 1** System model based on O-RAN architecture

from being maximum throughput-oriented to reaching a fair resource allocation scheme. Since non-RT RIC works with policies at a resolution greater than one second, it can be considered as a higher-level coordination mechanism. Non-RT RIC is connected to near-RT RIC through the A1 interface, defined by O-RAN. The aforementioned policies are transferred to near-RT RIC using the A1 interface. Additionally, the trained DRL agents can be transferred to near-RT RIC to influence scheduling operations in the O-DU. The xApp, central to this study, executes the model based on the policy received from the non-RT RIC. xApp operates on near-RT RIC, which is suitable for operations requiring a resolution between 10 ms and 1 second. The near-RT RIC interacts with the O-DU through the E1 interface, defined by O-RAN. Finally, the MAC scheduler in the O-DU allocates resources to users according to the actions determined by the DRL agents within the xApp.

The DRL agents studied in this chapter are referred to as DRL Agent-1 and DRL Agent-2. The first agent focuses solely on achieving the highest total uplink data rate, while the second agent considers both user throughput and fairness in resource allocation. Both agents are designed based on the deep Q-learning algorithm which is a model-free reinforcement learning algorithm.

Reinforcement learning is an interdisciplinary area of machine learning, where agents learn to make decisions by interacting with an environment and receiving feedback in the form of rewards. The overall objective is to maximize cumulative rewards over time. Deep Q-learning, a widely utilized method in RL, extends traditional Q-learning by leveraging deep neural networks (DNN) to approximate the Q-values for state-action pairs, allowing the algorithm to scale to more complex environments, in which classical Q-learning approach becomes impractical. Algorithm-1 provides pseudocode for deep Q-learning. At any given time step $t$, the agent observes state $s_t$, takes an action $a_t$, receives a reward $r_t$, and transitions to a new state $s_{t+1}$. These interactions are stored in replay memory $M$ for training. The weights of deep Q-network (DQN) are updated by minimizing a loss function, which is employed to measure the difference between the predicted Q-values and the target Q-values, also known as the Q-targets. $\gamma$ represents discount factor, which influences the balance between exploration and exploitation. Instead of updating the network after every single action, gradient descent is applied on batches of experience sampled from the replay memory, at regular intervals. This process helps break the correlation between consecutive actions and leads to more stable learning.

**Algorithm 1** Deep Q-learning algorithm

Initialize replay memory $M$
Initialize Q-network with random weights $\theta$
Initialize target $\hat{Q}$-network with weights $\theta^- = \theta$
**for** each episode **do**
  Reset environment
  Set initial state $s_t$
  **for** each step **do**
    Choose an action $a_t$
    Execute action $a_t$ in the environment
    Observe new state $s_{t+1}$
    Calculate reward $r_t$
    Store experience $(s_t, a_t, r_t, s_{t+1})$ in replay memory $M$
    Sample minibatch $(s_j, a_j, r_j, s_{j+1})$ from replay memory $M$
    set $y_{j+1} = \begin{cases} r_{j+1} & \text{final state} \\ r_{j+1} + \gamma \max_{a'} \hat{Q}(s_{j+1}, a'; \theta^-) & otherwise \end{cases}$
    Perform gradient descent on $(y_j - Q(s_j, a_j; \theta))^2$
    Every n steps, reset $\hat{Q} = Q$
  **end for**
**end for**

## 3 Proposed Method

In this section, DQN-based agents for uplink resource allocation are described, where the RL-based architecture is shown in Fig. 2. The resource allocation problem can be represented as a Markov decision process (MDP). In this context, S is defined as the state space, A is the set of actions and R represents the reward function.

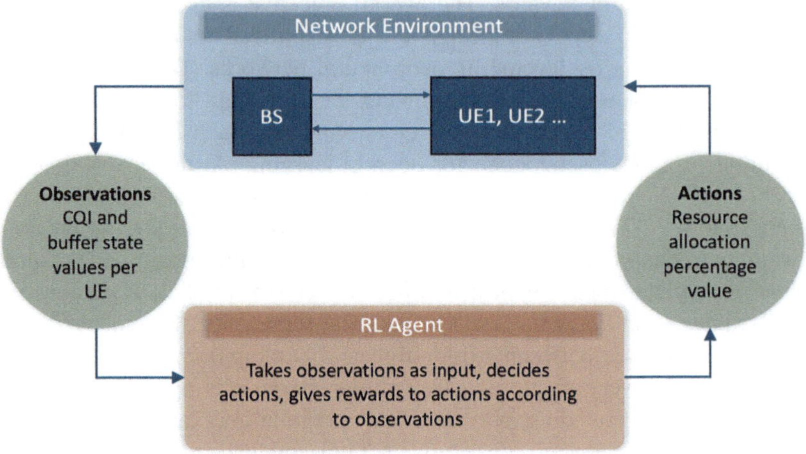


**Fig. 2** Simulation procedure of DRL agents

**State($S$)** The state function represents the current status of the environment. For both DRL agents, a 2 x $m$ matrix is used as the state, where m is the number of UEs. Both buffer status ($b$) and channel quality indicator ($c$) are considered part of the state, as resources should be allocated to a user only if that user has a packet to send and a high CQI to achieve higher total cell throughput.

$$S = \begin{bmatrix} b_{UE_1} & b_{UE_2} & \dots & b_{UE_m} \\ c_{UE_1} & c_{UE_2} & \dots & c_{UE_m} \end{bmatrix}. \tag{1}$$

**Action($A$)** The action is defined as the percentage of resource allocation to UEs, such as 10% for UE#1 and 90% for UE#2, etc. Given that the minimum resolution for near-RT RIC is 10 ms, the actions determined by the DRL agent in the xApp correspond to this minimum resolution. By representing actions as resource allocation percentages, the action space is reduced, simplifying the decision-making process.

$$A = \begin{bmatrix} a_{UE_1} & a_{UE_2} & \dots & a_{UE_m} \end{bmatrix}^T, where \sum_{k=1}^{m} a_{UE_k} = 1. \tag{2}$$

**Reward($R$)** The reward definition is a crucial component of the reinforcement learning agent design. For DRL Agent-1, the normalized value of the number of successful uplink transmitted bytes is used as the reward function. For DRL Agent-2, both the achieved data rate and the fairness of the allocation are jointly optimized. The reward function is calculated as follows, where $r$ is the normalized throughput metric and $f$ is the normalized fairness metric. $\alpha$ is taken as 1 for DRL Agent-1, and it is taken as 0.8 for DRL Agent-2.

$$R = \alpha \cdot r + (1 - \alpha) \cdot f. \tag{3}$$

The fairness ($f$) of resource allocation is evaluated using Jain's fairness index, calculated by the following formula, where $d$ represents throughput values:

$$f = \frac{(\sum_{k=1}^{m} d_{UE_k})^2}{m \cdot \sum_{k=1}^{m} d_{UE_k}^2}. \tag{4}$$

## 4 Simulation and Results

In this section, we conduct simulation experiments to evaluate the proposed DRL approaches and compare them with the conventional resource scheduling methods. For this purpose, MATLAB's 5G Toolbox and Reinforcement Learning Toolbox are utilized. The 5G Toolbox is used to simulate the network environment, while the Reinforcement Learning Toolbox is employed to create, train, and simulate the RL agents.

MATLAB's 5G Toolbox provides a flexible and 3GPP standard compliant realistic simulation environment. It can be customized to test and simulate a wide range of scenarios and algorithms. However, the 5G Toolbox is not integrated with the Reinforcement Learning Toolbox. In this chapter, we customized the resource allocation procedure within the 5G Toolbox to enable the execution of actions decided by the DRL agents. The parameter set for the environment created by the 5G Toolbox is provided in Table 1.

MATLAB's Reinforcement Learning Toolbox provides a platform for training and simulating RL agents. This toolbox includes built-in DQN, PPO, SAC, and DDPG agents, allowing users to design decision-making algorithms. In this chapter, we used the Reinforcement Learning Toolbox to create the RL agents, train them based on the designed inputs, and simulate the agents using the 5G environment provided by the 5G Toolbox. The parameters of the agents using the RL Toolbox are provided in Table 2.

**Table 1** 5G network environment parameters

| Parameter | Value |
|---|---|
| Simulation time | 1 frame (10 ms) |
| Subcarrier spacing | 15 KHz |
| Frequency band | 2.59 GHz |
| Bandwidth | 5 MHz |
| Duplex type | FDD |
| Number of RBs | 25 |
| Scheduling type | Slot-based |
| Traffic model | On-off type |
| Number of users | 2 |

**Table 2** RL agent parameters

| Parameter | Value |
|---|---|
| RL agent type | DQN |
| Number of hidden units | 256 |
| Discount factor | 0.99 |
| Initial epsilon | 1 |
| Epsilon decay | 0.005 |
| Epsilon min | 0.01 |
| Experience buffer length | $10^4$ |

**Table 3** Training parameters

| Parameter | Value |
|---|---|
| Number of episodes | 300 |
| Episode length | 5 |
| Average window length | 50 |
| Convergence metric | Max episode count |

## *4.1 Training*

The DRL agents are trained by using initial parameters in Table 3. The training procedure is shown in Fig. 3, where the process begins with the DRL agent taking random actions. Each action includes the information about the percentage of the available RBs assigned to each user. Afterward, the 5G simulation environment runs for one frame and executes the corresponding resource allocation action. The environment then provides the results of this run, such as the buffer status and data rates of the UEs. The DRL agent calculates rewards based on the resulting state. If the algorithm reaches the convergence criteria (i.e., the quantized throughput value reaches the objective value or the number of episodes reaches the maximum), the training procedure ends. Otherwise, it continues to the next step.

Although using resource percentage reduces the number of action-state-reward combinations, increasing the number of UEs and the resolution of the resource percentages requires more time and computational resources. There are 11 action alternatives for two UEs, 66 alternatives for three UEs, and 1001 alternatives when the number of UEs is 5. To minimize computational complexity, the algorithms are trained using the lower number of UEs.

DRL Agent-1 is trained for 300 episodes, with each episode consisting of up to five steps. As shown in Fig. 4, the agent is converged to a reward value after approximately 150 episodes. DRL Agent-2 is also trained for 300 episodes, with each episode having a maximum of five steps. As seen in Fig. 5, this agent is converged after nearly 150 episodes. DRL Agent-2 is trained with $\alpha = 0.8$, where $\alpha$ represents the weight of throughput in the reward and $(1 - \alpha)$ is the weight of fairness. While the xApp can select different weights for these criteria, in this chapter, the chosen values are considered baseline parameters, reflecting the agent's scheduling performance. The trained DRL Agent-1 prioritizes the UE with the best channel conditions and the highest likelihood of achieving a high data rate. For

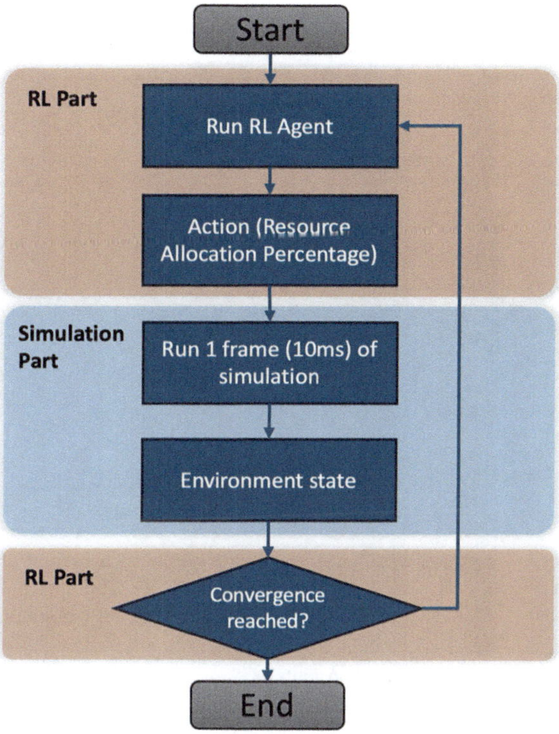


**Fig. 3** Training procedure of DRL agents

example, when one UE has a CQI of 10 and another has a CQI of 15, DRL Agent-1 assigns all resources to the UE with CQI 15. In contrast, DRL Agent-2, trained to balance throughput and fairness, may allocate 80% of resources to the UE with CQI 15 and 20% to the UE with CQI 10, reflecting its emphasis on throughput while still considering fairness.

## *4.2 Simulation Results*

The simulation phase for DRL agent-based resource allocation is shown in Fig. 2, where both DRL agents are simulated using this procedure. The simulations are run for 500 episodes, with the episode length set to 1, as each step begins with a randomly defined physical position. The UEs' positions are defined randomly between 100 and 500, values that directly impact the users' CQI in the 5G Toolbox. This allows the DRL agent to be tested under rapidly changing channel conditions.

The DRL agent-based resource allocation is compared with traditional Best-CQI and proportional fair resource allocation algorithms, as shown in Fig. 6. It can be observed that the DRL Agent-1 in the xApp nearly achieves the same throughput levels as the Best-CQI algorithm. It should be noted that the DRL agent has a

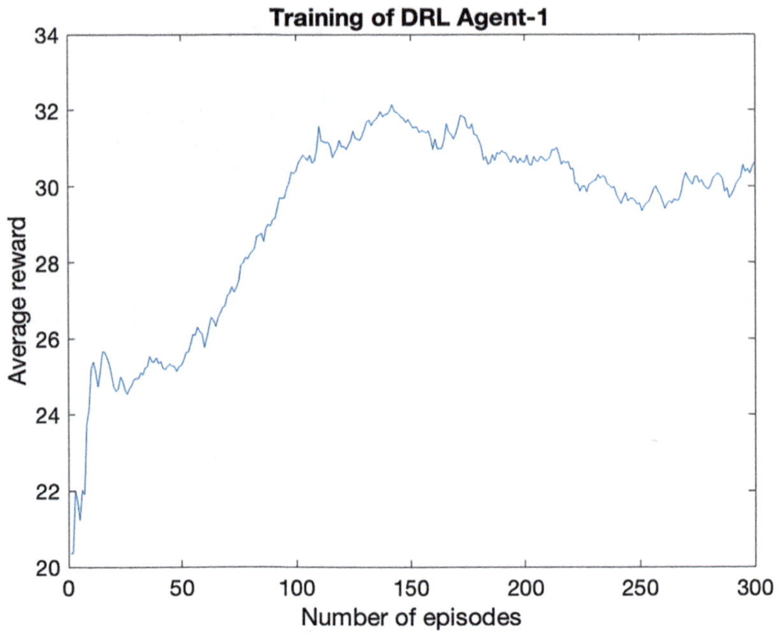


**Fig. 4** Training of DRL Agent-1

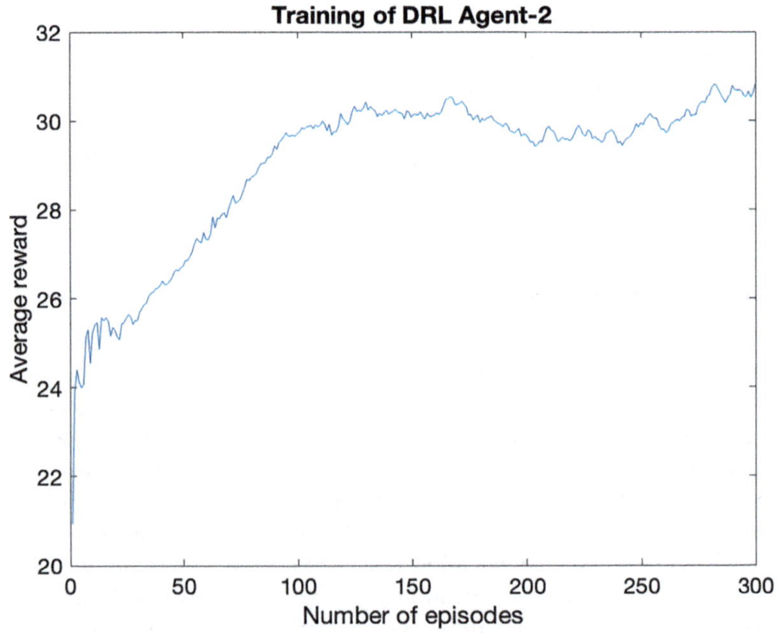


**Fig. 5** Training of DRL Agent-2

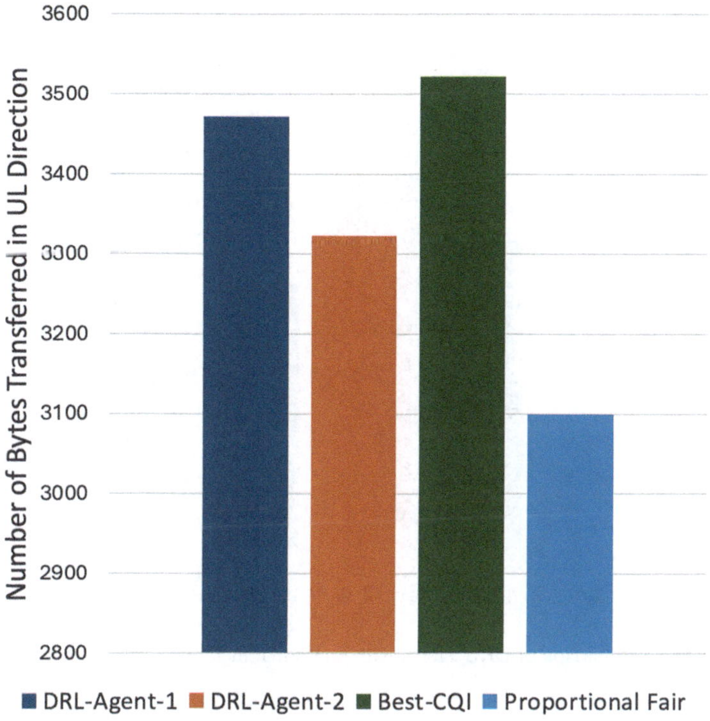


**Fig. 6** Cell throughput comparison of DRL agents with traditional methods

resource allocation resolution of 10 ms, whereas the Best-CQI algorithm schedules resources on a slot-based basis.

DRL Agent-2, which focuses on both throughput and fairness, is also compared with the Best-CQI and proportional fair agents. As shown in Fig. 7, DRL Agent-2 outperforms Best-CQI and DRL Agent-1 in terms of fairness. However, since DRL Agent-2 is configured with a policy that prioritizes throughput over fairness, its fairness level remains lower than that of the proportional fair agent.

Note that the selection of the DRL agent can be dynamically performed by rApp based on high-level network policies. By adjusting the weights of the throughput and fairness parameters, fine-tuning of the policy can also be achieved by rApp. Since the 5G Toolbox allows simulations to run for at least one frame, the simulations in this study are performed with the actions from the DRL agents taken every one frame. By using a simulation environment that permits more frequent resource allocation actions, this study will be extended to support more latency-sensitive applications in the future.

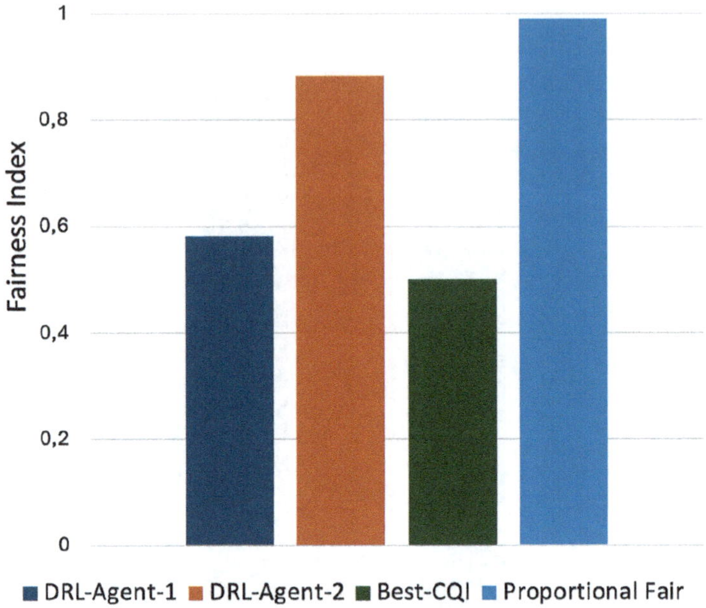


**Fig. 7** Fairness comparison of DRL agents with traditional methods

## 5 Training Results Across Different Configurations

Modifying the agent's parameters and environmental conditions can significantly impact the convergence time and performance of the DRL agents. In this section, we present the training results of agents under various configurations. Additionally, several scenarios with different simulation environment parameters are considered, and the corresponding training outcomes are discussed.

The parameters adjusted in the agent's configuration include the discount rate and learning rate, while the environmental parameter modified is the number of UEs. Changing the agents' discount rate did not result in a significant change in their training performance under the defined configurations. However, agents with higher learning rates failed to converge as effectively as the agent with a learning rate of 0.01, as illustrated in Fig. 8.

The training performance of different agent types also varied. MATLAB's Reinforcement Learning Toolbox supports three types of agents for discrete action spaces: DQN, PPO, and TRPO (trust region policy optimization). In this chapter, we compared the training performance of these agents in a resource allocation simulation environment. The agents were evaluated under the same configurations, including discount rate, learning rate, and the number of UEs. Among the alternatives, DQN agent was the only one to converge. The comparison is visualized in Fig. 9.

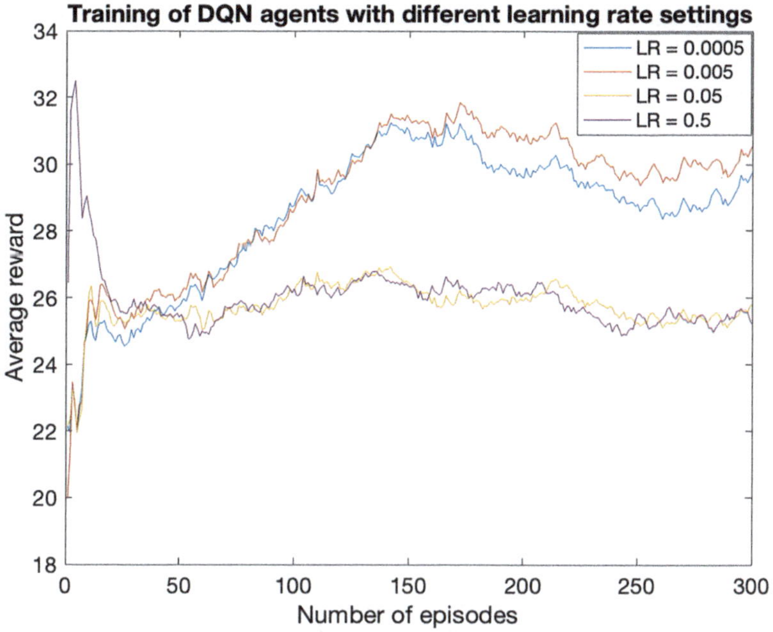


**Fig. 8** Training convergence performance for agents with different learning rates

As previously noted, increasing the number of UEs results in exponential growth in the action-state-reward space, which subsequently raises the computational resources required for training. The convergence performance of DRL Agent-1 for different numbers of UEs is compared in Fig. 10, where the maximum number of episodes is set to 300 and the maximum episode length is set to 5. It is evident that increasing the number of UEs prevents the agent from converging within the same number of episodes. Alternatively, running the training procedures on more powerful computational units, such as graphics processing units (GPUs), could help mitigate this issue. A key challenge in scaling DRL-based resource allocation to dynamic, real-world environments is the fluctuating number of UEs, as DRL agents are typically trained with a fixed UE count. Developing a solution for this variability is essential, and pointer networks may offer a viable approach [1].

## 6 Conclusion

This chapter proposes an uplink resource allocation scheme for 5G and beyond networks utilizing the RIC platforms within the O-RAN architecture. By leveraging DRL agents for resource allocation, the flexibility of the MAC scheduler is enhanced. Specifically, throughput-focused and throughput-fairness-focused DRL

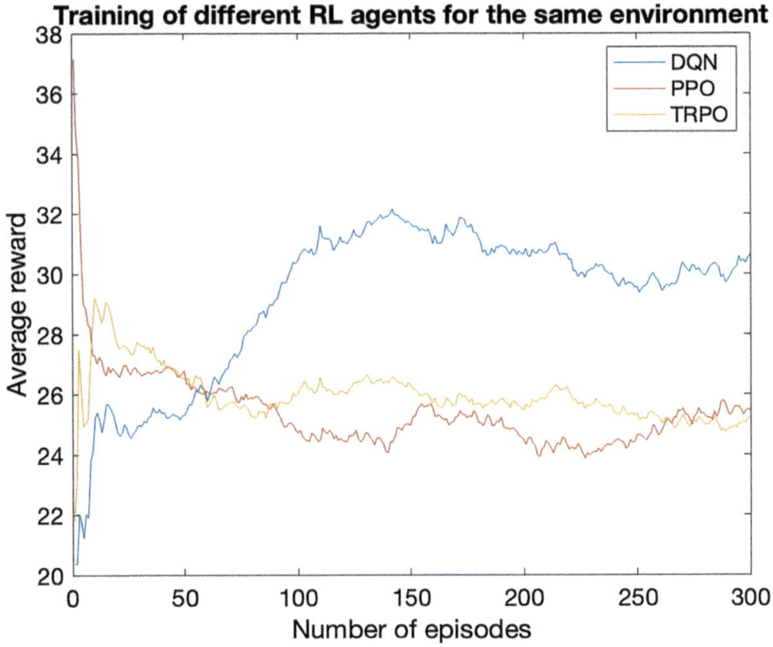


**Fig. 9** Training convergence performance for different types of RL agents

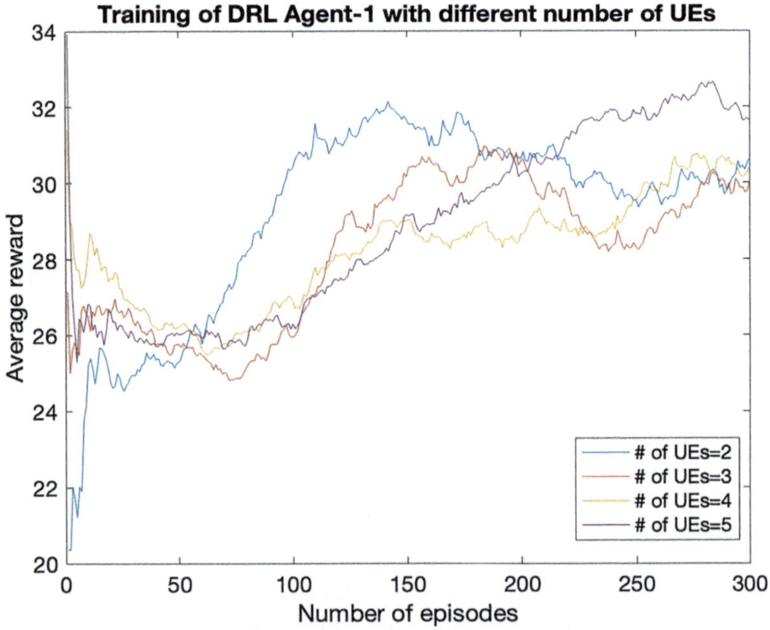


**Fig. 10** Training of DRL Agent-1 with a different number of UEs

agents were created, trained, and simulated for this purpose. The rApp, responsible for high-level policy decisions, plays a crucial role in conveying these policies to near-real-time units, such as the xApp, which influences near-real-time operations within the O-DU. Numerical results indicate that the DRL agents achieve comparable performance to traditional resource allocation methods. Additionally, training performances of different agent types and varying hyperparameters are compared and results are presented.

In future work, the proposed system will be extended to more realistic conditions, such as increasing the number of UEs. Additionally, creating a DRL agent with a primary focus on minimizing latency could be explored as a future research direction. Moreover, dynamically selecting the most suitable policy based on UE conditions, potentially facilitated by DRL agents, offers another promising area for future investigation.

## References


1. Seguin, M., Omer, A., Koosha, M., Malandra, F., Mastronarde, N.: Deep reinforcement learning for downlink scheduling in 5G and beyond networks: a review. In: IEEE International Symposium on Personal and Indoor Radio Mobile Communications (PIMRC), Toronto, pp. 1–6 (2023). https://doi.org/10.1109/PIMRC56721.2023.10293754
2. Akyildiz, H.A., Gemici, Ö.F., Hökelek, İ., Çirpan, H.A.: Hierarchical reinforcement learning based resource allocation for RAN slicing. IEEE Access **12**, 75818–75831 (2024). https://doi.org/10.1109/ACCESS.2024.3406949
3. Yan, D., Ng, B.K., Ke, W., Lam, C.-T.: Deep reinforcement learning based resource allocation for network slicing with massive MIMO. IEEE Access **11**, 75899–75911 (2023). https://doi.org/10.1109/ACCESS.2023.3296851
4. Qazzaz, M.M.H., et al.: Machine learning-based xApp for dynamic resource allocation in O-RAN networks. In: 2024 IEEE International Conference on Machine Learning for Communication and Networking (ICMLCN), pp. 492–497 (2024)
5. Al-Tam, F., Correia, N., Rodriguez, J.: Learn to schedule (LEASCH): a deep reinforcement learning approach for radio resource scheduling in the 5G MAC layer. IEEE Access **8**, 108088–108101 (2020). https://doi.org/10.1109/ACCESS.2020.3000893
6. AlQwider, W., Rahman, T.F., Marojevic, V.: Deep Q-network for 5G NR downlink scheduling. In: 2022 IEEE International Conference on Communications Workshops (ICC Workshops), pp. 312–317, Seoul (2022). https://doi.org/10.1109/ICCWorkshops53468.2022.9814547
7. Edelmann, E.-M., Kiyak, Y., Schmalen, L.: Deep reinforcement learning for uplink multi-carrier non-orthogonal multiple access resource allocation using buffer state information. arXiv preprint (2022). https://doi.org/10.48550/arXiv.2208.14689
8. Robaglia, B.-M., Coupechoux, M., Tsilimantos, D.: Deep reinforcement learning for uplink scheduling in NOMA-URLLC networks. IEEE Trans. Mach. Learn. Commun. Netw. **2**, 1142–1158 (2024). https://doi.org/10.1109/TMLCN.2024.3437351
9. Elsayem, M., Abou-zeid, H., Afana, A., Givigi, S.: Reinforcement learning-based dynamic resource allocation for grant-free access. In: GLOBECOM 2022 - 2022 IEEE Global Communications Conference, pp. 1091–1096, Rio de Janeiro (2022). https://doi.org/10.1109/GLOBECOM48099.2022.10001586
10. Arslan, H., Yilmaz, S., Şen, S.: Dynamic MAC scheduling in O-RAN using federated deep reinforcement learning. In: 2023 International Conference on Smart Applications, Communications and Networking, SmartNets 2023, pp. 1–6, İstanbul (2023). https://doi.org/10.1109/smartnets58706.2023.10216125

11. Hurtado Sánchez, J.A., Casilimas, K., Caicedo Rendon, O.M.: Mobility aware and energy-efficient federated deep reinforcement learning assisted resource allocation for 5G-RAN slicing. Sensors **22**(8), 3031 (2022). https://doi.org/10.3390/s22083031
12. D'Oro, S., Polese, M., Bonati, L., Cheng, H., Melodia, T.: dApps: distributed applications for real-time inference and control in O-RAN. IEEE Commun. Mag. **60**(11), 52–58 (2022). https://doi.org/10.1109/MCOM.001.2200401

# High-Frequency CPW-Fed Micro-Sensor Design and Optimization for Advanced Robotic Applications

**Djamel Sayad, Chemseddine Zebiri, Issa Elfergani, Kamil Karaçuha, Mouhamed Lamine Bouknia, Rami Zegadi, and Jonathan Rodriguez**

## 1 Introduction

The use of advanced micro-sensors is driving a significant shift in the robotics industry, with many robots now possessing this technology to enable greater autonomy and precise sensing capabilities necessary for navigation and control [1, 2]. Owing to their real-time data acquisition and processing, these highly sensitive and versatile sensors are revolutionizing a wide range of applications, including pressure detection and complex environmental monitoring, which require quick and accurate decision-making [3, 4]. Among the noteworthy developments in this area are micro-sensors that utilize coplanar waveguide (CPW) as their primary source of signal strength, have low footprint requirements, and can be easily integrated [5, 6]. The presence of these features is critical to the creation and advancement in modern robots, as accurate sensing greatly simplifies control algorithms. Also,

D. Sayad
Laboratoire d'Electrotechnique de Skikda (LES), Department of Electrical Engineering. University of 20 Aout 1955, Skikda, Skikda, Algeria
e-mail: d.sayad@univ-skikda.dz

C. Zebiri (✉) · M. L. Bouknia · R. Zegadi
Laboratoire d'Electronique de puissance et commande industrielle (LEPCI), Department of Electronics. University of Ferhat Abbas, Setif, Setif, Algeria
e-mail: czebiri@univ-setif.dz; ml.bouknia@univ-setif.dz; ramizegadi@univ-setif.dz

I. Elfergani · J. Rodriguez
Instituto de Telecomunicações, Campus Universitário de Santiago, Aveiro, Portugal
e-mail: i.t.e.elfergani@av.it.pt; jonathan@av.it.pt

K. Karaçuha
Electrical Engineering Department, Istanbul Technical University, Istanbul, Turkey
e-mail: karacuha17@itu.edu.tr

 
B. U. Töreyin et al. (eds.), *8th EAI International Conference on Robotic Sensor Networks*, EAI/Springer Innovations in Communication and Computing,
https://doi.org/10.1007/978-3-031-92143-8_8

the use of resonator-based sensors with millimeter-wave operating frequencies has become popular because it is capable of accurately measuring subtle changes in environment and physical behavior. The use of circular resonator designs and optimized superstrate materials in these sensors enhance their performance in dynamic and complex environments [7]. In applications where the movement of a vertical membrane is essential, such as in robotic hands and tactile interfaces, CPW-fed sensors are particularly useful for measuring force and pressure. High-frequency applications have seen a boost in sensor performance, with the refinement of geometric parameters such as feed line positioning, ground plane dimensions, and gap size being particularly useful [8, 9]. The use of CPW-fed sensors in miniaturized robotic platforms has made them an ideal choice for applications such as surgical robotics, industrial automation, and autonomous vehicles due to their robustness [10]. Significant progress has been achieved in designing resonator-based micro-sensors for real-time detection. These systems benefit from advancements in superstrate materials, geometric optimization of feed lines, and impedance matching. For instance, sensors using circular structures and vertically movable membranes exhibit enhanced sensitivity for pressure detection. The variation of geometric parameters, such as the feed lines position and ground plane width, allows fine adjustments to the resonance frequency, optimizing performance in dynamic environments [11, 12].

Herein, we present a 30 GHz circular-shaped CPW-fed micro-sensor optimized for high-frequency applications, integrating modifications to the resonator geometry and feed lines to enhance performance. The working principle of the micro-sensor was analyzed through circuit modeling, while its design and optimization were conducted using High-Frequency Structure Simulator (HFSS) software. The optimization process included a detailed parametric study of geometric factors such as feed line position, ground plane width, and gap dimensions, as well as the influence of a superstrate layer and vertical membrane displacement on the sensor's impedance matching and resonance frequency.

The results demonstrate that the optimized micro-sensor achieves a high sensitivity of approximately 1.2 GHz/$\mu$m, attributed to precise adjustments in permittivity and layer thickness. Additionally, the integration of advanced design elements ensures improved impedance matching and resonance control. The sensor's robustness and tunability make it suitable for precise pressure sensing and other high-frequency applications in challenging environments.

## 2 Proposed Micro-Sensor Design

Recent advances in superstrate layers with adjustable permittivity enhance sensor performance in extreme and hazardous environments, paving the way for more resilient and adaptable robotic systems [13]. The integration of high-frequency CPW-fed micro-sensors improves current robotic capabilities and opens new possibilities for future applications that require unprecedented precision and adaptability.

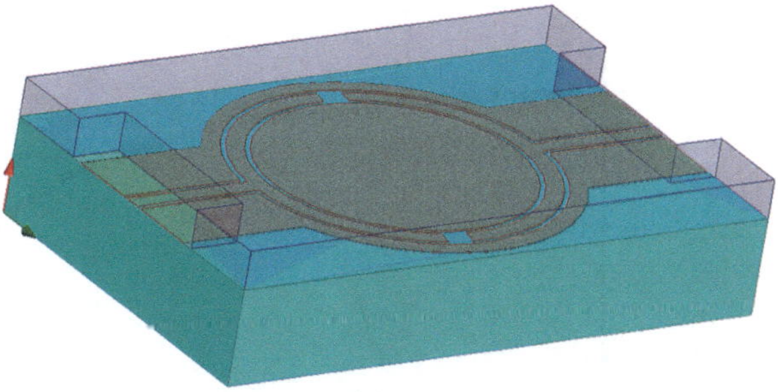

**Fig. 1** HFSS design of transducer micro-sensor

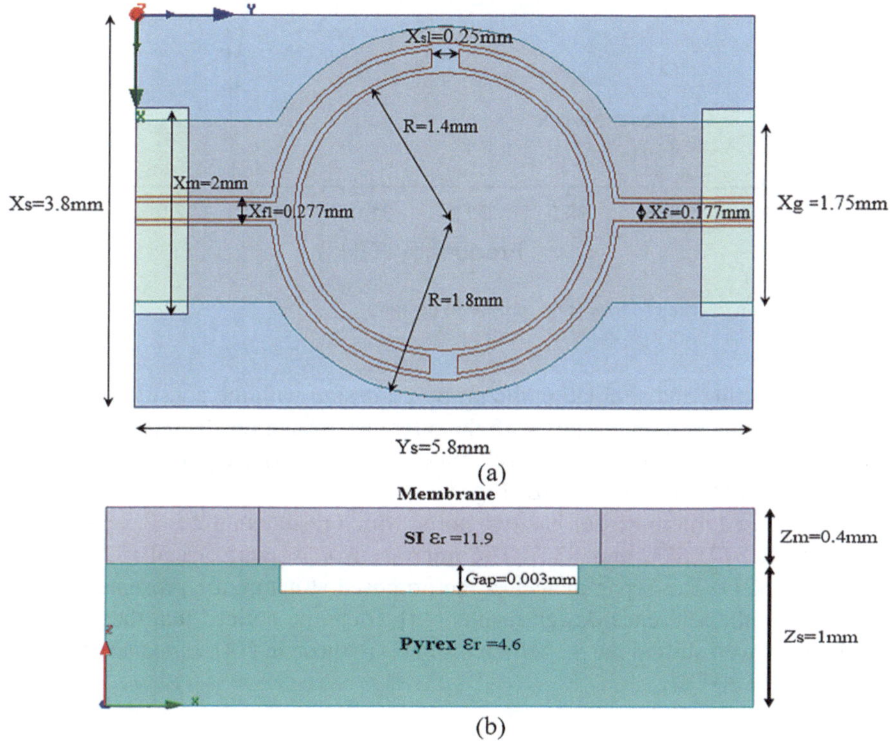


**Fig. 2** (**a**) Resonator dimensions and (**b**) Cross-sectional view of the micro-sensor

Our simulated design is based on the design presented in [14] (Fig. 1). The changes made concerned mainly the resonator shape and the excitation lines because the elliptical and circular shapes, studied in this case, converge quickly. The other dimensions and micro-sensor media are preserved as considered in [14].

The dimensions and media used are described in Fig. 2a, b, with a 1 mm Pyrex substrate, 400 μm high resistivity silicon superstrate, and between the two media resides a 3 μm air layer.

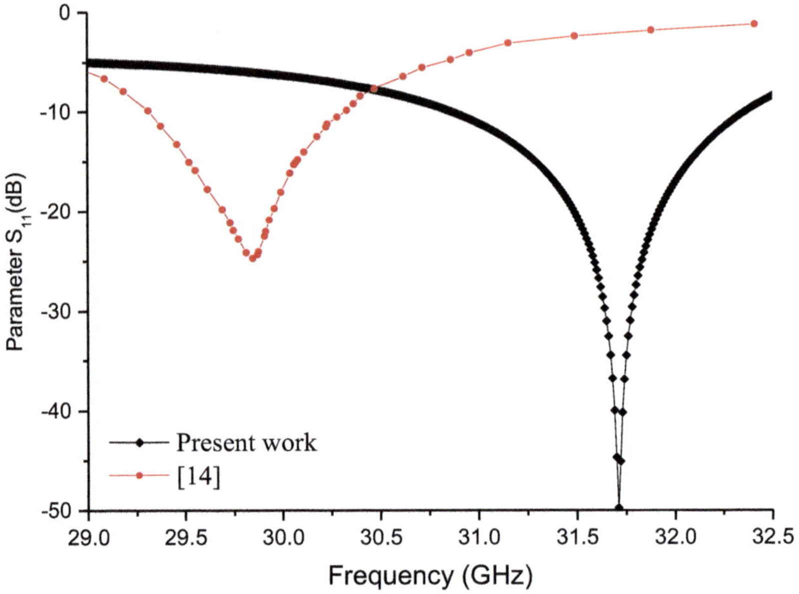


**Fig. 3** Reflection coefficient S11 of the proposed design

After modeling and simulating the proposed design with the considered modifications, we compared our results with the reference design results reported in [14]. It can be seen that our resonator is well matched to the frequency 31.71 GHz with a coefficient S11 lower than −50 dB (Fig. 3).

The proposed micro-sensor has two ports, which results in a 2 × 2 S-parameter matrix (S11, S12, S21, and S22). The micro-sensor is symmetrical (S11 = S22 and S12 = S21). The S-parameters of the proposed structure are presented in Fig. 4 along with the reference design results [14]. It can be noticed that the proposed structure has well shifted the S coefficients close to those in [14], especially for high frequencies at 30 GHz.

## 3 Parametric Study

For an optimized micro-sensor structure, a parametric study was conducted. We examined the effect of different geometric parameters on the resonance frequency and the S11 parameter of the micro-sensor, as well as the effect of a superstrate layer. For this, some geometric parameters of the micro-sensor were modified in order to evaluate their impact on performance, particularly in terms of adaptation over a narrow frequency band.

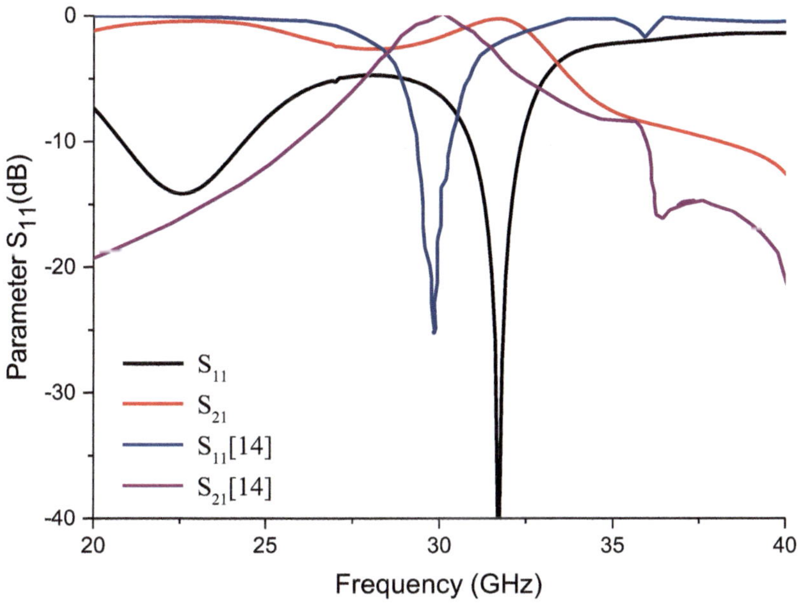


**Fig. 4** S11 and S21 parameters of the proposed design compared to the reference design

## 3.1 *Effect of the Position of the Feed Line (Fx)*

It is clearly observed from Fig. 5.b that the position of the excitation line with respect to the ground plane and the resonator has a significant impact on the impedance matching of our structure. A better matching is obtained for fx = 0.

## 3.2 *Effect of the Width of the Ground Plane of the CPW (Xg)*

The ground plane plays a major role on the different parameters of the CPW and the resonator structure. Therefore, we sought to demonstrate the influence of the ground plane and to determine its optimal dimensions. The impact of the ground plane width of the CPW (Xg) on the S11 parameter is illustrated in Fig. 6.

Figure 6 presents the S11 parameter for different ground plane geometries. A significant variation of the S11 coefficient is observed, as well as a shift in resonance frequency and bandwidth. For a better impedance matching and S11 coefficient, we opted for a ground plane width of Xg = 1.75 mm.

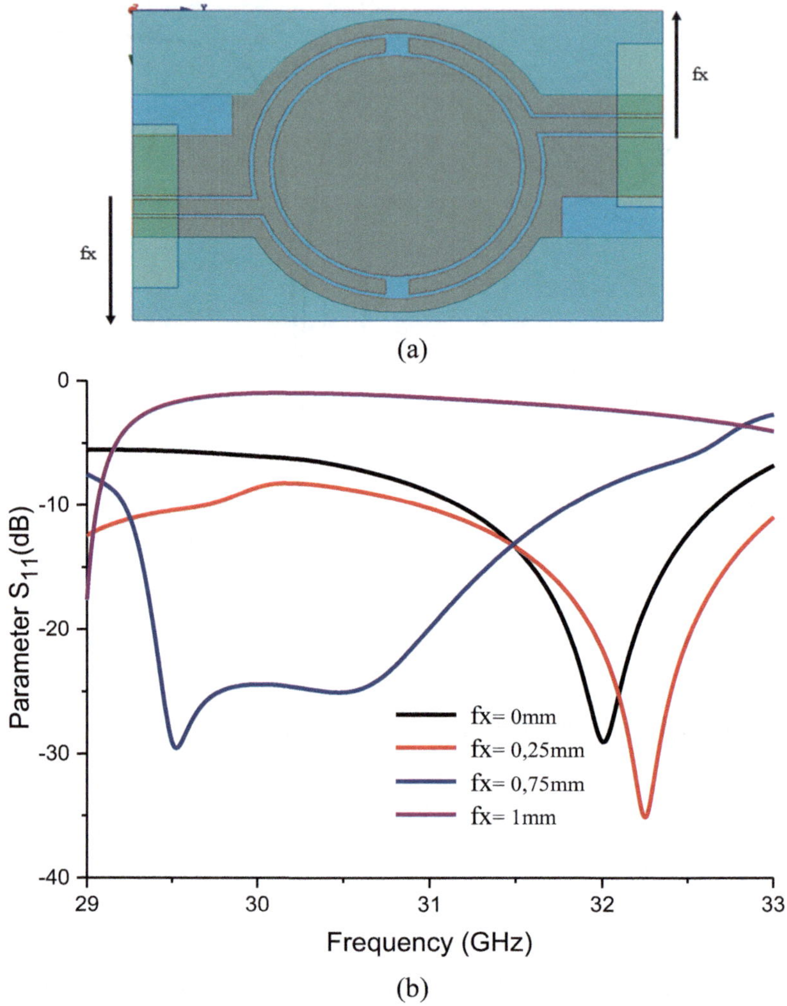


**Fig. 5** (**a**) Feed line geometry and (**b**) Simulated S11 for different feed positions

## 3.3 *Effect of the Gap Position of the Feed Lines (φ)*

In this subsection, the influence of the gap position between the excitation lines on the parameters of the micro-sensor is investigated, in particular on the resonance frequency and the S11 parameter. The gap position $\varphi$ of the CPW feed is varied. As illustrated in Fig. 7, it is observed that a better S11 parameter and a narrower bandwidth (less than $-10$ dB) are obtained for $\varphi = 90°$.

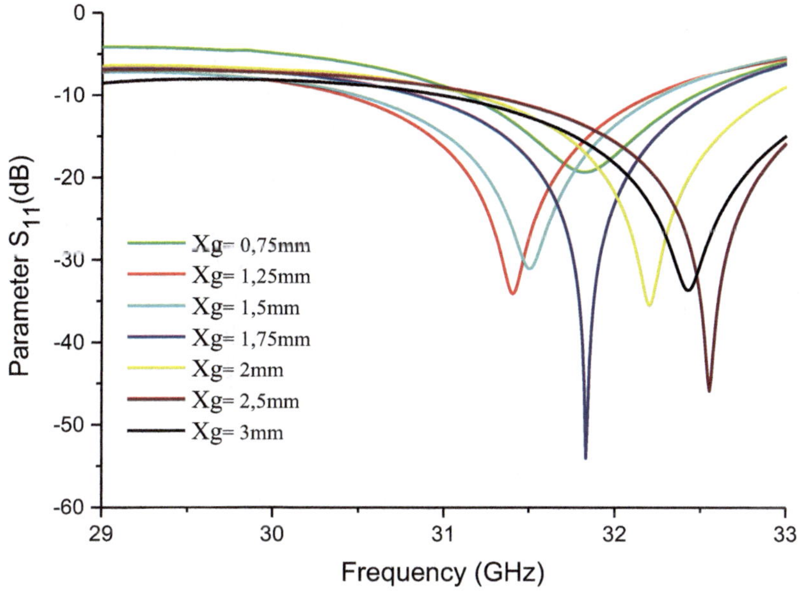


**Fig. 6** Resonance frequency for different ground plane widths

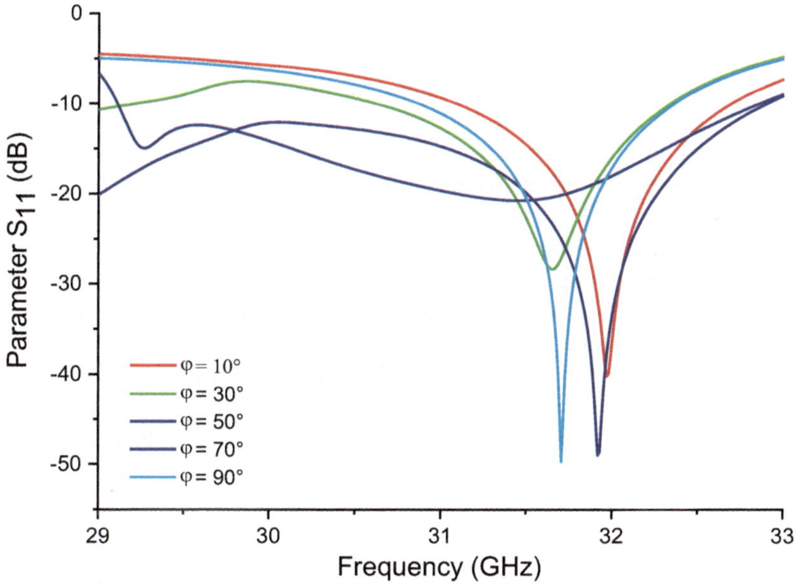


**Fig. 7** Simulated resonance frequency for different values of φ

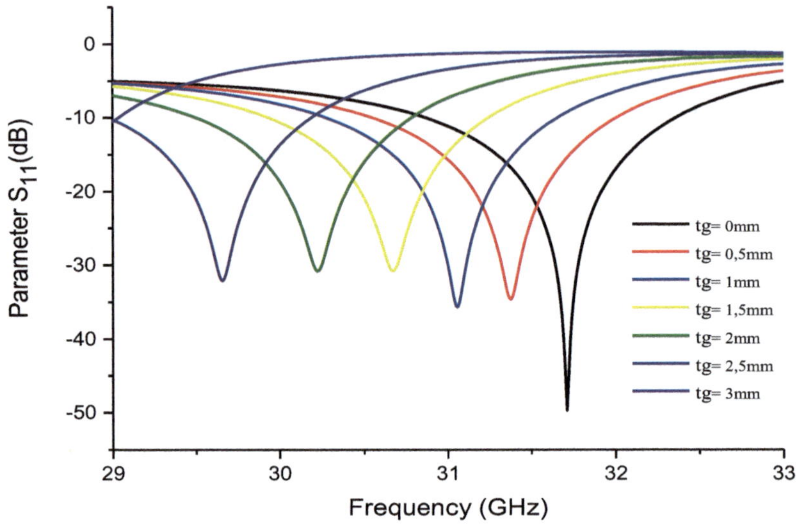


**Fig. 8** Simulated resonance frequency for different tg values

## 3.4 *Uniform Vertical Displacement Effect of the Membrane (Tg)*

After the parametric study and obtaining the different parameters of our microsensor, we try to express the displacement of the membrane, generated by the pressure, in terms of frequency variations. We now simulate the vertical and uniform movement of the membrane above the coplanar resonator.

Figure 8 presents the transmission coefficient obtained for different values of displacement of the superstrate under pressure, "tg", accompanied by a decrease in the thickness of the air layer. According to the results obtained, an increase in the displacement "tg" leads to a variation in the coefficient S11 as well as a shift in the resonator resonant frequency. By increasing "tg", the high resistivity silicon approaches the feeds and the resonator, which increases the effective permittivity of the medium. Since the resonant frequency is proportional to the inverse of the square root of the effective permittivity, the latter decreases as the effective permittivity increases.

## 3.5 *Control of the Resonance Frequency by Superstrate*

Substrate-superstrate structures are typically used to improve the performance of resonators. In addition, the superstrate also serves as protection against adverse weather conditions such as rain, fog, and snow, which can appear unpredictably and

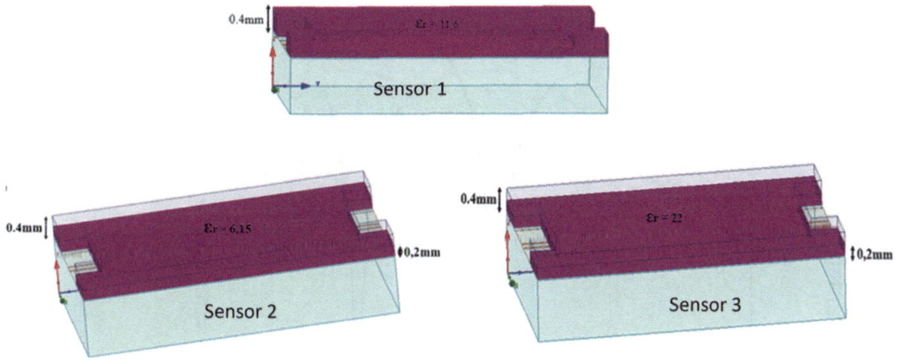


**Fig. 9** Different configurations of the designed sensor

affect the performance of printed resonators. Therefore, for optimal characterization and accurate design, these effects must be taken into account. In this subsection, we study the effect of the superstrate on the S parameters and how to adjust the resonance frequencies of this type of sensor. To do this, we replaced the single superstrate layer (400-μm silicon with a permittivity of 11.6) by two dielectric layers of different permittivities.

Two configurations were considered:

- Sensor 2: protected by a superstrate with a permittivity of 6.15 and a thickness of 200 μm, placed underneath a 200 μm silicon layer.
- Sensor 3: protected by a superstrate with a permittivity of 22 and a thickness of 200 μm, placed underneath a 200 μm silicon layer.

The three sensors modeled in the HFSS interface are shown in Fig. 9, and their corresponding transmission coefficients are illustrated in Fig. 10.

From Figs. 11 and 12, it is evident that the resonance frequency of the sensing design can be easily controlled by selecting the appropriate permittivity and thickness of the additional layer. Consequently, the thickness of the additional layer in Sensor 3 was set to 37 μm to achieve a resonance frequency of 30 GHz.

## 4 Evaluation of the Sensor Sensitivity

Figures 13 and 14 show the effect of vertical displacement of the membrane on the resonance frequency and the variation of the sensitivity of the three sensors respectively. Three different operating frequency zones of the three sensors (Sensor 1, 2, and 3) have been identified.

The first zone, corresponding to Sensor 3, is characterized by a linear variation of the resonance frequency as a function of pressure, with a sensitivity of approximately 1.2 GHz/μm. This zone is larger than those of the other sensors, and Sensor 3

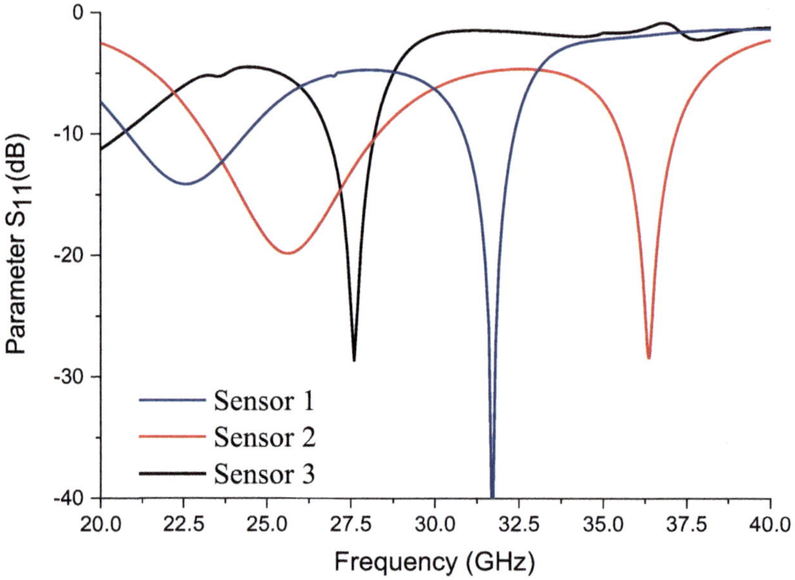


**Fig. 10** Effect of permittivity of the first superstrate layer on the reflection coefficient

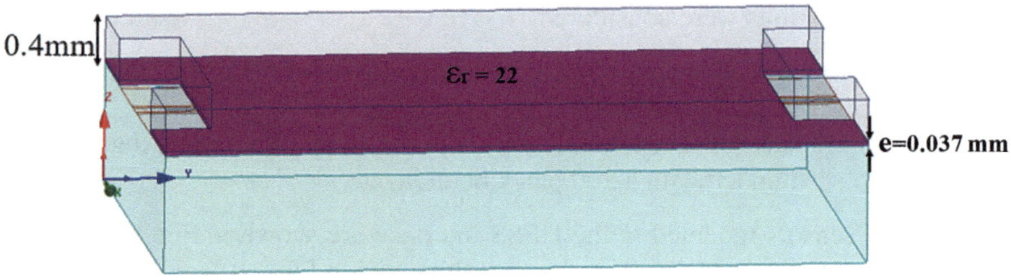


**Fig. 11** Sensor 3 with additional layer

exhibits significantly improved performance compared to the designs reported in the literature, in particular its increased sensitivity. In the second zone, corresponding to Sensor 1, the sensitivity of the sensor varies between 1 GHz/μm and 0.85 GHz/μm. This zone has an intermediate sensitivity. In the third zone, associated with Sensor 2, the resonance frequency and the effective permittivity undergo more significant variations as a function of the displacement of the membrane, resulting in a decrease in sensitivity to reach approximately 0.464 GHz/μm.

These results can play a key role in optimizing the operation of electromagnetic transducer sensors since linear behavior is always preferred to ensure optimal accuracy in the measurement units.

To better illustrate and understand the operation of the electromagnetic transducer sensor, we investigated the distribution of the electric field as shown in Fig. 15.

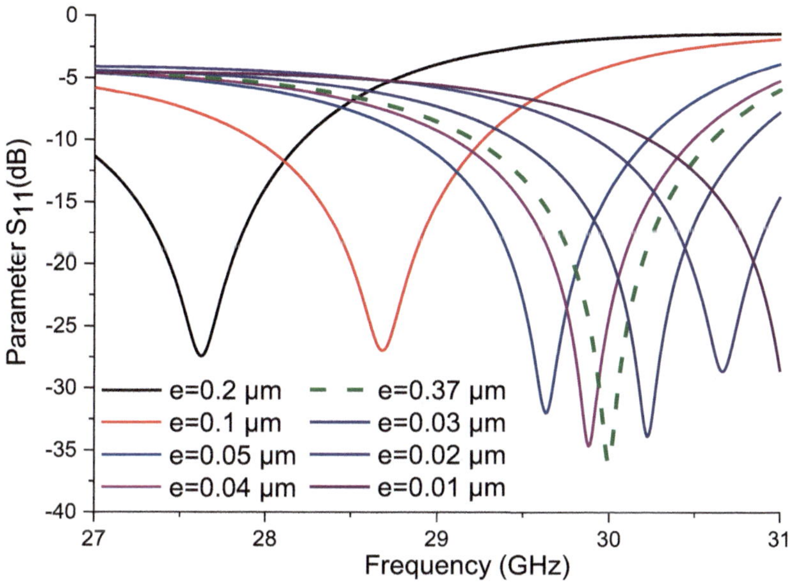


**Fig. 12** Effect of the superstrate additional layer thickness placed below the silicon of thickness 200 μm with a permittivity of 22

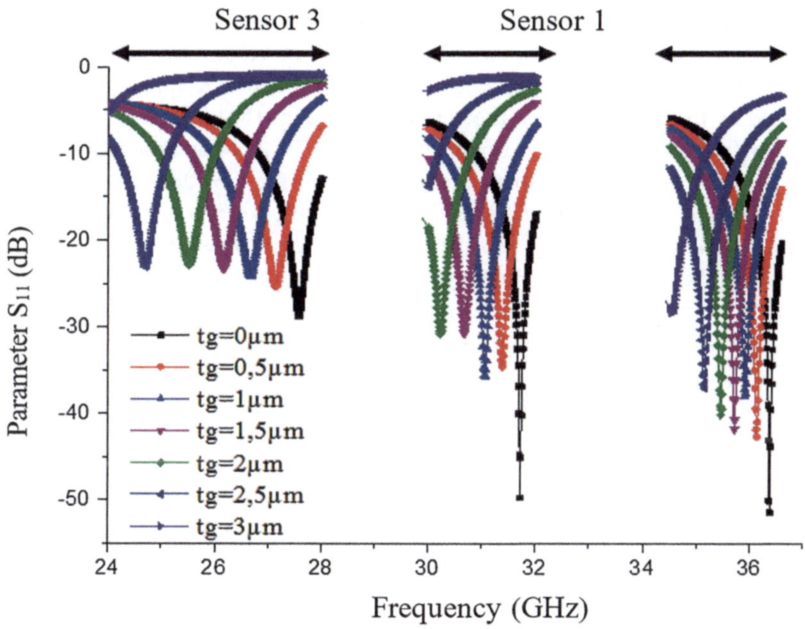


**Fig. 13** Influence of the membrane uniform vertical displacement (tg) on the resonance frequency of the three micro-sensors

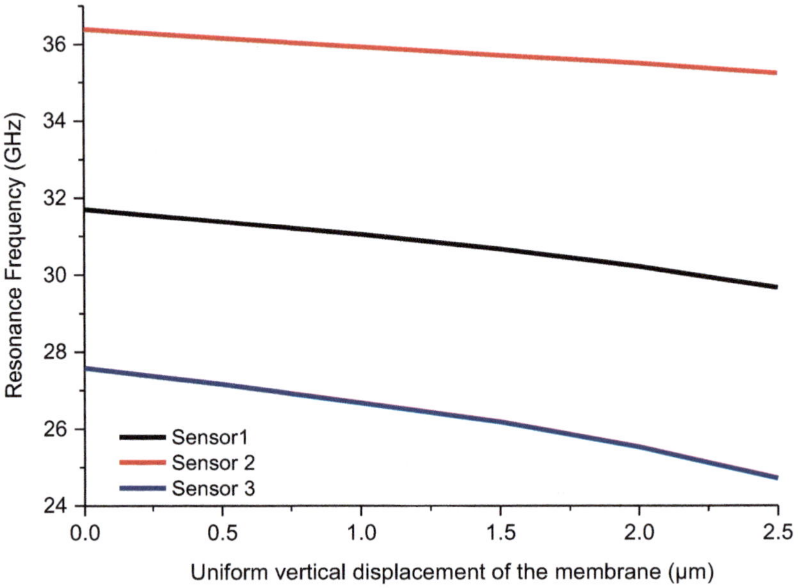


**Fig. 14** Determination of the sensitivity zones of the three micro-sensors

A pressure force applied to the sensor produces a vertical displacement of a thin membrane located above a circular planar millimeter-wave resonator. Such a displacement brings changes to the distribution of the electromagnetic field around the resonator accompanied by a change in the resonance frequency. This is well illustrated by the distribution of the electric field (Fig. 15a) for a zero displacement at 30 GHz (which presents the resonance frequency of this case) compared with that of a maximum membrane displacement (Fig. 15b) for the same frequency. Thus, the electromagnetic interaction responds appropriately to each pressure, causing the sensor's resonance frequency to shift according to its sensitivity by a few MHz around the initial resonance frequency, $f_0 = 30$ GHz, over a frequency range of 1 to 4 GHz.

## 5 Conclusion

This work outlines the development and optimization of a 30 GHz CPW-fed circular micro-sensor, aimed at enhancing its performance for high-frequency applications. By systematically modifying key parameters such as the resonator shape, feed line position, and superstrate configuration, we achieved significant improvements in impedance matching, resonance frequency control, and sensor sensitivity. The parametric study highlighted the importance of geometric factors and material properties, particularly the effect of vertical membrane displacement and superstrate

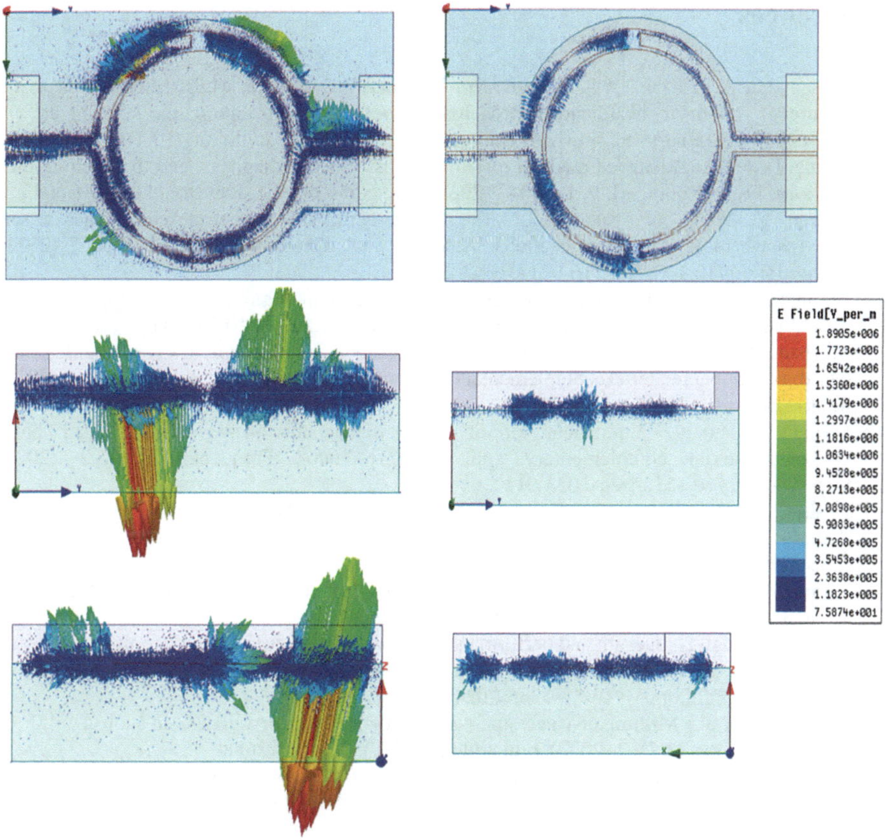


**Fig. 15** Electric field distribution in the micro-sensor

permittivity, in fine-tuning the sensor's response. The final optimized design demonstrates a high sensitivity of approximately 1.2 GHz/μm, with linear behavior over a wide range, making it a promising candidate for precise electromagnetic sensing applications. This approach can serve as a foundation for further sensor development, with potential applications in pressure sensing and environmental monitoring.

**Acknowledgments** This study was funded by the Scientific Research Unit (BAP) of Istanbul Technical University. This publication has received funding from the European Union's Horizon Europe research and innovation program under grant agreement HE-MSCA-SE-6G-TERAFIT-101131501. This work is also funded by the FCT/MEC through national funds and when applicable co-financed by the ERDF, under the PT2020 Partnership Agreement under the UID/EEA/50008/2020 and UIDP/50008/2020 project.

## References


1. Moshayedi, A.J., Khan, A.S., Davari, M., Mokhtari, T., Andani, M.E.: Micro robot as the feature of robotic in healthcare approach from design to application: the state of art and challenges. EAI Endorsed Trans. AI Robotics. **3**, 1 (2024). https://doi.org/10.4108/airo.5602
2. Zhang, D., et al.: Advanced medical micro-robotics for early diagnosis and therapeutic interventions. Front. Robot. AI. **9**, 1086043 (2023). https://doi.org/10.3389/frobt.2022.1086043
3. Cheng, Y., Wang, K., Xu, H., Li, T., Jin, Q., Cui, D.: Recent developments in sensors for wearable device applications. Anal. Bioanal. Chem. **413**(24), 6037–6057 (2021). https://doi.org/10.1007/s00216-021-03602-2
4. Borah, S.S., Khanal, A., Sundaravadivel, P.: Emerging technologies for automation in environmental sensing: review. Appl. Sci. **14**(8), 8 (2024). https://doi.org/10.3390/app14083531
5. Mosbah, S., et al.: Compact and highly sensitive bended microwave liquid sensor based on a metamaterial complementary Split-ring resonator. Appl. Sci. **12**, 4 (2022). https://doi.org/10.3390/app12042144
6. Mekki, S., et al.: Equivalent circuit of a planar microwave liquid sensor based on metamaterial complementary split ring resonator. Frequenz. **78**, 1–2 (2023). https://doi.org/10.1515/freq-2023-0111
7. Mishra, D.P., Behera, S.K.: Resonator based Chipless RFID: a frequency domain comprehensive review. IEEE Trans. Instrum. Meas. **72**, 1–16 (2023). https://doi.org/10.1109/TIM.2022.3225027
8. Ahmad, M.B.: Drawbacks of robotic cleaning or dry-cleaning robots. TMP Univ. J. Res. Rev. Arch. **3**, 1 (2024). https://doi.org/10.69557/ujrra.v3i1.68
9. Raj, A., Mandal, D.: Design and performance analysis of dielectric resonator antenna Array for 5G mm-wave ground-based navigation and wireless applications. Arab. J. Sci. Eng. **50**, 10611 (2024). https://doi.org/10.1007/s13369-024-09745-5
10. Jabbar, A., et al.: Millimeter-wave smart antenna solutions for URLLC in industry 4.0 and beyond. Sensors. **22**(7), 7 (2022). https://doi.org/10.3390/s22072688
11. Vinita, D.P., Islam, S.M., Singh, J.: Bulk acoustic wave resonators for sensing applications: a review. Sensors Actuators A Phys. **378**, 115839 (2024). https://doi.org/10.1016/j.sna.2024.115839
12. Kou, H., Tan, Q., Wang, Y., Zhang, G., Su, S., Xiong, J.: A wireless slot-antenna integrated temperature-pressure-humidity sensor loaded with CSRR for harsh-environment applications. Sensors Actuators B Chem. **311**, 127907 (2020). https://doi.org/10.1016/j.snb.2020.127907
13. Yuan, Z., Jiang, D., Chen, L., Zhu, Z.: The design and preparation of permittivity-adjustable FeNi@SrFe-MOF composite powders. Coatings. **14**(1), 1 (2024). https://doi.org/10.3390/coatings14010112
14. Jatlaoui, M.M., Chebila, F., Pons, P., Aubert, H.: Working principle description of the wireless passive EM transduction pressure sensor. Eur. Phys. J.Appl. Phys. **56**(1), 13702 (2011). https://doi.org/10.1051/epjap/2011100220

# Who Is Smoking? A Frame Analysis-Based Cigarette Detection Framework for Offline Videos

**Onur Can Koyun, Reyhan Kevser Keser, and Behçet Uğur Töreyin**

## 1 Introduction

Although cigarette detection has a wide range of applications, the number of studies on this problem in the literature is quite low. This problem holds critical importance

This research is supported in part by Turkcell Teknoloji A.Ş., partly by the Scientific Research Projects Coordination Department (BAP), Istanbul Technical University, under the project ITU-BAP MGA-2024-45372, and by the Scientific and Technological Research Council of Turkey (TUBITAK) 1515 Frontier R&D Laboratories Support Program for BTS Advanced AI Hub: BTS Autonomous Networks and Data Innovation Lab Project 5239903.

O. C. Koyun (✉)
Informatics Institute, İstanbul, Turkey

Signal Processing for Computational Intelligence Research Group (SP4CING), İstanbul, Turkey

Center for Artificial Intelligence Research, Wake Forest University School of Medicine, Winston-Salem, NC, USA
e-mail: okoyun@itu.edu.tr

R. K. Keser
Informatics Institute, İstanbul, Turkey

Signal Processing for Computational Intelligence Research Group (SP4CING), İstanbul, Turkey
e-mail: keserr@itu.edu.tr

B. U. Töreyin
Informatics Institute, İstanbul, Turkey

Signal Processing for Computational Intelligence Research Group (SP4CING), İstanbul, Turkey

Department of Artificial Intelligence and Data Engineering, İstanbul Technical University, İstanbul, Turkey
e-mail: toreyin@itu.edu.tr



B. U. Töreyin et al. (eds.), *8th EAI International Conference on Robotic Sensor Networks*, EAI/Springer Innovations in Communication and Computing,
https://doi.org/10.1007/978-3-031-92143-8_9

in various contexts, including monitoring smoking behaviors, ensuring compliance with smoking regulations, and promoting public health. Cigarette detection presents unique complexities owing to the small and inconspicuous nature of cigarettes, their diverse appearances, and the context in which they are typically found. This context often includes dynamic and cluttered scenes, making it exceptionally challenging to distinguish cigarettes from other objects and backgrounds.

In recent years, researchers have made significant strides in addressing the problem of cigarette detection using state-of-the-art object detection techniques [1–3]. Many studies have adapted well-established one-stage and two-stage object detectors to tackle this challenge, yielding promising results. However, notable obstacles remain.

A key challenge in cigarette detection is the scarcity of annotated datasets specifically designed for training accurate models. The limited availability of labeled data hinders the development of robust algorithms. Additionally, cigarettes are typically small and slender objects within frames, making their detection particularly challenging. Their inconspicuous appearance often evades computer vision systems. Further complicating matters, cigarettes frequently appear in dynamic and cluttered contexts, such as movies and TV shows, where distinguishing them from other objects and backgrounds becomes especially difficult.

To address these challenges, researchers have leveraged datasets originally designed for person and body part detection. Cigarettes are often associated with people, either being held or located nearby. This correlation allows for the use of preexisting datasets focused on human and body part detection to simplify the task of cigarette detection.

For example, the COCO dataset [4], which provides extensively labeled instances of people across diverse scenes, has been effective for person detection. Similarly, the COCO body part dataset [5], with detailed annotations of body parts such as hands and faces, has proven instrumental in body part detection. Detecting body parts like hands and mouths is particularly relevant for cigarette-related activities, as cigarettes are frequently located near these body parts during interactions.

By leveraging these datasets, researchers can effectively localize regions of interest within a frame—such as a person, hands, and mouth—thereby reducing the search space for cigarette detection. This hierarchical approach simplifies the detection process, particularly in complex real-world scenarios like movies and TV shows, where cigarettes are often held or interacted with by characters. By detecting humans and body parts first, we narrow down the locations in which cigarettes are likely to appear, improving both detection accuracy and computational efficiency.

In specific scenarios, such as when a cigarette is stationary or held by an individual, the use of person and body part detection datasets enables more focused analysis. This allows for better handling of cluttered scenes, dynamic backgrounds, or low-resolution frames where direct cigarette detection might otherwise be difficult or prone to errors. As a result, incorporating datasets like COCO and COCO body parts serves as a valuable tool in developing robust cigarette detection systems that can function effectively in complex environments.

In this chapter, we propose a comprehensive workflow for offline video-based cigarette detection. Our approach involves a multistep process that includes human detection, body part detection, and cigarette classification on hand and mouth crops. To achieve this, we have designed a framework consisting of two detection networks for person and body part detection. We employ the generalized focal loss (GFL) [6] as the base detector for each of these networks. The GFL detector is a single-stage detector that incorporates a ResNet50 [7] backbone and a feature pyramid network (FPN) [8] along with the detection head. MobileNetv2 [9] is utilized for classification tasks.

In the following sections of this chapter, we delve into the details of our proposed cigarette detection workflow, highlighting the key components and methodologies employed in each step. We also present experimental results to demonstrate the effectiveness of our approach in detecting cigarettes in offline videos. By addressing the challenges associated with cigarette detection, we aim to contribute to the advancement of computer vision techniques in the context of public health, regulatory compliance, and smoking behavior monitoring.

## 2 Related Work

Recent advances in object detection typically utilize powerful backbone networks, including ResNet [7], Hourglass [10], and ResNeXt [11]. For the neck model, architectures based on feature pyramid networks (FPN) [8] are commonly preferred. In the literature, multistage head models, such as Faster R-CNN [12], Mask R-CNN [13], and Cascade R-CNN [14], have gained prominence. Faster R-CNN generates region proposals using a region proposal network (RPN), while mask R-CNN extends faster R-CNN to perform both detection and segmentation simultaneously. Alternatively, single-stage detectors, such as YOLOv3 [15], SSD [16], GFL [6], and RetinaNet [17], omit the proposal stage and directly detect objects across densely sampled locations.

Recent two-stage methods integrate RPN-based modules [12] to generate region proposals. In object detection, the concept of anchors refers to predefined bounding boxes that are matched with the ground-truth boxes of objects. Several studies have proposed anchor-free approaches, aiming to reduce the computational cost associated with anchors [18]. A crucial component of deep learning-based object detection methods is the loss function, typically composed of two terms: regression loss and classification loss [12]. The regression branch predicts the localization of objects, while the classification branch identifies the object classes. Additionally, non-maximum suppression (NMS) and its variants [19] are essential in reducing redundant predictions produced by object detectors.

While significant progress has been achieved in general object detection [20], small object detection remains an area requiring further improvement [21]. Object detectors that perform well on medium and large objects often struggle with small objects, typically defined as objects occupying less than $32 \times 32$ pixels [4].

For example, it has been noted that DETR [22], despite outperforming faster R-CNN [12] on the MS COCO dataset [4], still does not achieve satisfactory performance on small object detection.

Most of the studies on cigarette detection adapt well-known one-stage and two-stage object detectors to the problem. For example, [23] propose to use faster R-CNN with inception ResNet v2 for cigarette detection. [24] introduce a driver action recognition method, where a deformable and dilated residual block is designed and employed in the faster R-CNN model to detect small and irregular-shaped objects such as cigarettes and cell phones. Another study that utilizes faster R-CNN [25] detects faces in images and cigarettes in face regions, respectively. Similarly, [26] employ YOLOv3, where a two-step workflow is proposed. First, hand and face regions are detected. After classifying those regions as smokers or non-smokers, regions that are classified as smokers are inputted into the cigarette detector model. As a previous study, YOLO [27]-based approaches are commonly used for this problem [28]. For instance, [29] utilize YOLOv5 to detect cigarettes in order to recognize smoking drivers, whereas SVD is used for compressing convolution layers of YOLOv5 to reduce the computational cost. Similarly, another study on smoking driver detection [3] uses YOLOv2, where a near-infrared camera is also employed to capture images of drivers. In that study, it is shown that the proposed approach performs well in day and night conditions.

It is observed that the attention mechanism is utilized to improve object detectors for this problem. For example, [30] employ an attention mechanism on the YOLOX model, while [31] use the SSD model by enhancing it with a feature fusion module and an attention mechanism. A recent study [32] proposes to combine the attention mechanism with YOLOv5 for real-time cigarette detection. That study utilizes Mosaic data augmentation [33] to expand the dataset, where the original dataset consists of only 600 images of cigarettes. Furthermore, [34] address the problem by concentrating on human-object interactions by using a transformer-based approach, where the aim is detecting smoking behavior.

## 3 Method

The cigarette detection problem poses several challenges due to its inherent limitations. One significant obstacle is the scarcity of available data specifically designed for training cigarette detection models. This shortage of annotated datasets hampers the development of accurate and robust algorithms. Furthermore, cigarettes are often small objects in real-world scenarios, which make their detection particularly challenging. Their slender and inconspicuous nature causes them to be easily overlooked by computer vision systems. Additionally, the context in which cigarettes appear, such as movies and TV shows, introduces further complexity, as they are often depicted in dynamic and cluttered scenes. Distinguishing cigarettes from other objects and backgrounds in these conditions becomes exceptionally difficult.

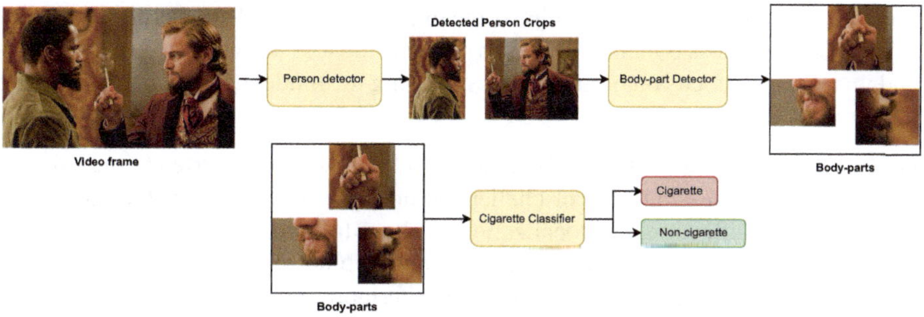


**Fig. 1** An illustration highlighting the main components of the proposed cigarette detection framework, which includes person detection, body part detection, and classification stages

Fortunately, the availability of datasets for person detection and body part detection can simplify the cigarette detection problem in specific scenarios. When a cigarette appears within a frame, it is typically associated with a person, either through interaction or by being stationary on an object. In scenarios involving direct interaction between a person and a cigarette, the utility of datasets such as COCO for person detection and the COCO body part dataset for body part detection becomes evident.

In this chapter, we propose a cigarette detection workflow for offline video processing. Our approach consists of three main steps: (1) person detection, (2) body part detection, and (3) cigarette classification using hand and mouth crops. These steps are illustrated in Fig. 1.

Our framework consists of two detection networks: one for person detection and another for body part detection. Both networks use the generalized focal loss (GFL) as the base detector. The GFL detector is a single-stage detector that consists of a ResNet-50 backbone, a feature pyramid network (FPN) as the neck, and a detection head.

*Focal Loss* The original focal loss (FL) [17] was introduced to address the challenge of class imbalance in one-stage object detection models. FL modifies the standard cross-entropy loss by applying a modulating factor that focuses training on hard examples. The formulation of FL is as follows:

$$\mathbf{FL}(p) = -(1 - p_t)^{\gamma} \log(p_t), \tag{1}$$

where

$$p_t = \begin{cases} p & \text{if } y = 1, \\ 1 - p & \text{if } y = 0. \end{cases} \tag{2}$$

Here, $y$ is the ground-truth class label (1 for positive, 0 for negative), and $p$ is the model's estimated probability for the positive class. The parameter $\gamma$ is a tunable focusing parameter that adjusts the importance of hard-to-classify examples. In essence, focal loss dynamically down-weights the loss assigned to easy examples and increases the focus on hard examples, helping the model handle class imbalance effectively.

*Generalized Focal Loss (GFL)* Generalized focal loss (GFL) extends the original focal loss by addressing critical challenges in object detection, including quality estimation, classification, and localization. GFL unifies quality estimation with class prediction, creating a joint representation that combines predicted class confidence with bounding box localization quality. Furthermore, GFL introduces continuous distributions for box locations, which enhances the model's ability to handle ambiguity and uncertainty in complex scenes. By resolving the inconsistencies between training and inference phases prevalent in traditional object detection models, GFL achieves a more accurate representation of the real data distribution. This makes GFL a highly effective optimization method, especially for challenging detection scenarios, such as identifying small or occluded objects.

*Network Architecture* In our approach, both the person detection and body part detection networks employ ResNet-50 as the backbone and a feature pyramid network (FPN) as the neck. ResNet-50 is chosen for its ability to extract hierarchical features, while the FPN enhances the model's ability to detect objects at multiple scales. For the detection head, we use a RetinaNet-based framework [17], which is known for its balance between detection accuracy and computational efficiency.

In the context of body part detection, which frequently involves detecting small objects like hands and mouths, we incorporate deformable convolutions into the backbone. Deformable convolutions allow the model to adapt its receptive field dynamically, focusing on critical details in small objects. This flexibility significantly improves detection performance for small and deformable objects, such as hands and mouths in complex scenes.

Finally, for cigarette classification, we employ a MobileNetv2 [9] model, which is a lightweight and efficient neural network pretrained on the ImageNet dataset [35]. After detecting hands and mouths from the body part detection model, we crop these regions and feed them into MobileNetv2 to classify the presence of a cigarette. This step ensures an accurate identification of cigarettes within the detected regions of interest, enabling our framework to perform well in both precision and computational efficiency.

Figure 2 shows examples of detections from the COCO dataset, featuring instances of people with bounding boxes around their hands and mouths, which correspond to body parts relevant to our detection pipeline.

In summary, our proposed framework integrates person detection, body part detection, and cigarette classification to address the challenges of cigarette detection in video frames. By leveraging advanced detection networks, including GFL and deformable convolutions, combined with a lightweight classifier for final cigarette recognition, our approach strikes a balance between accuracy and computational efficiency, making it well-suited for offline video analysis.

**Fig. 2** Illustrated examples are from the COCO dataset, featuring instances from the "person" class with bounding boxes shown in green, alongside corresponding mouth bounding boxes in red and hand bounding boxes in blue from the COCO body part dataset

# 4 Experiments and Results

## *4.1 Dataset*

In this chapter, we leveraged two distinct datasets to address the objectives of our research across three different tasks. Specifically, we employed the COCO dataset, renowned for its extensive utility in the realm of image recognition, catering to object detection, segmentation, and captioning tasks. This dataset comprises a vast repository of over 330,000 images, each annotated to encompass a rich diversity of 80 distinct object categories. Notably, each image in the COCO dataset is enriched with five detailed captions, providing invaluable contextual descriptions of the depicted scenes. Additionally, a designated subset of the COCO dataset, referred to as "COCO body part" [2], is tailored to accommodate annotations intended for the precise detection of human body parts. We have utilized these annotations to enable the training of our dedicated body part detection model.

The second dataset employed in this study is a cigarette classification dataset introduced by Khan et al. [36]. This dataset comprises a total of 2,400 raw images, evenly distributed between two categories: smoking and nonsmoking. The dataset was meticulously curated by scanning multiple search engines using a range of keywords related to cigarette smoking, smokers, various gestures, and activities such as coughing, using inhalers, talking on the phone, and drinking water. The inclusion of diverse images in both categories aimed to induce interclass confusion and enhance the model's training. Specifically, the smoking category contained images of smokers captured from various angles and with different gestures, while the nonsmoking category included images of nonsmokers engaged in activities resembling smoking gestures, such as drinking water, using an inhaler, and holding a mobile phone.

## 4.2 Setup

In our person detection task, we employed a GFL detector that was pretrained on the COCO dataset. Subsequently, for body part detection, we fine-tuned the same detector using the COCO body part dataset. Finally, we utilized the cigarette classification dataset to fine-tune a MobileNetv2 model that had been pretrained on the ImageNet dataset.

Regarding the fine-tuning of the detector, we conducted 24 epochs of training, utilizing the SGD optimizer with a weight decay value of 0.001, and we set the momentum parameter to 0.9. The initial learning rate was set at 0.01 and was reduced by half at training steps 16 and 22, respectively. In stage 3 and stage 4 of the backbone, deformable convolution is utilized.

For fine-tuning the MobileNetv2 classification model, we employed a training regimen spanning 50 epochs, implementing a cosine annealing learning rate schedule with an initial learning rate of 0.01.

## 4.3 Experiments

The overall framework was evaluated using episodes from a television series and a movie. Specifically, we conducted evaluations using the first two episodes of the Chernobyl TV series and the movie "Kiss Kiss Bang Bang" to gauge the effectiveness of the cigarette detection framework, where exemplary detection results are presented in Fig. 3.

**Fig. 3** Example cigarette detection results from the movie "Kiss Kiss Bang Bang" and TV show "Chernobyl"

**Table 1** Performance metrics for cigarette detection in movie and TV series. True positive (TP) denotes correctly identified instances, while false positive (FP) represents erroneous identifications. "# Detections" signifies the total count of frames positively detected for cigarettes among the processed frames, whereas "# Frames" denotes the overall number of frames processed

| Movie/Tv series | TP | FP | # Detections | # Frames | Precision |
|---|---|---|---|---|---|
| Chernobyl Episode-1 | 28 | 18 | 46 | 3360 | 0.608 |
| Chernobyl Episode-2 | 154 | 164 | 318 | 3735 | 0.484 |
| Kiss Kiss Bang Bang | 64 | 53 | 117 | 6162 | 0.547 |

**Table 2** Performance evaluation of models on different tasks: mean average precision (mAP) for GFL object detection and accuracy for MobileNetv2 cigarette classification on the cigarette classification dataset [36]

| Model | Dataset | mAP | Accuracy |
|---|---|---|---|
| GFL | COCO | 42.9 | – |
| GFL | COCO body part | 89.5 | – |
| MobileNetv2 | Cigarette Cls. | – | 93.4 |

In the case of movies and TV series, a distinct evaluation method was employed due to the absence of comprehensive annotations for these video sources. To assess the performance, each second of video footage was processed by our framework, while human annotators carefully reviewed each second of the videos. When a frame contained a cigarette, it was annotated as a positive instance; otherwise, it was marked as a negative frame. Subsequently, we computed the precision of our framework on these videos. The detailed results can be referenced in Table 1. Furthermore, the performance of the models was assessed individually for each task, as seen in Table 2.

We also compare the performance of our proposed model with the MolMo 1B model [37] on the task of detecting cigarettes within video frames of the "Chernobyl Episode-1 series." MolMo 1B is a state-of-the-art image-text model based on OpenAI's CLIP architecture and uses the fully open OLMoE-1B-7B mixture-of-experts large language model (LLM). For the MolMo model, we use the prompt *"Does this image contain a cigarette?"* to identify the presence of cigarettes in video frames.

This experiment consists of analyzing 100 video frames from the series, each of which contains a cigarette object. The total number of cigarettes was 50 across the frames. We report the true positives (TP), false positives (FP), precision, and frames per second (FPS) for each model in Table 3.

As shown in Table 3, the MolMo 1B model achieves a slightly higher precision (0.84) compared to our model's precision of 0.81. However, MolMo 1B processes frames at a significantly slower rate, achieving only 0.39 frames per second (FPS). In contrast, our model demonstrates a much higher efficiency, processing 11 frames per second, making it far more computationally efficient.

The trade-off between precision and computational efficiency is evident in this comparison. While MolMo 1B is more accurate in terms of precision, our model

**Table 3** Performance metrics for cigarette detection in 100 randomly selected frames from Chernobyl Episode-1. True positive (TP) denotes correctly identified instances, while false positive (FP) represents erroneous identifications. "# Cigarette" signifies the total count of frames where a cigarette or a smoker is present, whereas "# Frames" denotes the overall number of frames in this experiment. Frame per second (FPS) denotes the number of frames processed in a second for each model

| Chernobyl Episode-1 Series | TP | FP | # Cigarette | # Frames | Precision | FPS |
|---|---|---|---|---|---|---|
| MolMo 1B [37] | 84 | 16 | 50 | 100 | 0.84 | 0.39 |
| Ours | 81 | 19 | 50 | 100 | 0.81 | 11 |

achieves competitive results, while being 30 times faster, which is particularly advantageous for real-time applications. This suggests that our model is better suited for high-throughput tasks where rapid processing is crucial, such as live video analysis or applications in surveillance systems.

MolMo 1B is based on CLIP and incorporates the OLMoE-1B-7B mixture-of-experts model. It processes visual and textual inputs to answer the prompt *"Does this image contain an object of a cigarette?"*. However, this model, due to its large-scale architecture, trades off computational speed for higher precision.

Our model, on the other hand, is designed with efficiency in mind. Using dynamic neural network techniques, we reduce the number of parameters and layers used for low-frequency regions of the image, similar to the approach used in video compression algorithms. This allows us to achieve faster processing speeds, while maintaining a relatively high level of accuracy.

The results demonstrate that our model strikes a balance between speed and accuracy, which can be crucial for applications that require real-time processing. Furthermore, the lower computational load of our model means that it can be deployed on resource-constrained devices or systems without sacrificing much in terms of detection performance.

## 5 Conclusion

In conclusion, our proposed workflow for the effective detection of cigarettes in movies and television shows, which includes person detection, body part detection, and classification, presents a robust solution to a challenging problem. By segmenting the detection process into these sequential stages, we mitigate the occurrence of false positive detections. This approach is particularly advantageous given the limited availability of cigarette detection data and the inherent diversity across a wide spectrum of movies and TV shows. Our framework not only improves the accuracy of detection, but also offers a versatile solution that can be adapted to various cinematic contexts, contributing to the advancement of automated cigarette detection in visual media.

## References


1. Dhanwal, S., Bhaskar, V., Agarwal, T.: Automated censoring of cigarettes in videos using deep learning techniques. Decision Analytics Applications in Industry, pp. 339–348 (2020)
2. Ma, Y., Yang, J., Li, Z., Ma, Z.: Yolo-cigarette: an effective yolo network for outdoor smoking real-time object detection. In: 2021 Ninth International Conference on Advanced Cloud and Big Data (CBD), pp. 121–126. IEEE, Piscataway (2022)
3. Chien, T.-C., Lin, C.-C., Fan, C.-P.: Deep learning based driver smoking behavior detection for driving safety. J. Image and Graph. 8(1), 15–20 (2020)
4. Lin, T.-Y., Maire, M., Belongie, S., Hays, J., Perona, P., Ramanan, D., Dollár, P., Zitnick, C.L.: Microsoft COCO: common objects in context. In: European Conference on Computer Vision, pp. 740–755. Springer, Berlin (2014)
5. Li, X., Yang, L., Song, Q., Zhou, F.: Detector-in-detector: multi-level analysis for human-parts. In: Computer Vision–ACCV 2018: 14th Asian Conference on Computer Vision, Perth, December 2–6, 2018, Revised Selected Papers, Part II 14, pp. 228–240. Springer, Berlin (2019)
6. Li, X., Wang, W., Wu, L., Chen, S., Hu, X., Li, J., Tang, J., Yang, J.: Generalized focal loss: learning qualified and distributed bounding boxes for dense object detection. Preprint. arXiv:2006.04388 (2020)
7. He, K., Zhang, X., Ren, S., Sun, J.: Deep residual learning for image recognition. In: Proceedings of the IEEE Conference on Computer Vision and Pattern Recognition, pp. 770–778 (2016)
8. Lin, T.-Y., Dollár, P., Girshick, R., He, K., Hariharan, B., Belongie, S.: Feature pyramid networks for object detection. In: Proceedings of the IEEE Conference on Computer Vision and Pattern Recognition, pp. 2117–2125 (2017)
9. Sandler, M., Howard, A., Zhu, M., Zhmoginov, A., Chen, L.-C.: Mobilenetv2: inverted residuals and linear bottlenecks. In: Proceedings of the IEEE Conference on Computer Vision and Pattern Recognition, pp. 4510–4520 (2018)
10. Newell, A., Yang, K., Deng, J.: Stacked hourglass networks for human pose estimation. In: European Conference on Computer Vision, pp. 483–499. Springer, Berlin (2016)
11. Xie, S., Girshick, R., Dollár, P., Tu, Z., He, K.: Aggregated residual transformations for deep neural networks. In: Proceedings of the IEEE Conference on Computer Vision and Pattern Recognition, pp. 1492–1500 (2017)
12. Ren, S., He, K., Girshick, R., Sun, J.: Faster R-CNN: towards real-time object detection with region proposal networks. In: Advances in Neural Information Processing Systems, pp. 91–99 (2015)
13. He, K., Gkioxari, G., Dollár, P., Girshick, R.: Mask R-CNN. In: Proceedings of the IEEE International Conference on Computer Vision, pp. 2961–2969 (2017)
14. Cai, Z., Vasconcelos, N.: Cascade R-CNN: delving into high quality object detection. In: Proceedings of the IEEE Conference on Computer Vision and Pattern Recognition, pp. 6154–6162 (2018)
15. Redmon, J., Farhadi, A.: Yolov3: an incremental improvement. Preprint. arXiv:1804.02767 (2018)
16. Liu, W., Anguelov, D., Erhan, D., Szegedy, C., Reed, S., Fu, C.-Y., Berg, A.C.: SSD: single shot multibox detector. In: European Conference on Computer Vision, pp. 21–37. Springer, Berlin (2016)
17. Lin, T.-Y., Goyal, P., Girshick, R., He, K., Dollár, P.: Focal loss for dense object detection. In: Proceedings of the IEEE International Conference on Computer Vision, pp. 2980–2988 (2017)
18. Tian, Z., Shen, C., Chen, H., He, T.: FCOS: fully convolutional one-stage object detection. In: Proceedings of the IEEE/CVF International Conference on Computer Vision, pp. 9627–9636 (2019)
19. Bodla, N., Singh, B., Chellappa, R., Davis, L.S.: Soft-NMS–improving object detection with one line of code. In: Proceedings of the IEEE International Conference on Computer Vision, pp. 5561–5569 (2017)

20. Dai, X., Chen, Y., Xiao, B., Chen, D., Liu, M., Yuan, L., Zhang, L.: Dynamic head: unifying object detection heads with attentions. In: Proceedings of the IEEE/CVF Conference on Computer Vision and Pattern Recognition, pp. 7373–7382 (2021)
21. Liu, Y., Sun, P., Wergeles, N., Shang, Y.: A survey and performance evaluation of deep learning methods for small object detection. Expert Syst. Appl. **172**, 114602 (2021)
22. Carion, N., Massa, F., Synnaeve, G., Usunier, N., Kirillov, A., Zagoruyko, S.: End-to-end object detection with transformers. In: European Conference on Computer Vision, pp. 213–229. Springer, Berlin (2020)
23. Poonam, G., Shashank, B.N., Rao Athri, G.: Development of framework for detecting smoking scene in video clips. Indones. J. Eng. Comput. Sci. **13**(1), 22–26 (2019)
24. Lu, M., Hu, Y., Lu, X.: Driver action recognition using deformable and dilated faster R-CNN with optimized region proposals. Appl. Intell. **50**, 1100–1111 (2020)
25. Han, G., Li, Q., Zhou, Y., Duan, J.: Rapid cigarette detection based on faster R-CNN. In: 2019 IEEE Symposium Series on Computational Intelligence (SSCI), pp. 2759–2765. IEEE, Piscataway (2019)
26. Pundhir, A., Verma, D., Kumar, P., Raman, B.: Region extraction based approach for cigarette usage classification using deep learning. In: International Conference on Computer Vision and Image Processing, pp. 378–390. Springer, Berlin (2021)
27. Redmon, J., Divvala, S., Girshick, R., Farhadi, A.: You only look once: unified, real-time object detection. In: Proceedings of the IEEE Conference on Computer Vision and Pattern Recognition, pp. 779–788 (2016)
28. Arief, L., Tantowi, A.Z., Novani, N.P., Sundara, T.A.: Implementation of yolo and smoke sensor for automating public service announcement of cigarette's hazard in public facilities. In: 2020 International Conference on Information Technology Systems and Innovation (ICITSI), pp. 101–107. IEEE, Piscataway (2020)
29. Shi, F., Zhou, H., Ye, C., Mai, J.: Faster detection method of driver smoking based on decomposed YOLOv5. In: Journal of Physics: Conference Series, vol. 1993, p. 012035. IOP Publishing, Bristol (2021)
30. She, Y.-h., Zhang, X.: Improved YOLOX method for small target smoking detection algorithm. In: International Conference on Cyber Security, Artificial Intelligence, and Digital Economy (CSAIDE 2023), vol. 12718, pp. 452–460. SPIE, Bellingham (2023)
31. Han, L., Rong, L., Li, Y., Qin, Z., Xu, Y.: CA-SSD-based real-time smoking target detection algorithm. In: 2021 5th International Conference on Digital Signal Processing, pp. 283–288 (2021)
32. Zhao, Z., Zhao, Y.: YOLOv5-transformer: improved YOLOv5 network for real-time detection of cigarette smoking based on image processing. In: 2023 4th International Seminar on Artificial Intelligence, Networking and Information Technology (AINIT), pp. 672–675. IEEE, Piscataway (2023)
33. Bochkovskiy, A., Wang, C.-Y., Liao, H.-Y.M.: YOLOv4: optimal speed and accuracy of object detection. Preprint. arXiv:2004.10934 (2020)
34. Ling, H.-B., Huang, D., Cui, J., Wang, C.-D.: HOLT-Net: detecting smokers via human–object interaction with lite transformer network. Eng. Appl. Artif. Intell. **126**, 106919 (2023)
35. Deng, J., Dong, W., Socher, R., Li, L.-J., Li, K., Fei-Fei, L.: ImageNet: a large-scale hierarchical image database. In: 2009 IEEE Conference on Computer Vision and Pattern Recognition, pp. 248–255. IEEE, Piscataway (2009)
36. Khan, A.: Dataset containing smoking and not-smoking images (smoker vs non-smoker). Mendeley Data 1 (2020)
37. Deitke, M., Clark, C., Lee, S., Tripathi, R., Yang, Y., Park, J.S., Salehi, M., Muennighoff, N., Lo, K., Soldaini, L., et al.: Molmo and PixMo: open weights and open data for state-of-the-art multimodal models. Preprint. arXiv:2409.17146 (2024)

# Establishing Visualization-to-Hardware Communication Through Middleware for Digital Twin with ROS-Compatible Microcontrollers Latency Tests

**Diana Belolipetskaja, Anton Rassõlkin, Hadi Ashraf Raja, Pāvels Maksimkins, and Andrejs Stupāns**

## 1 Introduction

### *1.1 Digital Twin Technology*

Digital Twin (DT) technology has emerged as a transformative concept at the nexus of academia and industry, revolutionizing complex systems' design, monitoring, and optimization. DTs are a sophisticated set of virtual models that accurately replicate physical entities' characteristics, actions, and behaviors, providing a detailed, functional, and comprehensive representation that operates simultaneously with real systems. This duality enables enhanced simulations and optimizations, leading to significant advancements across various industries, as presented in [1–4].

DT technology is paving the way for more energy-efficient and environmentally friendly vehicles in the automotive industry. By creating virtual replicas of vehicles and their components, DTs facilitate advanced simulations that could reduce fuel consumption, improve performance, and enhance overall sustainability [5]. Applications of DTs in this sector span from predictive maintenance [6] and battery management [7] to autonomous navigation, driver assistance systems [8], and safety of the vehicle [9], showcasing their potential to revolutionize automotive operations and contribute to sustainable transportation solutions.

D. Belolipetskaja (✉) · A. Rassõlkin · H. A. Raja
Tallinn University of Technology, Tallinn, Estonia
e-mail: diana.belolipetskaja@taltech.ee; anton.rassolkin@taltech.ee; hadi.raja@taltech.ee

P. Maksimkins · A. Stupāns
Riga Technical University, Riga, Latvia
e-mail: pavels.maksimkins@rtu.lv; andrejs.stupans@rtu.lv

 
B. U. Töreyin et al. (eds.), *8th EAI International Conference on Robotic Sensor Networks*, EAI/Springer Innovations in Communication and Computing,
https://doi.org/10.1007/978-3-031-92143-8_10

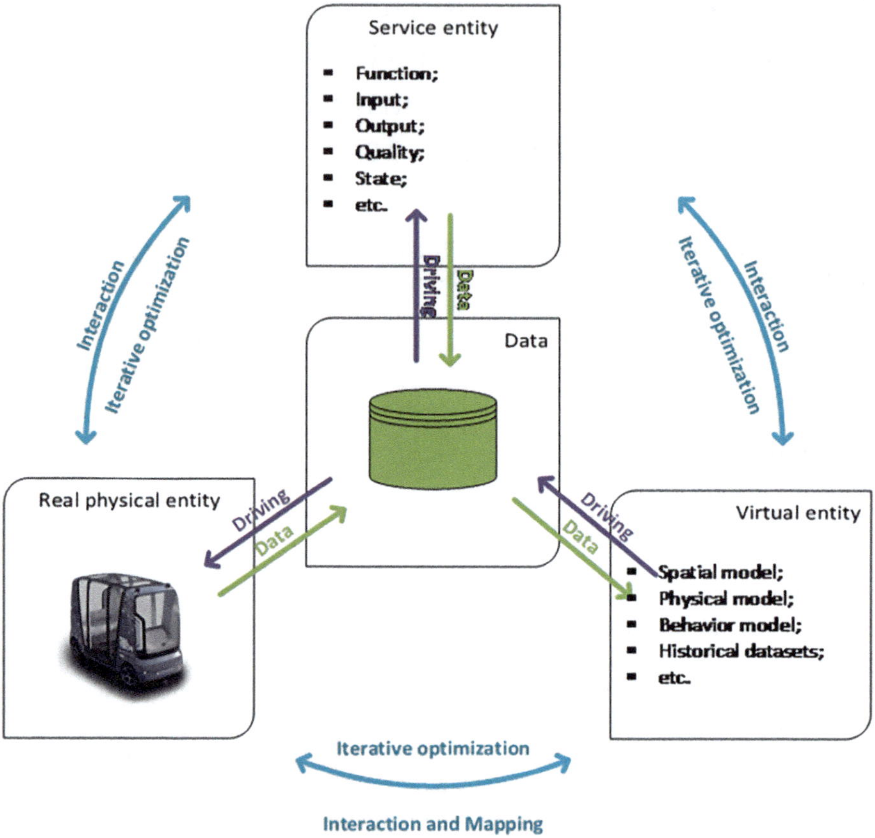


**Fig. 1** Flowchart showing how the service, physical, and virtual entities of DT interact through data. The service entity handles functions, inputs, and outputs; the physical entity is represented as a real vehicle; and the virtual entity includes models and datasets. Arrows labeled "Driving Data," "Interaction," and "Iterative Optimization" illustrate continuous data exchange between them, all connected through a central data storage which is depicted as a green cylinder [6]

The three main components of DT technology are the physical entity, which represents the real-world system and interacts with the external environment; the virtual entity, which is the digital counterpart of the physical entity; and the connection between the physical and virtual entities, which manages the synchronized data flow and communication. Additionally, collected data and service systems that represent an integrated service platform responding to the demands of both physical and virtual entities are usually added to the DT [6]. All of these components are crucial for creating a DT, and the interconnection of the components is shown in "Fig. 1."

The most critical part of the DT is real-time communication, which enables a constant flow of data between the physical and virtual entities. This real-time data exchange is essential for performing predictive maintenance, where potential issues are identified and addressed before they lead to costly failures. By continuously monitoring the physical system's performance, the DT can detect anomalies, predict when maintenance is needed, and suggest optimal times for intervention. Additionally, real-time communication allows for immediate action when required, ensuring that the system can be adjusted or controlled in a timely manner to avoid disruptions, improve efficiency, and enhance overall performance. This responsiveness is key to maximizing uptime, reducing operational risks, and optimizing processes, making real-time communication a fundamental element of a fully functioning DT. The main focus of this research study is to establish a connection between virtual and physical entities through a dedicated middleware solution. To achieve this, the selection of appropriate middleware and compatible controllers is discussed. Moreover, the process of selecting controllers based on latency tests is also presented.

## *1.2 Middleware in Digital Twins*

Middleware simplifies the connections between various sensors, actuators, and applications, optimizing the integration and communication within the system. Otherwise, the attempt to connect all the hardware parts with different software could become really complicated, as shown in Fig. 2a. In different DT solutions [4, 10, 11], the middleware layer serves as an intermediary between hardware and software layers, managing data flow and ensuring interoperability, as shown in Fig. 2b. In the context of DT projects, particularly in automotive applications, the choice of middleware is crucial for seamless and real-time operation. Suitable middleware ensures efficient communication and data exchange between the DT and physical components, enabling accurate monitoring, control, and synchronization. From a

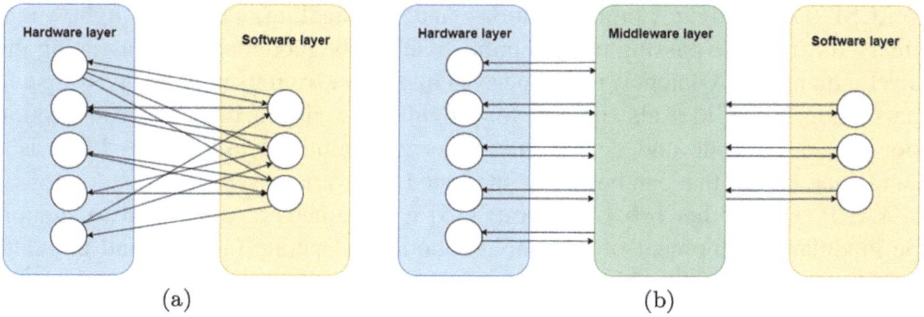


**Fig. 2** Diagram comparing two system architecture flowcharts: (**a**) A direct connection between hardware and software layers, with multiple arrows showing interactions. (**b**) A version with a middleware layer between hardware and software, illustrating a more organized and modular communication structure

safety perspective, the vehicle must react instantly to obstacles and pedestrians on the road to prevent accidents and protect lives. The middleware plays a key role in processing sensor data and executing real-time commands that allow the vehicle to detect and respond to unexpected situations, such as sudden obstacles or pedestrians crossing, ensuring a safe and reliable driving experience.

Many middleware solutions are currently suitable for data transfer between sensors, actuators, and software applications. Some commonly used middleware platforms are ach, LCM, YARP, ROS, and ROS2, see "Table 1." Each platform has its strengths and ideal use cases, and choosing the right middleware depends on factors such as data throughput, latency requirements, scalability, and specific use cases.

There is no systematic way of comparing different middleware frameworks. However, in the study by Cardoso et al. [12], a set of qualitative criteria for Internet of Things (IoT) frameworks were suggested, and therefore, we will define our set of criteria for evaluating and comparing the available middleware as follows:

- Area of application (web development, embedded programming, etc.)
- Support for the desired communication model (publish-subscribe, request-response, etc.)
- Availability and clarity of the documentation, as well as available tutorials
- Quality of the support and livelihood of developer communities

The communication model should preferably be modular and of publisher-subscriber type since, at any point of operation, it should be possible to get and study the data of a single component. This can help create systems where elements may need to get the data from another shared element. Likewise, controlling subscriptions and publishers is possible, either by stopping any publishing or publishing simulated data to observe the system's reaction.

Ach [13] provides a message bus or publish-subscribe style of communication between multiple writers and multiple readers. A real-time system has multiple Ach channels for publishing individual data samples. Ach was created to communicate in real-time systems that sample data from physical processes.

LCM (Lightweight Communications and Marshalling) [14] is a lightweight library for message-passing using a publish-subscribe model aimed at assisting the development of low-latency, real-time systems. Apart from its message-passing and marshaling, the LCM is also notable for providing a real-time deep traffic inspection tool that can decode and display messages with minimal user effort. LCM is a standalone library that can be easily integrated into various systems.

YARP (yet another robot platform) [15] is a framework developed to support the modular development of humanoid robotics to operate in harsh and nonideal conditions. YARP is highly modular and is compatible with other frameworks if needed. YARP includes a transport-neutral communication model so that data flow is decoupled from the details of the underlying networks and protocols in use. Importantly for the long term, YARP is designed to play well with other architectures [16].

**Table 1** Comparison of available middleware

| Platform | Release | Type | Messaging type | Advantages | Disadvantages |
|---|---|---|---|---|---|
| ach | 2013 | Interprocess communication mechanism | Message bus publish-subscribe | + Real-time support<br>+ Solved head-of-line problem for accessing the newest message<br>+ Extensive documentation | – Inactive community<br>– Development discontinued<br>– No ready software packages |
| YARP | 2002 | Robotics middleware | Publish-subscribe | + Extensible family of connection types<br>+ Extensive documentation<br>+ Active community<br>+ QoS policies | – Limited real-time support<br>– No ready software packages |
| LCM | 2006 | Libraries and tools for message-passing and data marshalling, targeted at real-time systems | Publish-subscribe | + Distributed network topology<br>+ Low-latency interprocess communication<br>+ Large support of programming languages | – No ready software packages<br>– Development stalled<br>– Weak documentation<br>– Inactive community |
| ROS | 2007 | Robotics middleware | Publish-subscribe | + Extensive collection of ready-to-use packages<br>+Extensive documentation<br>+ Active community | – Limited real-time support<br>– Has a master server through which all connections are handled<br>– Support ends in 2025 |

(continued)

**Table 1** (continued)

| Platform | Release | Type | Messaging type | Advantages | Disadvantages |
|---|---|---|---|---|---|
| ROS2 | 2017 | Robotics middleware, successor of ROS | Publish-subscribe | + Real-time support<br>+ Distributed network topology<br>+ Native embedded support<br>+ Based on a standard<br>+ Active community<br>+ Extensive documentation<br>+ QoS policies | – Development is still ongoing<br>– Documentation is aimed more on ROS users<br>– Some of ROS ready packages are still being ported |

**Table 2** Comparison of ROS1 and ROS2 systems

| | ROS1 | ROS2 |
|---|---|---|
| Supported OS platforms | Ubuntu | Ubuntu, macOS, Windows, Red Hat Enterprise Linux (RHEL) |
| Supported languages | Python 2 and 3; C++ 98, 11, 14 | Python 3; C++ 11, 14, 17 |
| Real-time communication support | No | Yes |
| Development | Not developing anymore, the last distribution will reach the end of a lifetime in May 2025 | The new distribution of ROS2 is released every year |

ROS is a framework for robotics development that consists of a pub-sub mechanism that exchanges ROS messages over a TCP/UDP network. It is widely used in academia and research for its rich documentation and available ROS packages—already-developed software components for complex tasks (navigation, localization, computer vision, etc.) that are open source and available to everybody.

ROS2 is the successor of ROS. ROS2 is currently being actively developed to provide the following features: support of the data distribution service (DDS) standard, industry-grade support and the deprecation of original ROS in 2025, transfer of ROS core libraries to use the C++11 standard, real-time operating systems (RTOS), and microcontrollers native support [17].

ROS is a highly suitable middleware solution for DT projects due to its robust publish-subscribe messaging architecture, real-time support, and extensive documentation. Originally developed to facilitate communication and integration in robotic systems, ROS provides a standardized framework that enhances the development and deployment of DTs. Its functionalities, such as message-passing, hardware abstraction, and package management, make it a powerful tool for integrating diverse components in complex environments. More detailed differences between ROS1 and ROS2 are presented in Table 2.

ROS2, in particular, supports newer programming languages, operates on a wider range of operating systems, enables real-time communication, and continues to be actively developed by the community. Compared to ROS1, ROS2 is a more suitable choice for modern robotic systems, especially for DT development, due to its real-time communication capabilities, which are essential for handling dynamic and complex scenarios.

Considering these advantages and the prior experience of using the ROS platform in the ISEAUTO self-driving vehicle designed for research and educational purposes [18], it is logical to adopt ROS2 for the current DT project. Since the ISEAUTO propulsion drive system serves as the foundation for this DT project and with future plans to test the DT not only on a scaled demonstrator but also on the actual vehicle, using the same system is both practical and efficient. This approach ensures smoother integration of hardware and software components, facilitating seamless operation and scalability in future developments.

### *1.3 Microcontrollers*

A microcontroller is a compact integrated circuit designed to control a specific function or a set of functions in an embedded system. A single chip typically contains a processor, memory, and input/output (I/O) peripherals. Microcontrollers are widely used in various electronic devices, from home appliances to automotive systems, due to their ability to perform real-time control tasks with low power consumption. In DT technology, microcontrollers can be used to collect real-time data from physical systems, such as sensors or machines, and send this data to the virtual entity for analysis and simulation. This allows the DT to monitor, predict, and optimize the physical system's performance, making microcontrollers essential for creating an accurate and responsive DT.

### *1.4 Research Gap*

The current study is focused on the DT for the propulsion drive system of an autonomous electric vehicle that uses ROS2 as the middleware. The main aim of this work is to connect the visualization software of DT, represented by Unity, and the loading motors of the DT scaled demonstrator, which simulate the road load on the propulsion drive system. Microcontrollers should be utilized effectively to control the hardware components of a DT or to retrieve data from sensors. ROS2 includes the micro-ROS concept, designed to bridge the gap between resource-constrained microcontrollers and larger processors in ROS-based robotic applications. Besides creating a connection, this chapter aims to evaluate the latency of different microcontrollers that could be implemented with the ROS2 system, testing their performance to identify the most efficient solutions for seamless data exchange and optimal functioning of DT.

By comparing the performance of different microcontrollers, this study serves as a guide for future research involving the implementation of microcontrollers for communication in ROS2-based projects. The findings will aid in selecting appropriate message transmission frequencies to ensure reliable and efficient data exchange in diverse applications.

## 2 Methodology

Microcontrollers were tested to select the fastest board to establish a seamless connection between the scaled demonstrator loading motors, which simulate the road load experienced by a bus during travel, and the Unity visualization environment. Users can select various driving conditions in this virtual setting to test motor behavior. The system is designed to automatically control the torque the loading

**Table 3** Comparison of microcontrollers used for the latency test

| | ESP32 DevKitC-32E | RaspberryPi Pico 2040RP | Teensy 4.0 |
|---|---|---|---|
| CPU | dual-core Xtensa LX6 | ARM Cortex-M0+ | ARM Cortex-M7 |
| CPU max speed | 240 MHz | 133 MHz | 600 MHz |
| Supported interfaces | SPI, I2C, UART, CAN, Ethernet, IR, PWM, DAC, ADC, USB | SPI, I2C, UART, PWM, ADC, USB | SPI, I2C, UART, CAN bus, USB |
| Dimensions (mm) | 28 × 54 | 21 × 51 | 18 × 36 |

motors produce based on user-defined parameters in the visualization environment and provide a user with feedback. The ROS2 Humble Hawksbill distribution was selected for this task due to its extended support until May 2027. Although ROS2 supports various operating systems, Ubuntu 22.04 OS was specifically chosen for its stability and compatibility with ROS2 Humble Hawksbill. ROS2 supports different DDS vendors, and for this work, ROS2's default DDS vendor—eProsim's Fast DDS.

## 2.1 Selection of Microcontrollers

An essential criterion for selecting suitable microcontrollers for the DT solution is their capability to connect with ROS2, facilitated by the micro-ROS framework. Only specific 32-bit microcontrollers support micro-ROS. For the latency test were chosen next microcontrollers from all the officially supported boards by the micro-ROS: Raspberry Pi Pico RP2040, ESP32-DevKitC-32E, and Teensy 4.0. To identify the most efficient options for seamless data exchange and optimal DT functioning, these microcontrollers were chosen based on their compact size and cost-effectiveness, ensuring practical implementation within DT projects and real vehicles. The tests focus on USB communication protocols to ensure consistent performance across all selected boards.

Table 3 compares the three microcontrollers, focusing on their physical dimensions and power consumption. Such a comparison is crucial for selecting the appropriate microcontroller based on design constraints, particularly when considering applications that prioritize compact form factors or energy efficiency.

## 2.2 Latency Test

The primary criterion of the DT project is the response time, or latency, which is crucial for effectively monitoring electric vehicle propulsion drive system power consumption and detecting issues in the system. Real-time communication, requir-

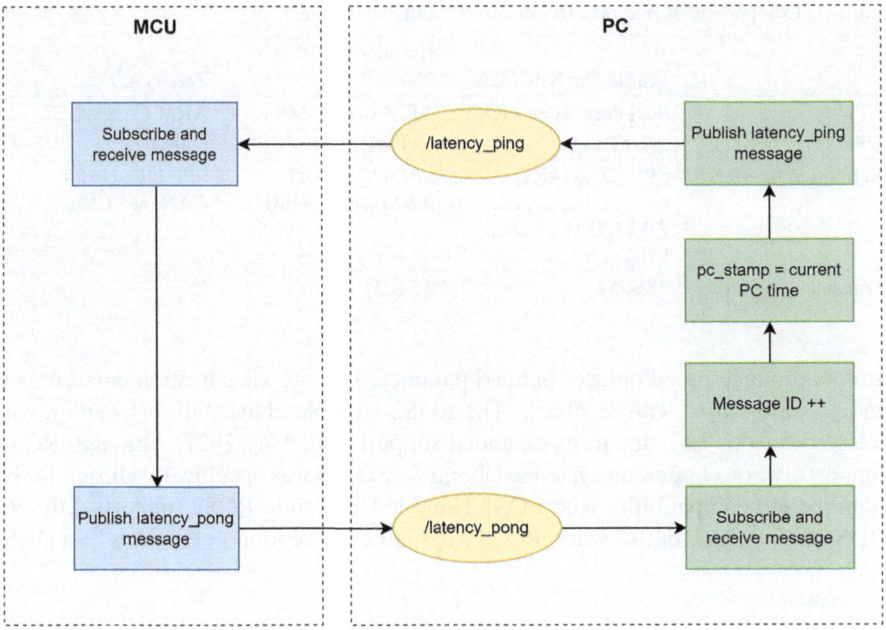


**Fig. 3** Flowchart illustrating the round-trip test principle between an MCU and a PC. The process shows how the PC sends a latency_ping message to the MCU, which receives it and responds with a latency_pong message back to the PC, completing a full communication loop. By incrementing the message ID on each cycle and recording the PC timestamp, it becomes possible to calculate the time each message takes to travel from the PC to the MCU and back

ing minimal latency for accurate and timely data transmission, is a key project criterion. A latency test was conducted to evaluate the latency of each microcontroller. A latency test was performed using the round-trip test (RTT) to see how long it takes one ROS message to be sent to the microcontroller and back. The data transmission distance was also taken into account, while it could influence the latency of the microcontroller.

RTTs were done using different cable lengths to see their influence on latency results. The main components of the RTT are the PC that runs ROS2 and three different microcontrollers. PC published a message containing the PC time stamp (builtin_interfaces/Time message type) and message ID number (int32 message type) to the ROS topic to which the microcontroller was subscribed. After receiving the message, the microcontroller published the same message to a different topic to which the PC is subscribed. Therefore, the PC updated the time stamp and message ID, and the loop continued. "Figure 3" shows a simplified RTT code working principle scheme.

To prevent clock desynchronization and avoid further complications, only the PC time stamp was used. This process was repeated until the PC and the microcontroller exchanged 100,000 messages. The RTT was conducted with all the selected microcontrollers using USB cable lengths of 5.0 m, 2.0 m, 1.0 m, and 0.2 m.

## 3 Results

### 3.1 Latency Test Results

"Figure 4" shows the distribution of how long it took for each message to do a round trip. The results are shown for each board and with different cable lengths.

Among the tested boards, the Teensy 4.0 consistently demonstrated the shortest average message transmission time, averaging approximately 0.66 ms across all cable lengths, resulting in a message-passing frequency of 1515.15 Hz. The Raspberry Pi Pico RP2040 ranked second, with an average transmission time of approximately 3.67 ms and a message-passing frequency of about 272.48 Hz, both consistent across varying cable lengths. Despite the ESP32-DevKitC-32E having a higher CPU clock speed than the Raspberry Pi Pico RP2040, it exhibited the

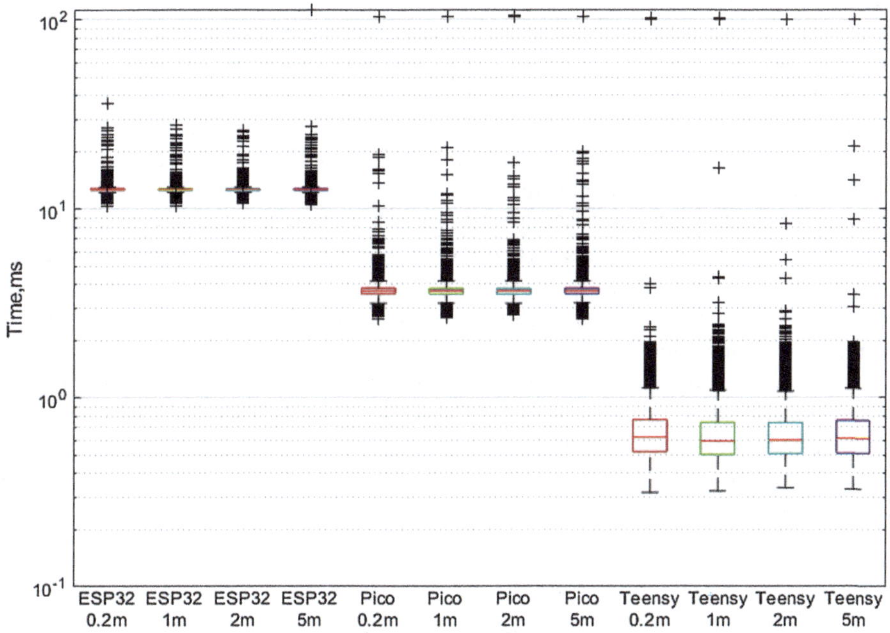


**Fig. 4** Box plot comparing time in milliseconds for three devices: ESP32, Pico, and Teensy, across four distances: 0.2m, 1m, 2m, and 5m. The y-axis is logarithmic, ranging from 0.1 to 100 ms. Each device shows varying performance, with Teensy generally having lower median times. Outliers are marked with plus signs

**Table 4** Additional results of latency test for each microcontroller

| Microcontroller | Cable length, in m | Min time, in ms | Max time, in ms | Total time for 100,000 messages, in min |
|---|---|---|---|---|
| ESP32 DevKitC-32E | 0.2 | 10.490 | 36.427 | 21.204 |
| | 1.0 | 10.439 | 27.737 | 21.182 |
| | 2.0 | 10.799 | 26.204 | 21.176 |
| | 5.0 | 10.653 | 114.047 | 21.181 |
| Raspberry Pi Pico RP2040 | 0.2 | 2.609 | 104.592 | 6.126 |
| | 1.0 | 2.659 | 104.403 | 6.132 |
| | 2.0 | 2.740 | 105.455 | 6.130 |
| | 5.0 | 2.599 | 104.697 | 6.136 |
| Teensy 4.0 | 0.2 | 0.316 | 101.891 | 1.125 |
| | 1.0 | 0.322 | 102.073 | 1.112 |
| | 2.0 | 0.335 | 101.461 | 1.095 |
| | 5.0 | 0.329 | 101.633 | 1.098 |

longest transmission time, averaging approximately 12.70 ms, corresponding to a message-passing frequency of 78.74 Hz, regardless of cable length. These results indicate that the Teensy 4.0 is the fastest performer in terms of message-passing on average, followed by the Raspberry Pi Pico RP2040, and the ESP32-DevKitC-32E is the slowest out of three. Although the ESP32-DevKitC-32E showed the more consistent message-passing time with only one message being passed with time over 100 ms. Other boards could not show the same results as it appeared more frequently the messages that were passed over 100 ms. This problem needs more research to find the reason for such delays and possible solutions. Additional parameters gained during the latency tests are presented in "Table 4," along with the minimum and maximum times for one message exchange and the total amount of time it took to send 100,000 messages.

Although Teensy 4.0 message-passing time results on average were the fastest, the wider spread of results indicates potential issues under specific conditions (such as handling data bursts) leading to delays. In real-time systems, where predictability is as critical as speed, this could limit the Teensy's effectiveness despite its low average latency.

## 3.2 *Practical Implementation*

The principal diagram for the setup is illustrated in "Fig. 5a," and it serves as a foundational reference for understanding the overall architecture and operational workflow, with insights into the relationships between various scaled demonstrator

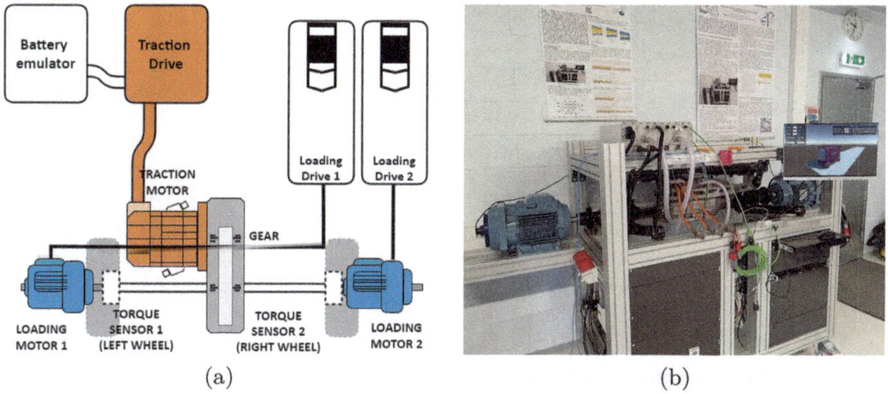


**Fig. 5** Diagram and photograph of a traction drive system. Panel (**a**) is a flow chart illustrating the components of a traction drive system, including a battery emulator, traction motor, gear, torque sensors for left and right wheels, and loading motors. Panel (**b**) is a photograph of the physical setup, showing motors, wiring, and a monitor displaying system data

subsystems and their functionalities. The loading motor(s) is controlled via a frequency converter(s), which manages various motor parameters at the hardware control level. The converter was set up so that the motor's torque value could be controlled by the analog input signal. The signal range is set from 0 to 3.3 V, where 0 V is the motor's maximum torque in one direction, 3.3 V is the motor's maximum torque in the opposite direction, and 1.65 V is the zero point.

To test and validate the communication, the scaled demonstrator for DT of the propulsion drive system of the electric vehicle was utilized. The microcontroller was connected to the frequency converters that control the loading motors. "Figure 5b" shows the scaled demonstrator used for the project and communication test.

The microcontroller is subscribed to the ROS topic where the road conditions are published. The road load is calculated on the microcontroller based on the received data, and the corresponding analog signal is sent to the motor frequency converter to produce proper torque. Additionally, torque value is also sent to other ROS topics to reach the user.

### 3.3 *Implementation*

To seamlessly integrate the Unity visualization environment with the DT system, establishing a connection with ROS2 is essential. Unity offers a dedicated solution for this integration through the ROS TCP Connector package, which serves as a bridge between Unity and ROS using TCP connections. This setup involves deploying the ROS TCP Connector on the Unity side and the ROS TCP Endpoint on the ROS2 side. The ROS TCP Endpoint functions as a node within ROS2 and

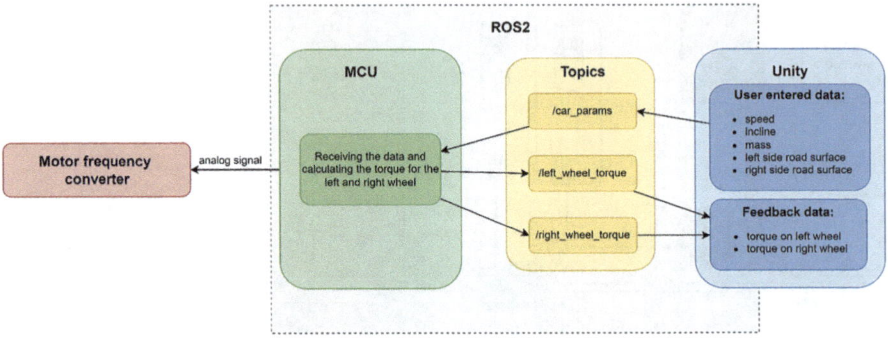


**Fig. 6** Flowchart illustrating a system using ROS2. The process starts with the user entering road condition data in Unity, such as speed, incline, mass, and surface type. The MCU subscribes to the topic containing these parameters, calculates the required torque, and sends it to the motor frequency converter, which controls the loading motors of the test bench. Afterward, the MCU sends load data back to Unity, providing real-time feedback on the torque applied to the left and right wheels

establishes a TCP connection to facilitate message exchange between Unity and ROS2. Since Unity predominantly uses C#, while ROS uses C + + or Python, the ROS TCP Connector handles the critical task of message serialization and deserialization, ensuring smooth communication between the two environments. Using the TCP protocol for communication allows the ROS2 system and Unity to run on one computer or, if needed, on separate computers to divide the system's load.

A C# script was developed to enable Unity to subscribe to ROS2 topics. This script establishes a ROS2 connection, managing both a publisher and a subscriber. The publisher transmits user-defined parameters such as speed, mass, incline, left-wheel road surface, and right-wheel road surface to the `car_param` topic. The subscriber listens to the `torque_left_wheel` and `torque_right_wheel` topics, receiving torque values calculated by the microcontroller for each wheel and displaying them to the user in real time on the main visualization screen. This setup ensures efficient collaboration and interaction within the integrated system, providing real-time feedback on the system's operation. The full connection scheme is shown in "Fig. 6."

## 4 Conclusions

DT technology is rapidly gaining traction in the automotive industry, offering precise and cost-effective simulation and testing environments, as well as advanced platforms for predictive maintenance. Autonomous electric vehicles present a unique case due to their extensive sensor arrays and the vast amounts of data they generate, allowing for the creation of highly detailed models and testing environments. To manage this extensive data effectively, a robust and reliable connection is crucial for accurate analysis and maintenance predictions.

In this research work, latency tests were conducted for several microcontroller units using different cable lengths to assess how these variations impact communication speed. The Teensy 4.0 exhibited the best performance, maintaining an average latency of 0.66 ms regardless of cable length. In comparison, the Raspberry Pi Pico RP2040 and the ESP32-DevKitC 32E displayed average latencies of approximately 3.67 ms and 12.70 ms, respectively. However, further research is needed to investigate the factors affecting Teensy 4.0's performance consistency, as occasional delays and variability in latency were observed.

The successful integration of the ROS2 middleware enhances the contributions of this research, facilitating seamless bidirectional communication between the DT's motors and the Unity visualization platform. This implementation allows for real-time data exchange between hardware and software, creating a flexible system that can adapt to various hardware configurations.

Future work will focus on refining communication protocols and optimizing system responsiveness to further improve scalability and robustness. Additional testing will involve evaluating microcontrollers by transmitting messages in one direction, while varying message frequency and other parameters to understand their impact on performance. Investigating the effect of potential factors such as microcontroller heating on performance will also be a priority. Moreover, comparisons of different communication protocols will be conducted to identify optimal solutions for specific applications.

For the DT, the next steps involve integrating additional hardware and software components to create a fully functional system. This research provides a solid foundation for advancing DT technology, paving the way for high-performance and flexible architectures that can be applied across various industries.

**Acknowledgments** The research has been supported by the Estonian Research Council under grant PSG453 "Digital twin for propulsion drive of autonomous electric vehicle."

Research team would like to express sincere gratitude to Sergei Jegorov for his contribution to this paper. His insights and expertise in middleware technologies greatly enhanced the quality of our comparative analysis.

## References

1. Alazab, M., Khan, L.U., Koppu, S., et al.: Digital twins for healthcare 4.0—recent advances, architecture, and open challenges. IEEE Consum. Electron. Mag. **12**(6), 29–37 (2023). https://doi.org/10.1109/MCE.2022.3208986
2. Dembski, F., Wössner, U., Letzgus, M., Ruddat, M., Yamu, C.: Urban digital twins for smart cities and citizens: the case study of Herrenberg, Germany. Sustainability **12**(6) (2020). ISSN: 2071–1050. https://doi.org/10.3390/su12062307 [Online]. Available: https://www.mdpi.com/2071-1050/12/6/2307
3. Purcell, W., Neubauer, T.: Digital Twins in Agriculture: A State-of-the-Art Review (2023). https://doi.org/10.1016/j.atech.202.100094 [Online]. Available: https://www.sciencedirect.com/science/article/pii/S2772375522000594

4. Sueldo, C.S., Colo, I.P., Paula, M.D., Villar, S.A., Acosta, G.G.: ROS-based architecture for fast digital twin development of smart manufacturing robotized systems. Ann. Oper. Res. **322**, 75–99 (2023). ISSN: 15729338. https://doi.org/10.1007/s10479-022-04759-4
5. Bhatti, G., Mohan, H., Singh, R.R.: Towards the future of smart electric vehicles: digital twin technology (2021). https://doi.org/10.1016/j.rser.2021.110801
6. Rassõlkin, A., Rjabtšikov, V., Vaimann, T., Kallaste, A., Kuts, V.: Concept of the test bench for electrical vehicle propulsion drive data acquisition. In: 2020 XI International Conference on Electrical Power Drive Systems (ICEPDS), pp. 1–8 (2020). https://doi.org/10.1109/ICEPDS47235.2020.9249078
7. Eaty, N.D.K.M., Bagade, P.: Digital twin for electric vehicle battery management with incremental learning. Exp. Syst. Appl. **229** (2023). ISSN: 09574174. https://doi.org/10.1016/j.eswa.2023.120444
8. Liao, X., Zhao, X., Wang, Z., et al.: Driver digital twin for online prediction of personalized lane-change behavior. IEEE Internet Things J. **10**, 13,235–13,246 (2023). ISSN: 23274662. https://doi.org/10.1109/JIOT.2023.3262484
9. Almeaibed, S., Al-Rubaye, S., Tsourdos, A., Avdelidis, N.P.: Digital twin analysis to promote safety and security in autonomous vehicles. IEEE Commun. Stand. Mag. **5**, 40–46 (2021). ISSN: 24712825. https://doi.org/10.1109/MCOMSTD.011.2100004
10. Kuts, V., Modoni, G.E., Otto, T., et al.: Synchronizing physical factory and its digital twin through an IIoT middleware: a case study. Proc. Est. Acad. Sci. **68**, 364–370 (2019). ISSN: 17367530. https://doi.org/10.3176/proc.2019.4.03
11. Cakir, L.V., Bilen, T., Özdem, M., Canberk, B.: Digital twin middleware for smart farm IoT networks. In: 2023 International Balkan Conference on Communications and Networking, BalkanCom 2023, Institute of Electrical and Electronics Engineers Inc. (2023). ISBN: 9798350339109. https://doi.org/10.1109/BalkanCom58402.2023.10167962
12. Cardoso, J., Pereira, C., Aguiar, A., Morla, R.: Benchmarking IoT middleware platforms, pp. 1–7 (2017). https://doi.org/10.1109/WoWMoM.2017.7974339
13. Dantam, N.T., Lofaro, D.M., Hereid, A., Oh, P.Y., Ames, A.D., Stilman, M.: The ACH library: a new framework for real-time communication. IEEE Robot. Autom. Mag. **22**(1), 76–85 (2015). ISSN: 1070-9932. https://doi.org/10.1109/MRA.2014.2356937
14. Huang, A.S., Olson, E., Moore, D.C.: LCM: lightweight communications and marshalling, pp. 4057–4062 (2010). https://doi.org/10.1109/IROS.2010.5649358
15. Metta, G., Fitzpatrick, P., Natale, L.: YARP: yet another robot platform. Int. J. Adv. Robot. Syst. **3**(1), 8 (2006). https://doi.org/10.5772/5761
16. Fitzpatrick, P., Metta, G., Natale, L.: Towards long-lived robot genes. Robot. Auton. Syst. **56**(1), 29–45 (2008). https://doi.org/10.1016/j.robot.2007.09.014
17. Thomas, D.: Changes between ROS 1 and ROS 2, ROS 2 design [Online]. Available: http://design.ros2.org/articles/changes.html
18. Rassõlkin, A., Sell, R., Leier, M.: Development case study of first estonian self-driving car ISEAUTO. Electr. Control. Commun. Eng. **14**(1), 81–88 (2018)

# Morphing Detected Human Intention Via Human-Computer Interactions

**Cevdet Şencan, Aydan Müşerref Erkmen, and Aygün Ertuğrul**

## 1 Introduction

Human-Computer Interaction (HCI) provides significant positive contributions to many areas of our lives, from improving the living conditions of disabled individuals to security applications by mining human intentions. A particularly intriguing area of research in this field is the identification of intentions through human facial expressions to affect and reshape those intentions. The Theory of Mind (ToM) posits that emotions conveyed through facial expressions serve as crucial indicators of an individual's underlying intentions. ToM emphasizes the capacity to comprehend the beliefs, knowledge, and intentions of others by interpreting their behaviors [1]. Paul Ekman asserts that emotions are fundamental in communicating and understanding others [2]. As a result, the attempt to detect human facial expressions and emotional states, thereby inferring intentions, using image processing technology has emerged

C. Şencan (✉)
Department of Neuroscience and Neurotechnology, Graduate School of Natural and Applied Sciences, Middle East Technical University, Ankara, Turkey
e-mail: sencan.cevdet@metu.edu.tr

A. M. Erkmen
Department of Neuroscience and Neurotechnology, Graduate School of Natural and Applied Sciences, Middle East Technical University, Ankara, Turkey

Department of Electrical and Electronics Engineering, Faculty of Engineering, Middle East Technical University, Ankara, Turkey

A. Ertuğrul
Department of Neuroscience and Neurotechnology, Graduate School of Natural and Applied Sciences, Middle East Technical University, Ankara, Turkey

Department of Psychiatry, Faculty of Medicine, Hacettepe University, Ankara, Turkey

 
B. U. Töreyin et al. (eds.), *8th EAI International Conference on Robotic Sensor Networks*, EAI/Springer Innovations in Communication and Computing,
https://doi.org/10.1007/978-3-031-92143-8_11

as a significant research area encompassing both technological [3] and psychological dimensions [2].

Human intention recognition can be identified not just through facial expressions but also by analyzing voice, gaze [4], posture [5], and behaviors or movements. The assessment of intentions in the context of human-robot interactions is emerging as a significant area of interest, especially as robots become increasingly integrated into everyday life [6–8]. One of the pioneering studies in this field, conducted by Durdu et al., extends the focus beyond merely predicting human intentions in human-robot interaction (HRI) [9, 32] by also seeking to reshape human intentions through robotic engagement. This perspective is relatively novel, and research in this area remains limited [10].

Görür and Erkmen [11] describe intention reshaping as the process of "controllably altering individuals' goals or intentions by making pertinent modifications in the environment that influence their perception of the physical surroundings and their psychological state." Several studies suggest that deep intentions can be evoked and converted into behavioral intentions [12]. Er and Erkmen [13] characterize deep intention as impulses that individuals often express in their daily lives under specific conditions, lying dormant without needing to manifest frequently in behavior. Findings from this emerging area of research suggest that altered intentions can lead to behavioral changes [10, 12, 14].

Another critical aspect is the accurate and sensitive recognition of intentions that are to be reshaped. The Convolutional Neural Network (CNN) method is among the most effective approaches for deep learning, demonstrating its efficiency in numerous studies for feature extraction and mining [15–17]. It excels in managing large datasets for tasks such as pattern recognition, image classification, and interpretation.

In this paper, we present an innovative approach to human-computer interaction by (1) developing emotional triggers that uncover the underlying intentions of human participants as they emerge in real time through diverse proactive stimuli, and (2) guiding the current identified intention toward newly established intentions that the computer autonomously defines. We select the bluff card game as an ideal setting to showcase the effectiveness of our proactive HCI system for intention reshaping, focusing on the changes in facial expressions that reveal the player's underlying move intentions. The computer system interacts with the human player to bring their next move intention to light, employing a CNN-based training method to learn how to recognize deep intentions. This trained model is developed using real-time recordings from our experimental process (specifically, the bluff game) conducted with participants in a natural environment.

This paper presents a performance analysis regarding how our HCI system facilitates the transformation of players' intentions, specifically, those identified as being about to bluff (surfaced intention), into alternative actions (intention reshaping) suggested subtly (rather than overtly directed) by our computer interaction system through suitable proactive stimuli. To achieve this, we design a variety of visuals that serve as proactive psychological stimuli during the interaction. These psychological elements set our study apart from the limited existing approaches to intention reshaping within the realm of HRI [9–11, 13].

# 2 Methodology

## *2.1 Experimental Configuration*

We regard the bluff card game as the ideal experimental environment for observing variations in people's facial expressions and linking these changes to behavioral intentions during gameplay. Additionally, we employ a controlled real play session for model verification and training of our computer interaction system, which will proactively reshape the player's intentions by validating the cards a player discards when bluffing. Given that the primary objective of the bluff card game is to bluff effectively and win, it is a logical game that necessitates psychological analysis between moves. The player who can make decisions grounded in logic and persuade others through their demeanor and facial expressions, without revealing the true state of their cards on their face, stands the best chance of winning. However, Paul Ekman [2] notes that individuals can display microexpressions in certain emotional states, which are often imperceptible to the person they are interacting with. Microexpressions are brief facial expressions that occur involuntarily and last for about 1/25 of a second. Since these microexpressions emerge automatically, they cannot be entirely suppressed or concealed. A person may exhibit these microexpressions based on their emotional state while bluffing. Consequently, our proactive HCI aims to detect these microexpressions by initially learning from the extensive data present in actual game videos.

Therefore, in this study, during the controlled phase of the experiment that will train our system, we capture recordings of the players' faces, the cards in their hands, and the cards on the table using cameras. This setup is designed to identify potential connections between facial expressions, the emotional states of the participants playing the bluff card game, and the cards held by the players, as well as their actions (behavioral intentions). We then simultaneously observe the facial expressions and cards on the same time scale to identify and label significant facial expressions.

### 2.1.1 Training Set Up and Data Acquisition of Our System

The learning process for playing the bluff card game is straightforward, quick, and easy. In this study, the bluff card game is conducted with four participants, each receiving 13 cards. The players are seated around a square table, making a total of four participants for the game. We record the entire session using four cameras, with one positioned behind each player. Each camera is angled to capture the cards in the player's hand and the facial expressions of the opposing player across the table. As illustrated in Fig. 1, which presents both a sketch and the actual setup of the experimental layout, our experiment includes 4 players, 4 cameras, and playing cards.

We utilize the "Splitcam (version 10.5.80)" [18] application to transmit images from the four cameras to the computer undergoing training. The training process

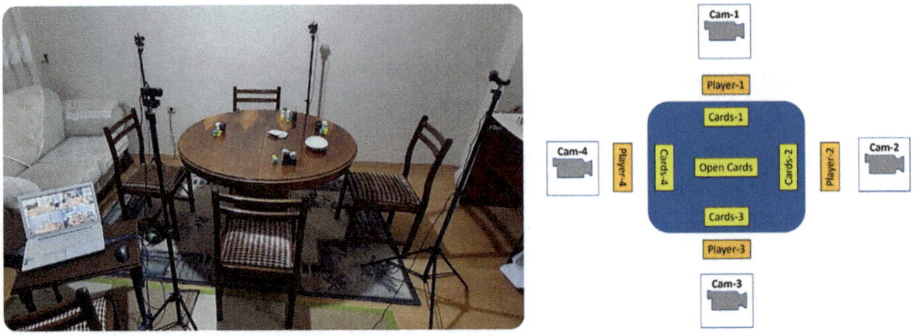


**Fig. 1** Actual experiment arrangement and conceptual design of the experiment

incorporates a total of 64 bluff card games played by 10 volunteers. It is important to highlight that the participants are not briefed and are completely unaware of the study's purpose; however, the game does include a reward mechanism as recommended in [19, 20]. If the cards that the player places on the table are detected while being separated from the cards in their hand, bluff labeling can be performed with high accuracy by generating an intention matrix. We create all intention matrices based on the recordings from all the games.

According to Ekman, microexpressions such as a slight smile, a rapid eyebrow raise, and involuntary eye movements manifest similarly across all individuals [2]. The key to finding a solution lies in detecting expressions that deviate from a person's typical behavior and do not align with their natural conduct. Additionally, the theory of mind (ToM), which encompasses the perception of others' behaviors, expressions, gestures, and other non-verbal signals, highlights the significance of perceptual processes in comprehending other people's mental states [21]. In light of these perspectives, we monitor the microexpressions that players may exhibit while bluffing in the experimental setting. Accordingly, we analyze the moves in the intention matrix that have been previously identified as bluffs. As a result, we record a total of 1345 bluffing facial expressions, as shown in Fig. 2.

It is important to highlight that during the recording process, we conduct the initial dimension reduction by cropping only the facial area from the 1280x720p images containing the bluff facial expressions. By eliminating this large, unnecessary portion of the image, we significantly decrease both the computational load and processing time. We utilize the recorded dataset of 1345 bluffing facial expressions to develop a model that recognizes microexpressions associated with bluffing.

## 2.2 Training Stage

We develop a model by processing the labeled bluff facial expression images using the deep learning algorithm based on CNN. For this study, we use Python version

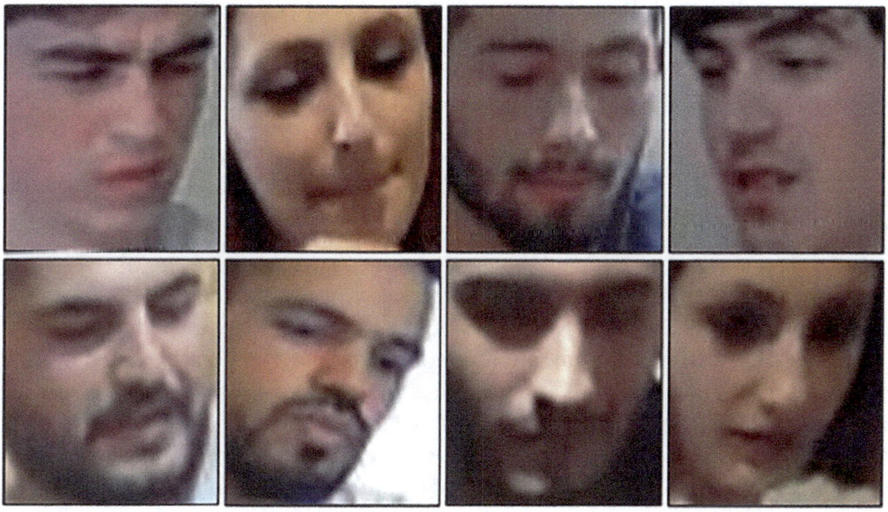

**Fig. 2** Instances of facial expressions classified as bluff

3.9 (64-bit). To enhance the efficiency of the algorithm, facilitate quick responses to results and errors and minimize computational time, we utilize Google Colab and the Turkish National Science e-Infrastructure (TRUBA) environments [22]. Using transfer learning from [23], we create the CNN algorithm which comprises four convolutional layers and three fully connected layers, similar to the approach in [24]. Each convolutional layer consists of varying numbers of "3x3" convolution filters, along with the ReLU activation, batch normalization, max pooling (2x2), and dropout function. Each fully connected layer contains different numbers of fully connected neurons, the ReLU, batch normalization, and dropout function. The output layer comprises two neurons with a softmax activation function. To mitigate overfitting [25] and because test accuracy improves as the training set expands [26], we increased the existing dataset by "imagedatagenerator" function from the tensorflow library, which is frequently used in deep learning and computer vision. We divide the preprocessed data such that 80% is allocated for training and 20% for validation. In the model compilation phase, we utilize the Adam optimizer along with a loss function, setting the learning rate to 0.0001. Additionally, we record the training accuracy and loss, as well as the validation accuracy and loss values to enhance the models based on the outcomes.

## 2.3 *Intention Recognition*

Our HCI system carries out the intention recognition process using the intention recognition algorithm, we developed with the best-trained model. At this stage, we utilize two different bluff card game videos that we set aside for testing (not used in training or validation). These test games include a total of 21 distinct bluff moves. The intention recognition algorithm we developed is employed to identify these moves. The detection performance results will be examined in Subsection 3.2.

## 2.4 Intention Reshaping

We apply the stimuli to one target player at a time to effectively demonstrate our intention reshaping HCI system. Since the intervention directed at the target player, designed to reshape intentions through computer-generated visual stimuli, will impact all players in a real face-to-face bluff game, we strive for a controlled experimental environment. We create a game approach using laptops to ensure that the proactive visuals employed as stimuli affect only one target player at a time among the players. Thus, to facilitate more effective and efficient reshaping, we implement the human-computer interaction method using visual cards on the laptops instead of the actual cards held in each player's hand. Furthermore, to enhance the reshaping effect, we ensure that the proactive visuals are presented on the PC screen that captures the players' attention, specifically targeting the player whose intentions will be reshaped through the stimulus. In the controlled experimental environment depicted in Fig. 3, we facilitate the players to engage in the same game within a virtual setting using computers, while the participants remain unaware of the proactive reshaping objective of the HCI.

In a bluff game played by four participants around a table using their laptops, we execute the reshaping algorithm we developed on the target player's computer to recognize their intentions. As illustrated in Fig. 3, the internal webcam of the target player's computer captures images to provide input to our HCI system. Additionally, an external webcam captures the target player's facial images in real time simultaneously. We record the player's computer screen using Splitcam, ensuring that the target player's facial image is displayed in the bottom right corner of the screen.

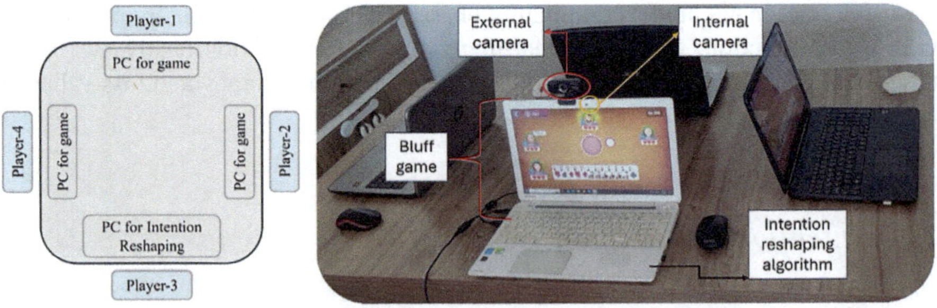


**Fig. 3** Experimental configuration for intention reshaping

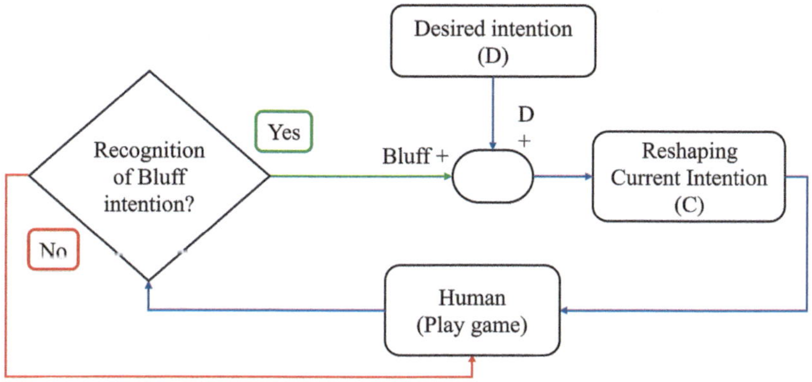


**Fig. 4** Block diagram of our reshaping system

As a result, we record which moves belong to each player and the corresponding facial expressions they display while making those moves. When our algorithm within the proactive HCI system identifies that it is the target player's turn, it begins scanning for bluff recognition. If the target player intends to bluff and the HCI subsystem detects this intention, a proactive visual stimulus (predetermined by us to prevent recurrence) representing a new, computer-generated desired intention is displayed on the player's screen to reshape their current bluff intention to the new one. The block diagram illustrating the system for reshaping intentions is presented in Fig. 4. The performance analysis of intention reshaping will be debated in Subsection 3.3.

## *2.5 Proactive Stimulus Impact Approach*

In intention reshaping, human psychology plays a significant role, as the willingness to abandon a current intention is influenced by the player's psychological state. Additionally, the characteristics of the stimuli used for intention reshaping are critical factors that directly impact the process. We develop an approach to examine the effects on the reshaping of human intentions.

Research in the literature indicates that verbal communication constitutes 7%, paralinguistic elements account for 38%, and emotional facial expressions make up 55% of face-to-face interpersonal communication [27]. In today's context, where communication increasingly occurs in social media environments rather than face-to-face interactions, these emotional facial expressions have often been replaced by emojis [28]. Consequently, it is proposed that the most prevalent method for expressing emotional states, particularly among younger generations, is through the use of emojis [29]. Therefore, by utilizing these emojis as proactive visuals, the stimuli presented to the player can be more easily understood, especially since

**Table 1** Clarified intentions and developed proactive visuals

| Nu. | Intentions | Yellow Emojis |
|---|---|---|
| D1 | Open |  |
| D2 | Bluff |  |
| D3 | Not bluff |  |
| D4 | Pass |   |

our HCI system does not rely on verbal communication. Moreover, we assess that emojis are more effective in triggering changes in the target player's emotions and, consequently, in prompting behavioral changes in their moves related to intention shifts.

Conversely, it is understood that the perception of an event occurs through the association with past events and experiences [30]. This tendency may enhance the likelihood of recognizing familiar yellow emojis. Therefore, we select yellow as the color for the emojis.

Building on these psychological studies, we design four distinct yellow emojis and align them with the four different intentions outlined in Table 1. Through these proactive visuals, we aim to influence the player prior to their move and alter their behavioral intention. In this manner, the proactive HCI establishes a foundation for two innovative approaches in the intention reshaping process:

- Creating emotional triggers that will uncover the deep intentions of individuals (surfacing intentions that are typically buried due to habituation).
- Guiding them toward new intentions that the HCI has predetermined in advance (reshaping intention).

The player whose turn it is to take action in the bluff game has four options: they can call the previous hand by asserting that a bluff was made, bluff while placing cards in the center (bluff), place cards in the center without bluffing (not bluff), or pass without taking any action, thereby allowing the turn to move to the next player (pass). These four options correspond to the four intentions defined in the demonstration experiment for our HCI system. Consequently, the proactive HCI guides the player whose turn it is toward the targeted intention by alerting them with proactive visual stimuli. The four intentions and the proactive visuals we developed to reshape these intentions are presented in Table 1. The desired intentions that the HCI system seeks to impart to the target player through proactive visuals are represented as Di (with i: 1, 2, 3, 4). Meanwhile, the player's current intention, as detected by the HCI system in a video frame, is referred to as Ci (with i: 1, 2, 3, 4).

All of this information indicates that the quality of the stimulus plays a crucial role in reshaping intentions, as our innovative approach offers improvements in this process. Consequently, we have developed this approach specifically for the purpose of reshaping intentions.

The rationale behind placing the bluff (D2) image here can be viewed as a means of boosting the player's confidence to continue the game. This visual enhances the player's self-assurance; without it, the player might lose interest in the game. Our approach with this visual can be described as "fishing" to engage the player more deeply. By doing so, we instill confidence in the player without them realizing that their intention has shifted. In other words, to enhance engagement in the game, we provide the player with what they desire to do.

# 3 Results and Discussion

This section concentrates on the sensitivity analysis of our proactive HCI utilized in this paper for recognizing bluffs and reshaping the recognized intentions based on the data collected from the actual bluff card game.

## *3.1 Model Development Results*

We implement a range of optimizations on the parameters and hyperparameters that influence the performance of CNN-based learning within our proactive HCI.

Twelve different results compile the CNN performance sensitivity analysis using output function = softmax, lr = 0.0001, two different epoch values (20 and 40), three different batch sizes (32, 64, and 128), and an additional augmented dataset to the original dataset. When we compare the results in Table 2, the best result is found to be achieved at 40 epochs using a batch size of 64, similar to Lin's work [24], in both performance criteria, validation loss and accuracy with 0.146 and 0.958, respectively.

Upon comparative analysis, improved results are observed with increased data across all parameters, except for the 128 batch size, where meaningful and reliable outcomes could not be achieved. This is attributed to a reduction in learning efficiency as the batch size increases, as seen with the value of 128. This aligns with the widely recognized property that smaller batch sizes tend to yield more stable and reliable results during training [31].

Finally, we observe how increasing the learning rate to lr = 0.001 will affect the result. When we increase the learning rate while keeping all other parameters constant, we understand that the results do not improve as accuracy = 0.94, loss = 0.179.

**Table 2** Comparison of augmented and original data across various batch sizes and epoch values

| Percentage of accuracy | | | | |
|---|---|---|---|---|
| Batch size | Original data | | Augmented data | |
| | 20 | 40 | 20 | 40 |
| 32 | 0.885 | 0.878 | 0.941 | 0.952 |
| 64 | 0.866 | 0.929 | 0.925 | 0.958 |
| 128 | 0.680 | 0.825 | 0.695 | 0.847 |
| Percentage of loss | | | | |
| Batch size | Original data | | Augmented data | |
| | 20 | 40 | 20 | 40 |
| 32 | 0.244 | 0.249 | 0.162 | 0.186 |
| 64 | 0.261 | 0.195 | 0.200 | 0.146 |
| 128 | 0.614 | 0.381 | 0.607 | 0.364 |

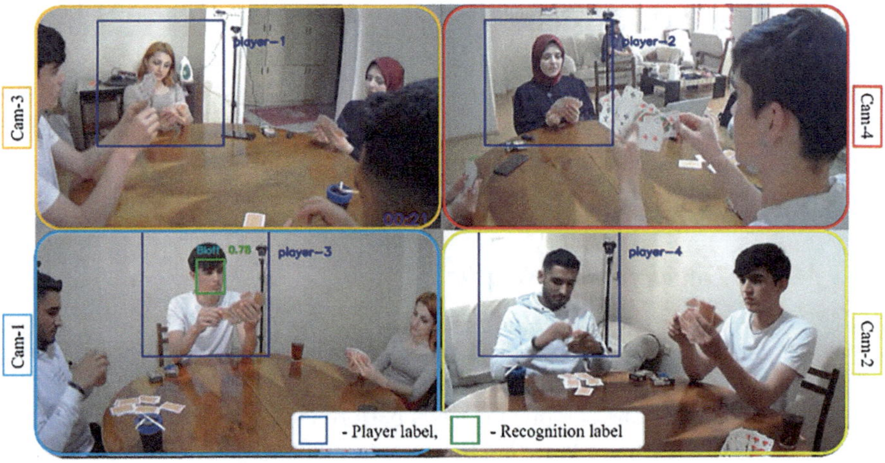


**Fig. 5** Instances of bluff detection utilizing the trained model

## 3.2 *Intention Recognition Results*

In this subsection, we evaluate the performance of the intention recognition unit of our Proactive HCI, trained with the optimal parameters from Subsection 3.1. The intention recognition process is tested on two games, recorded during the real bluff card game but excluded from the training phase. Thanks to our camera transmitting 25 frames per second, our algorithm can process each frame and capture microexpressions that occur for a very short time. Our system successfully recognizes 14 bluffing moves in the game. Instance of intentions identified as bluff (highlighted with a green frame when correctly detected) is shown in Fig. 5.

We also evaluate our system using nine non-bluff moves. The results obtained, along with the confusion matrix, are displayed in Fig. 6. As illustrated in Fig. 6, 14 out of the 21 intentions that are genuinely labeled as bluffs are accurately identified,

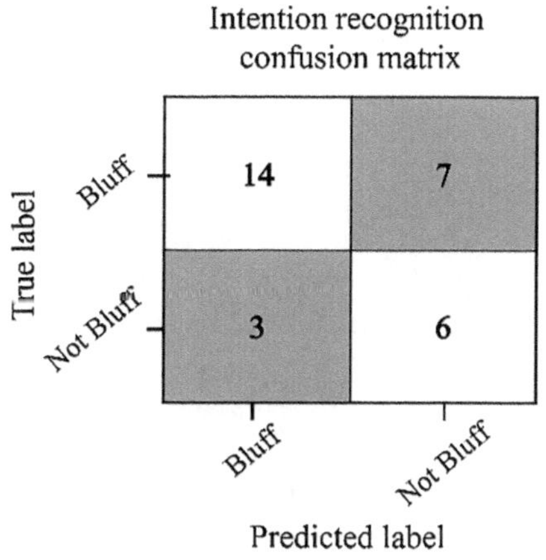


**Fig. 6** The confusion matrix of intention recognition

while 7 are incorrectly classified as bluffs (incorrect recognition). Conversely, 3 out of the 9 intentions that are not actually bluffs are mistakenly identified as bluffs (false positives). Based on the data from the confusion matrix in Fig. 6, we derive the following performance metrics: accuracy = 66.6%, precision = 0.66, recall = 0.823, and F1-score = 0.736. The F1-score is computed here due to the imbalance between bluff and non-bluff categories (21 and 9).

### *3.3 Reshaping Human Intention Results*

This subsection emphasizes the performance evaluation of intention reshaping via human-computer interaction involving 12 participants. An illustrative visual of this process is presented in Fig. 7.

In Fig. 7a, the target player's turn is highlighted, and we are monitoring their intention, which we aim to reshape. The current claim in the game indicates a "9." In Fig. 7b, our system has successfully identified the player's intention to "bluff," and it auto-generates a proactive visual stimulus emoji representing the desired intention (D4). In Fig. 7c, the player is shown engaging with our proactive visual, which responds to her facial expression. Figure 7d illustrates that the player's initial intention to bluff has been reshaped into a "pass" (new current intention: C4) as a result of the influence of our proactive visual. Moreover, when comparing this frame with the previous one, it is evident that the player discards the "Q" card. Consequently, the player who initially intended to bluff now opts to pass when our HCI system is activated.

It is crucial to highlight that participants are not made aware of the proactive visuals we implement: no information is provided to any of the players regarding the purpose of these visuals or the intentions behind each one. The experimental outcome depicted in Fig. 7 indicates that the proactive HCI effectively reshapes

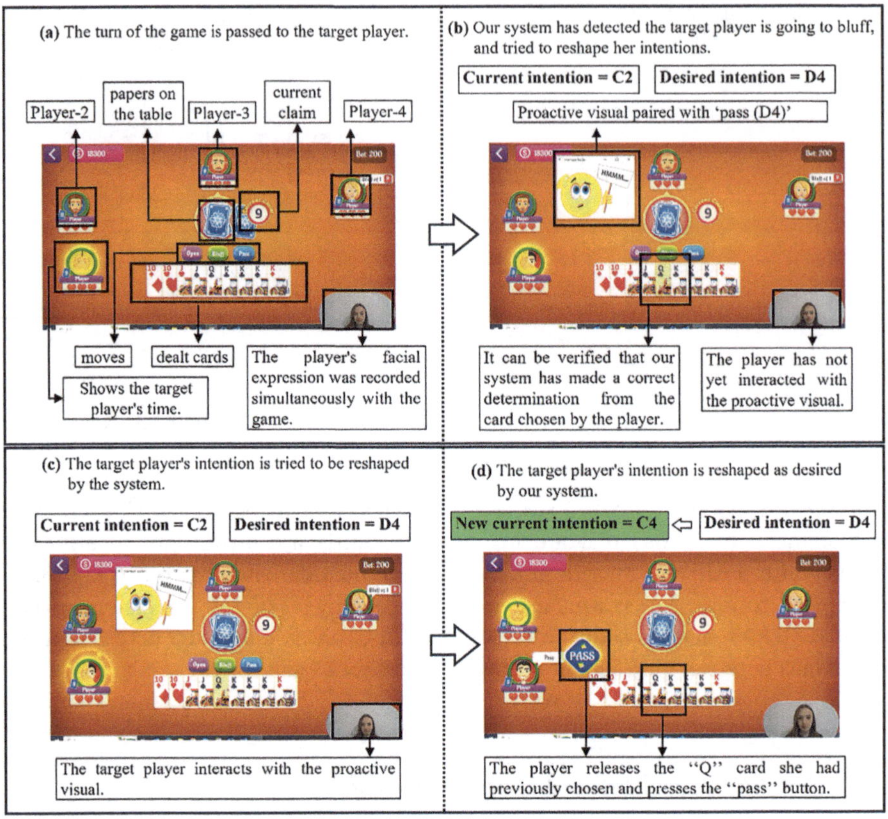


**Fig. 7** Reshaping human intention during the gameplay process

the intention of this particular target player though the results vary among different players. In this test run, we conduct intention reshaping for 12 distinct players across a total of 12 different games, following the proactive stimulus impact approach.

Since the participants have no knowledge of the experiment's purpose, the initial appearance of proactive visuals during the game captures their attention, regardless of which target player is involved, suggesting a potential to influence human intentions. Figure 8 illustrates the effectiveness of our scenario aimed at reshaping human intentions as each target player engages with proactive visuals.

## 4 Conclusion

In this paper, we introduce and demonstrate that reshaping of human intentions into auto created ones without verbal cues can be realized by interaction of our two-stage

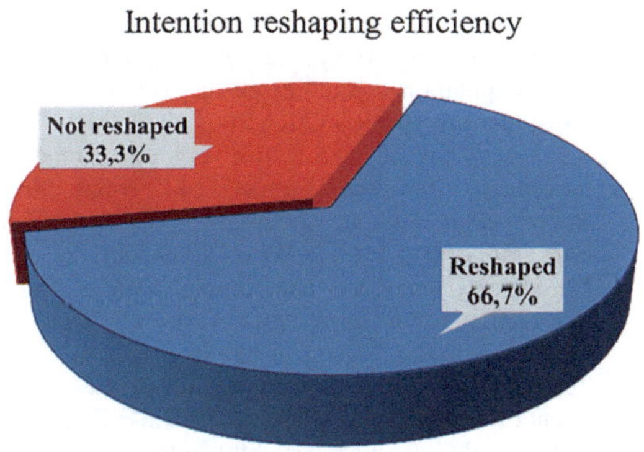


**Fig. 8** Effectiveness of human intention reshaping

proactive HCI system. For this purpose, in the first stage of the HCI system, we use our CNN algorithm, which we train with the facial expressions of people playing games in a real environment, to detect the player's instantaneous deep intention as bluff with 66.6% success. In our research on the reshaping of human intentions using our proactive HCI system, we present that our novel approach which contains characteristics of the stimuli can significantly affect the success of reshaping. These results introduce that our HCI system achieves 66.7% success in reshaping players' intentions using various proactive stimuli.

**Acknowledgments** We appreciate TUBITAK ULAKBIM, the High-Performance and Grid Computing Center (TRUBA resources), for partially offering computational assistance.

## References

1. Navarro, E.: What is theory of mind? A psychometric study of theory of mind and intelligence. Cogn. Psychol. **136**, 101495 (2022). https://doi.org/10.1016/j.cogpsych.2022.101495
2. Ekman, P., Oster, H.: Facial expressions of emotion. Annu. Rev. Psychol. **30**, 527–554 (1979). https://doi.org/10.1146/annurev.ps.30.020179.002523
3. Anwar, S., Milanova, M., Bigazzi, A., Bocchi, L., Guazzini, A.: Real time intention recognition. In: IECON 2016-42nd Annual Conference of the IEEE Industrial Electronics Society, pp. 1021–1024 (2016, October). https://doi.org/10.1109/IECON.2016.7794016
4. Singh, R., Miller, T., Newn, J., Sonenberg, L., Velloso, E., Vetere, F.: Combining planning with gaze for online human intention recognition. In: In Proceedings of the 17th International Conference on Autonomous Agents and Multiagent Systems (AAMAS 2018), pp. 488–496 (2018, July). https://doi.org/10.5555/3237383
5. Li, S., Zhang, L., Diao, X.: Deep-learning-based human intention prediction using rgb images and optical flow. J. Intell. Robot. Syst. **97**, 95–107 (2020). https://doi.org/10.1007/s10846-019-01049-3
6. Nakauchi, Y., Noguchi, K., Somwong, P., Matsubara, T., Namatame, A.: Vivid room: human intention detection and activity support environment for ubiquitous autonomy. In: Conference on Intelligent Robots and Systems, pp. 773–778 (2003, October). https://doi.org/10.1109/IROS.2003

7. Noguchi, K., Somwong, P., Matsubara, T., Nakauchi, Y.: Human intention detection and activity support system for ubiquitous autonomy. In: IEEE International Symposium on Computational Intelligence in Robotics and Automation, pp. 906–911 (2003, July). https://doi.org/10.1109/CIRA.2003.1222300
8. Koo, S., Kwon, D.S.: Recognizing human intentional actions from the relative movements between human and robot. In: IEEE International Symposium on Robot and Human Interactive Communication, pp. 939–944 (2009, October). https://doi.org/10.1109/ROMAN.2009.5326127
9. Durdu, A., Erkmen, I., Erkmen, A.M.: Estimating and reshaping human intention via human–robot interaction. Turk. J. Electr. Eng. Comput. Sci. **24**, 88–104 (2016). https://doi.org/10.3906/elk-1306-85
10. Terada, K., Shamoto, T., Mei, H., Ito, A.: Reactive movements of non-humanoid robots cause intention attribution in humans. In: IEEE Conference on Intelligent Robots and Systems, pp. 3715–3720 (2007, October). https://doi.org/10.1109/IROS.2007.4399429
11. Görür, O.C., Erkmen, A.M.: Elastic networks in reshaping human intentions by proactive social robot moves. In: The 23rd IEEE International Symposium on Robot and Human Interactive Communication, pp. 1012–1017 (2014August). https://doi.org/10.1109/ROMAN.2014.6926385
12. Webb, T.L., Sheeran, P.: Does changing behavioral intentions engender behavior change? A meta-analysis of the experimental evidence. Psychol. Bull. **132**, 249–268 (2006). https://doi.org/10.1037/0033-2909.132.2.249
13. Er, N.B., Erkmen, A.M.: Mining deep intention by human-robot interactions in a proactive social environment. In: Robotics conference of Turkey, ToRK 2016 Proceedings Book, pp. 16–19 (2016) (in Turkish). https://www.gedik.edu.tr/wp-content/uploads/tork-2016-bildiriler.pdf
14. Meltzoff, A.N.: Understanding the intentions of others: reenactment of intended acts by 18-month-old children. Dev. Psychol. **31**, 1–16 (1995). https://doi.org/10.1037/2F0012-1649.31.5.838
15. He, K., Zhang, X., Ren, S., Sun, J.: Deep residual learning for image recognition. In: Proceedings of IEEE Conference on Computer Vision and Pattern Recognition, pp. 770–778 (2015 June). https://doi.org/10.1109/CVPR.2016.90
16. Fernando, C., Banarse, D., Reynolds, M., Besse, F., Pfau, D., Jaderberg, M., Lanctot, M., Wierstra, D.: Convolution by evolution: differentiable pattern producing networks. In: Proceedings of the Genetic and Evolutionary Computation Conference, Association for Computing Machinery, pp. 109–116 (2016). https://doi.org/10.1145/2908812.2908890
17. Sheng, M., Jiang, J., Su, B., Tang, Q., Yahya, A.A., Wang, G.: Short-time activity recognition with wearable sensors using convolutional neural network. In: Proceedings of the 15th ACM SIGGRAPH Conference on Virtual-Reality Continuum and Its Applications in Industry, pp. 413–416 (2016, December). https://doi.org/10.1145/3013971.3014016
18. SplitCam. (2022). https://splitcam.com/. Accessed January 20, 2022
19. Lorenz, R.C., Gleich, T., Gallinat, J., Kühn, S.: Video game training and the reward system. Front. Hum. Neurosci. **9**, 40 (2015). https://doi.org/10.3389/fnhum.2015.00040
20. Green, C.S., Bavelier, D.: Learning, attentional control, and action video games. Curr. Biol. **22**, R197–R206 (2012). https://doi.org/10.1016/2Fj.cub.2012.02.012
21. Byom, L.J., Mutlu, B.: Theory of mind: mechanisms, methods, and new directions. Front. Hum. Neurosci. **2013**, 413 (2013). https://doi.org/10.3389/fnhum.2013.00413
22. Truba explanation document. (2022). https://docs.truba.gov.tr/TRUBA/TRUBAgenel.html. Accessed January 01, 2022
23. Algorithm for transfer learning. (2022). https://www.kaggle.com/code/ akshitmadan/emotion-classification-cnn-using-keras. Accessed February 05, 2022
24. Lin, R.: Analysis on the selection of the appropriate batch size in CNN neural network. In: 2022 International Conference on Machine Learning and Knowledge Engineering (MLKE), pp. 106–109 (2022). https://doi.org/10.1109/MLKE55170.2022.00026
25. Han, D., Liu, Q., Fan, W.: A new image classification method using CNN transfer learning and web data augmentation. Expert Syst. Appl. **95**, 43–56 (2018). https://doi.org/10.1016/j.eswa.2017.11.028

26. Luo, C., Li, X., Wang, L., He, J., Li, D., Zhou, J.: How does the data set affect CNN-based image classification performance? In: 2018 5th International Conference on Systems and Informatics (ICSAI), pp. 361–366 (2018). https://doi.org/10.1109/ICSAI.2018.8599448
27. Mehrabian, A., Wiener, M.: Decoding of inconsistent communications. J. Pers. Soc. Psychol. **6**(1), 109–114 (1967). https://doi.org/10.1037/h0024532
28. Erol, S., Çalışır, G.: Using emoji in interpersonal communication: a review on whatsapp. Gumushane Univ. J. Faculty Commun. **11**(2), 1568–1599. (in Turkish) (2023). https://doi.org/10.19145/e-gifder.1283938
29. Harn, R. W. J.: The Visual Language of Emojis: a Study on College Students' Social Support Communication in Online Social Networks. University of Kansas, Ph.D. Thesis. (2017). https://api.semanticscholar.org/CorpusID:148766162
30. Özdemir, A.: The criteria affecting color selection in design. Çukurova Univ. Soc. Sci. Inst. J. **14**(2), 391–401 (2005) (in Turkish). https://dergipark.org.tr/tr/download/article-file/50244
31. Masters, D., & Luschi, C.: Revisiting small batch training for deep neural networks. ArXiv. (2018) https://doi.org/10.48550/arXiv.1804.07612
32. Durdu, A., Erkmen, I., Erkmen, A.M., Yilmaz, A.: Morphing estimated human intention via human-robot interactions. In: Proceedings of the World Congress on Engineering and Computer Science, vol. 1. WCECS, San Francisco (2011, October)

# Long Short-Term Memory Algorithm in Intrusion Detection: A Deep Learning Approach to Time Series Data

**Azizjon Meliboev, Elbek Asqarov, Jasurbek Sulaymonov, and Axrorjon Yo'ldashev**

## 1 Introduction

An Intrusion Detection System (IDS) [1, 2] is a vital cybersecurity tool used to identify various security breaches within a network. There are two primary types of IDS: host-based intrusion detection system [3] and network-based intrusion detection system [4]. Host-based intrusion detection system monitors the behavior of a single host, analyzing elements such as sensors, system logs, software logs, file systems, disk resources, and network traffic. Network-based intrusion detection system, on the other hand, inspects all incoming packets to detect suspicious patterns. When an intrusion is detected, an IDS can either trigger an alert or block the source IP address from accessing the network. This study focuses on Network-based intrusion detection system.

Deep learning, a subset of machine learning characterized by artificial neural networks with multiple layers, has revolutionized the way complex patterns and representations are learned from data. By hierarchically extracting features through successive layers of abstraction, deep learning models have demonstrated remarkable success in domains such as computer vision [5], natural language processing [6], and speech recognition [7]. This success is largely attributed to their ability to automatically learn intricate patterns directly from raw data without the need for manual feature engineering.

A. Meliboev (✉) · E. Asqarov · J. Sulaymonov · A. Yo'ldashev
Kokand University, Kokand, Uzbekistan
e-mail: a.meliboyev@kokanduni.uz; e.e.askarov@kokanduni.uz; j.sulaymonov@kokanduni.uz; m.a.yuldashev@kokanduni.uz



B. U. Töreyin et al. (eds.), *8th EAI International Conference on Robotic Sensor Networks*, EAI/Springer Innovations in Communication and Computing,
https://doi.org/10.1007/978-3-031-92143-8_12

In recent years, the proliferation of sophisticated cyber threats has posed significant challenges to traditional intrusion detection systems (IDS). Conventional IDS approaches, often reliant on signature-based or heuristic methods, struggle to detect novel or evolving attacks, particularly zero-day exploits and advanced persistent threats. These limitations underscore the need for more adaptable and intelligent detection mechanisms capable of learning from the dynamic nature of network traffic.

Deep learning offers promising solutions to these challenges by leveraging its powerful representation learning capabilities. Specifically, Long Short-Term Memory (LSTMs) [8, 9], designed to handle sequential data, are adept at modeling temporal dependencies inherent in network traffic. LSTMs maintain internal memory states that capture information about previous inputs, making them suitable for analyzing time series data where the sequence and timing of events are critical.

This study explores the application of deep learning, with a focus on LSTMs, to develop an accurate and scalable network-based intrusion detection system. By transforming Transmission Control Protocol/Internet Protocol (TCP/IP) [10] packet data into time series representations within fixed intervals, we enable the LSTM to learn patterns associated with both normal and anomalous network behaviors. Supervised learning techniques are employed to train the model using labeled datasets, allowing it to distinguish between legitimate traffic and potential security breaches.

To evaluate the effectiveness of the proposed approach, we utilize the UNSW-NB15 [11] dataset, a comprehensive benchmark dataset that captures a wide range of network intrusion activities. These models are evaluated based on their performance over multiple training epochs, learning rates, and data balancing techniques to ensure a rigorous assessment.

The findings indicate that LSTM-based models outperform conventional classifiers in detecting intrusions within network traffic. This superiority is attributed to the LSTM's ability to extract high-level, abstract features from low-level network data, effectively capturing the complex temporal dynamics associated with malicious activities. The incorporation of deep learning into IDS not only enhances detection accuracy but also provides a scalable solution capable of adapting to the ever-evolving landscape of cyber threats.

This study contributes to the growing body of research advocating for the integration of deep learning methodologies in cybersecurity applications. By demonstrating the practical advantages of LSTMs in intrusion detection, we highlight the potential for these models to form the backbone of next-generation IDS solutions, ultimately strengthening network defense mechanisms against sophisticated attacks.

In an LSTM, the input is fed to a layer of the network, and it outputs a prediction. Then, a second input is fed to another layer, and the network outputs a prediction. However, it also uses information computed from the previous layer. The process is repeated until a final output is obtained. Therefore, a typical LSTM is shown in Fig. 1.

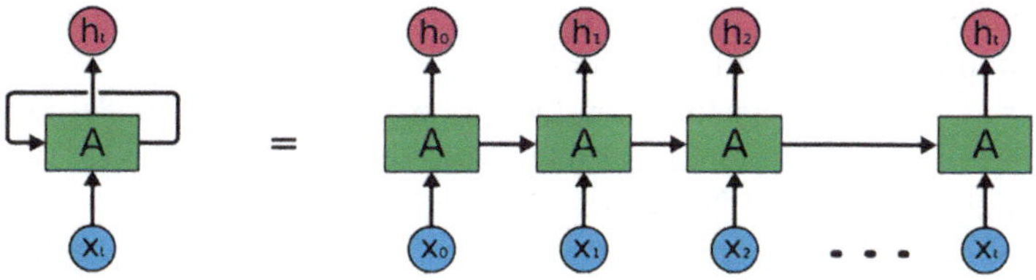


**Fig. 1** The architecture of LSTM

## 2 Related Works

The application of deep learning techniques, particularly Recurrent Neural Networks (RNNs) and their variants like Long Short-Term Memory (LSTM) networks, has gained significant traction in the field of IDS due to their ability to model sequential and temporal data inherent in network traffic. Sara et al. [12] use the CIDDS dataset and employ a deep learning technique, Long Short-Term Memory (LSTM), to develop an intrusion detection system. The approach attains a satisfactory accuracy of 85%.

Fatima et al. [13] developed deep learning methods for detecting attacks using Long Short-Term Memory (LSTM) networks. To enhance the model, we applied Principal Component Analysis (PCA) for dimensionality reduction and Mutual Information (MI) for feature selection. The authors evaluated proposed approach on the benchmark dataset KDD99. The experimental results demonstrate that models incorporating PCA achieved the highest accuracy in both training and testing stages, for both binary and multiclass classification scenarios.

Vinayakumar et al. [14] applied an RNN-based deep learning algorithm for IDS, creating hybrid models like RNN-RNN, RNN-LSTM, and RNN-GRU. As a test process, they utilize the established benchmark dataset from the Defense Advanced Research Projects Agency (DARPA) and Knowledge Discovery and Data Mining (KDD) Cup 99 intrusion detection contest to illustrate the effectiveness of these RNN architectures. All deep learning experiments are conducted for up to 1000 epochs with learning rates ranging from 0.01 to 0.5 on GPU-enabled TensorFlow. For traditional machine learning algorithms, experiments are carried out using Scikit-learn. The results show that the different families of RNN architectures achieve a low false positive rate compared to traditional machine learning classifiers.

Javaid Ahmad [15] introduced a deep learning approach for developing an effective and adaptable Intrusion Detection System (IDS). They employed a self-taught learning (STL) algorithm that combines a sparse autoencoder with SoftMax regression. Their method was tested on the NSL-KDD dataset, achieving an accuracy of 88.38% for binary classification and 79.10% for five-class classification.

Shone et al. [16] proposed a Non-Symmetric Deep Autoencoder (NDAE) for unsupervised feature learning, implemented in TensorFlow on a GPU. Their model was evaluated using the KDD Cup '99 and NSL-KDD datasets, and it achieved

85.42% accuracy in five-class classification and 89.22% accuracy in 13-class classification on the NSL-KDD dataset.

Kim et al. [17] build an Intrusion Detection System (IDS) model using a deep learning approach. They employ a Long Short-Term Memory (LSTM) architecture within a Recurrent Neural Network (RNN) and train the model on the KDD Cup 1999 dataset. The model performance tests confirm that the deep learning method is effective for IDS.

Nayyar et al. [18] focuses on utilizing a Machine Learning model based on Long Short-Term Memory (LSTM) networks to detect anomalies in network traffic. The approach involves redirecting all malicious requests to a honeypot acting as a black hole server. The algorithm was trained and tested on the CICIDS2017 dataset, which includes a variety of attacks such as Patator attacks, web-based brute force attacks, DoS Slowloris, DoS Slowhttptest, DoS Hulk, DoS GoldenEye, DDoS LOIT, among other categorical attacks.

Yin et al. [19] suggested using a Recurrent Neural Network (RNN) for IDS, with 122 input features and two output nodes for binary classification. The authors introduce a deep learning method for intrusion detection utilizing recurrent neural networks (RNN-IDS). They investigate the model's performance in both binary and multiclass classification, assessing how varying the number of neurons and different learning rates impacts its effectiveness. The experimental results demonstrate that RNN-IDS is highly effective for building a classification model with high accuracy, outperforming traditional machine learning methods in both binary and multiclass classification tasks. Their approach achieved 83.28% accuracy for binary classification and 81.29% accuracy for five-class classification on the NSL-KDD dataset.

Building upon the aforementioned works, our study utilizes the dataset to implement an intrusion detection system using LSTM networks. By leveraging the temporal dependencies in network traffic data, our approach aims to improve detection accuracy and address the limitations of prior methods that relied on outdated datasets. Our LSTM-based IDS achieves a satisfactory performance, indicating its potential effectiveness in real-world applications.

## 3 Proposed Method

In this study, we propose a deep learning approach utilizing LSTMs to enhance intrusion detection systems (IDS) by effectively modeling network traffic as time series data. The method involves preprocessing network traffic data, transforming it into a suitable format for LSTM input, designing an LSTM architecture tailored for intrusion detection, and evaluating its performance against traditional machine learning classifiers.

## 3.1 Dataset Description

We employ the UNSW-NB15 [11] dataset, a comprehensive benchmark for network intrusion detection research. This dataset includes a diverse range of modern attack types and normal activities, capturing realistic network traffic scenarios. The attack_cat feature represents the category of the attack in the dataset. It is a nominal type feature that classifies the type of malicious behavior or network attack associated with a particular record as shown Table 1. It consists of 49 features extracted from raw network packets using the Argus and Bro-IDS tools, with additional features generated using twelve algorithms but we used 44 features due to duplication.

## 3.2 Data Preprocessing

To ensure data quality, we perform the following preprocessing techniques that Handling Missing Values, Feature Scaling, Categorical Encoding using Scikit-learn

**Table 1** Features of UNSW_NB15 dataset

| № | Feature | Types | № | Feature | Types |
|---|---|---|---|---|---|
| 1 | id | Nominal | 23 | dtcpb | dtcpb |
| 2 | dur | float | 24 | dwin | dwin |
| 3 | proto | Nominal | 25 | tcprtt | tcprtt |
| 4 | service | Nominal | 26 | synack | Float |
| 5 | state | Nominal | 27 | ackdat | Float |
| 6 | spkts | Integer | 28 | smean | Integer |
| 7 | dpkts | Integer | 29 | dmean | Integer |
| 8 | sbytes | Integer | 30 | trans_depth | Integer |
| 9 | dbytes | Integer | 31 | response_body_len | Ineger |
| 10 | rate | Integer | 32 | ct_srv_src | Integer |
| 11 | sttl | Integer | 33 | ct_state_ttl | Integer |
| 12 | dttl | Integer | 34 | ct_dst_ltm | Integer |
| 13 | sload | Float | 35 | ct_src_dport_ltm | Integer |
| 14 | dload | Float | 36 | ct_dst_sport_ltm | Integer |
| 15 | sloss | Integer | 37 | ct_dst_src_ltm | Integer |
| 16 | dloss | Integer | 38 | is_ftp_login | Binary |
| 17 | sinpkt | Integer | 39 | ct_ftp_cmd | Integer |
| 18 | dinpkt | Integer | 40 | ct_flw_http_mthd | Integer |
| 19 | sjit | Float | 41 | ct_src_ltm | Integer |
| 20 | djit | Float | 42 | ct_srv_dst | Integer |
| 21 | swin | Integer | 43 | is_sm_ips_ports | Binary |
| 22 | stcpb | Integer | 44 | attack_cat | Nominal |

[20]. Scikit-learn is one of the best machine learning library for the data preprocess. It provides simple and efficient tools for data analysis and modeling, including classification, regression, clustering, and dimensionality reduction.

With handling missing values by imputing the mean value is preprocessing technique. This method replaces missing values in a numerical feature with the mean value of that feature.

Feature Scaling deals with continuous features for normalizing by using min-max scaling to a range between 0 and 1. It helps machine learning models to train faster and better.

$$x_i = \frac{x_i - \mathrm{Min}}{\mathrm{Max} + \mathrm{Min}} \quad (1)$$

Categorical Encoding techniques for nominal features can execute converting into numerical representations using one-hot encoding. We applied one-hot-encoding technique for the features dataset that proto, service, state.

### *3.3 LSTM Model Architecture*

The proposed LSTM model is designed to capture temporal patterns in network traffic as shown in Fig. 2. We developed and evaluate the model using Keras framework [21, 22]. Keras is an open-source, high-level deep learning framework that allows for the easy and fast implementation of neural networks. Keras provides to build, train, and evaluate our LSTM model.

Input layer accepts our data that sequences of network traffic features transformed into time series format. In our case, our training set data contains 42 features and 175,342 instances as a input.

LSTM is architecture that is particularly effective for learning from sequential data like time series, natural language, and video frames. LSTM layers intent to address the vanishing gradient problem and effectively learn long-term dependencies. In our proposed architecture, we used two LSTM layers with 64 hidden units.

Dropout [23] technique can help prevent overfitting by randomly setting a fraction of input units to 0 at each update during training. In our model, we used two dropout layers after LSTM layer.

A dense layer for the fully connected that maps the learned representations to the output space.

Output layer uses a SoftMax activation function [24] for binary classification, providing probability distributions over the attack, normal classes.

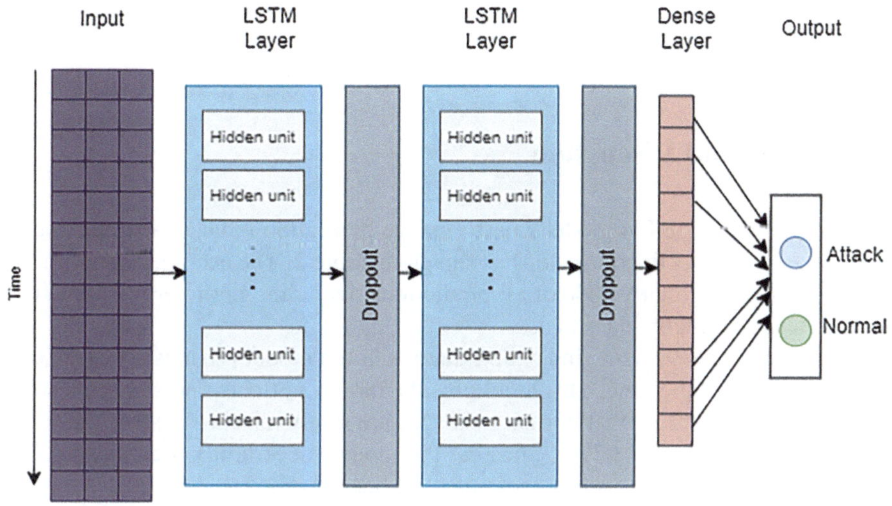


**Fig. 2** The proposed architecture of LSTM

## *3.4 Model Evaluation*

In our model, the evaluation performance metric used to evaluate the effectiveness of the LSTM model for intrusion detection is Accuracy (AC) [25, 26]. In addition to accuracy, we also account for the detection rate and the false positive rate. True Positive (TP) represents correctly identified attacks, indicating the number of attack records accurately classified as such. False Positive (FP) refers to regular records misclassified as attacks. True Negative (TN) corresponds to correctly identify normal records, while False Negative (FN) reflects instances where attacks are incorrectly classified as normal. The accuracy is calculated as the proportion of correctly classified records out of the total records, as shown in Eq. (2).

$$\mathrm{AC} = \frac{\mathrm{TP} + \mathrm{TN}}{\mathrm{TP} + \mathrm{TN} + \mathrm{FP} + \mathrm{FN}} \tag{2}$$

True Positive Rate (TPR) that shows the percentage of the number of records identified correctly over the total number of anomaly records, as shown in Eq. (3).

$$\mathrm{TRP} = \frac{\mathrm{TP}}{\mathrm{TP} + \mathrm{FN}} \tag{3}$$

False Positive Rate (FPR) where the percentage of the number of records rejected incorrectly is divided by the total number of normal records, as shown in Eq. (4).

$$\mathrm{FRP} = \frac{\mathrm{FP}}{\mathrm{FP} + \mathrm{TN}} \tag{4}$$

## 4 Results and Discussion

The results of our LSTM model's performance evaluation indicate a strong overall performance in detecting intrusions as shown in Table 2. The accuracy of the model is 91.3%, meaning that 91.3% of all predictions, including both normal and attack instances, are correctly classified.

The precision is 94.91%, indicating that when the model predicts an attack, it is correct 94.91% of the time. This highlights the model's effectiveness in minimizing false positives, ensuring that normal instances are rarely misclassified as attacks.

The recall stands at 88.79%, meaning the model successfully identifies 88.79% of actual attacks, though some attacks are still missed (false negatives).

Finally, the F-score of 91.74% represents the harmonic mean of precision and recall, reflecting a balanced performance between correctly identifying attacks and minimizing false positives. These results demonstrate that the model is reliable and effective in accurately detecting intrusions with a good balance between precision and recall.

Figure 3 shows the accuracy curves of LSTM model during training and testing across 200 epochs. The training accuracy starts low in the initial epochs but increases rapidly, reaching around 0.95 within the first 50 epochs. The test accuracy also starts relatively low but increases rapidly in the first 20 epochs, reaching around 0.90. The training accuracy steadily improves and converges close to 0.96, while the test accuracy stabilizes around 0.92.

Fig. 4 is illustrated a confusion matrix to evaluate the performance of a classification model. According to plot, the model has an overall accuracy of 91.3%. This means that 91.3% of the total predictions made by the model are correct. The model is highly effective at correctly identifying attacks, with a 95% success rate.

## 5 Conclusion

In this study, we present the development and evaluation of a deep learning model for time series classification, particularly focusing on distinguishing between normal and attack behaviors in UNSW-NB15 dataset. Through the implementation

**Table 2** The LSTM performance

| Algorithm | Accuracy (%) | Precision (%) | Recall (%) | F-score (%) |
|---|---|---|---|---|
| LSTM | 91.3 | 94.91 | 88.79 | 91.74 |

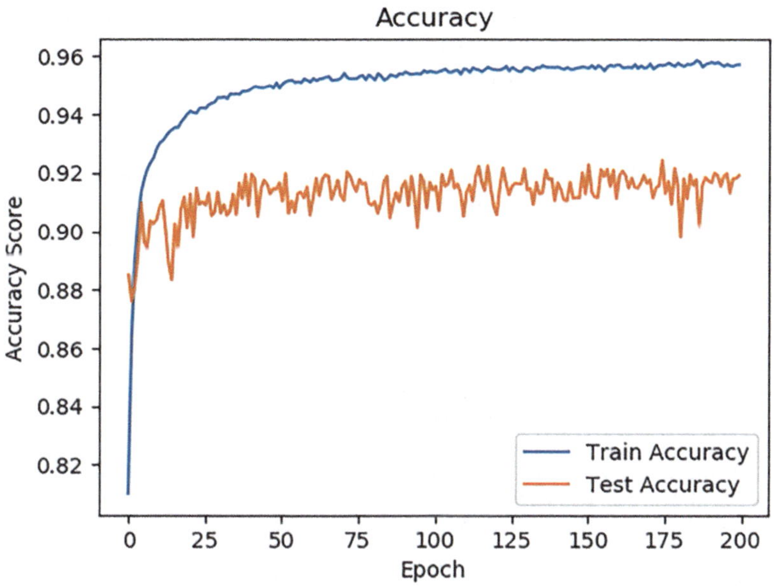


**Fig. 3** Accuracy curve of LSTM

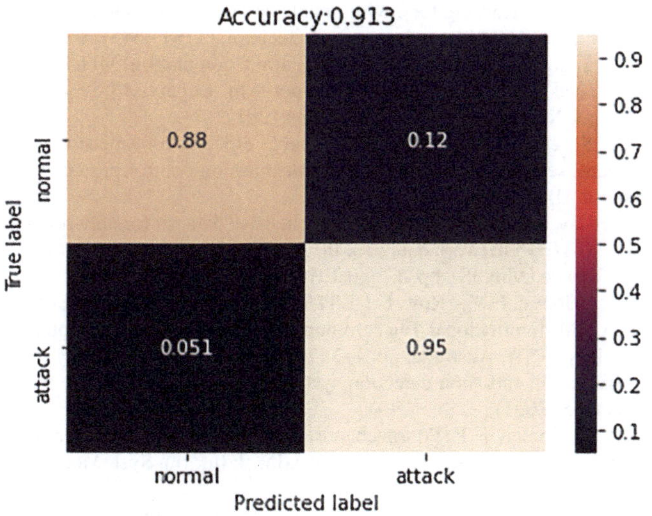


**Fig. 4** Confusion matrix of LSTM

of a Recurrent Neural Network (RNN) architecture with LSTM layers, the model demonstrates strong classification capabilities, achieving an overall accuracy of 91.3%. The confusion matrix analysis reveals that the model performs well in identifying attack instances with a high true positive rate of 95%, while correctly classifying 88% of normal instances.

Overall, this study demonstrates the potential of using LSTM for time series classification, particularly in intrusion detection system application. The model provides a strong foundation for further research and development, where future work can focus on refining its classification accuracy and generalization capability to improve its practical applicability in real-world scenarios.

## References

1. Bace, R.G.: Intrusion detection. Sams Publishing (2000)
2. Liao, H.-J., et al.: Intrusion detection system: a comprehensive review. J. Netw. Comput. Appl. **36**(1), 16–24 (2013)
3. Vokorokos, L., Baláž, A.: Host-based intrusion detection system. In: 2010 IEEE 14th International Conference on Intelligent Engineering Systems. IEEE (2010)
4. Vigna, G., Kemmerer, R.A.: NetSTAT: a network-based intrusion detection system. J. Comput. Secur. **7**(1), 37–71 (1999)
5. Hassaballah, M., Awad, A.I. (eds.): Deep Learning in Computer Vision: Principles and Applications. CRC Press (2020)
6. Otter, D.W., Medina, J.R., Kalita, J.K.: A survey of the usages of deep learning for natural language processing. IEEE Trans. Neural Networks Learn. Syst. **32**(2), 604–624 (2020)
7. Zhang, Z., et al.: Deep learning for environmentally robust speech recognition: an overview of recent developments. ACM Trans. Intell. Syst. Technol. (TIST). **9**(5), 1–28 (2018)
8. Hochreiter, S.: Long short-term memory. In: Neural Computation. MIT-Press (1997)
9. Graves, A., Graves, A.: Long short-term memory. In: Supervised Sequence Labelling with Recurrent Neural Networks, pp. 37–45. Springer (2012)
10. Mahmoodi, K.S., et al.: An intelligent sustainable efficient transmission internet protocol to switch between user datagram protocol and transmission control protocol in IoT computing. Expert. Syst. **40**(5), e13129 (2023)
11. Moustafa, N., Slay, J.: UNSW-NB15: a comprehensive data set for network intrusion detection systems (UNSW-NB15 network data set). In: 2015 Military Communications and Information Systems Conference (MilCIS), pp. 1–6. IEEE (2015)
12. Althubiti, S.A., Jones, E.M., Roy, K.: LSTM for anomaly-based network intrusion detection. In: 2018 28th International Telecommunication Networks and Applications Conference (ITNAC), Sydney, NSW, Australia, pp. 1–3 (2018)
13. Laghrissi, F.E., et al.: Intrusion detection systems using long short-term memory (LSTM). J. Big Data. **8**(1), 65 (2021)
14. Vinayakumar, R., Soman, K.P., Poornachandran, P.: Evaluation of recurrent neural network and its variants for intrusion detection system (IDS). Int. J. Inform. Syst. Model. Design (IJISMD). **8**(3), 43–63 (2017)
15. Javaid, A., et al.: A deep learning approach for network intrusion detection system. In: Proceedings of the 9th EAI International Conference on Bio-Inspired Information and Communications Technologies (Formerly BIONETICS), pp. 21–26 (2016)
16. Shone, N., et al.: A deep learning approach to network intrusion detection. IEEE Trans. Emerg. Top. Comput. Intell. **2**(1), 41–50 (2018)
17. Kim, J., Kim, J., Thu, H.L., Kim, H.: Long short term memory recurrent neural network classifier for intrusion detection. In: 2016 International Conference on Platform Technology and Service (PlatCon), pp. 1–5. IEEE (2016)

18. Nayyar, S., Arora, S., Singh, M.: Recurrent neural network based intrusion detection system. In: 2020 International Conference on Communication and Signal Processing (ICCSP), pp. 0136–0140. IEEE (2020)
19. Yin, C., et al.: A deep learning approach for intrusion detection using recurrent neural networks. IEEE Access. **5**, 21954–21961 (2017)
20. Kramer, O., Kramer, O.: Scikit-learn. In: Machine learning for evolution strategies, pp. 45–53 (2016)
21. Manaswi, N.K., Manaswi, N.K.: Understanding and working with Keras. In: Deep Learning with Applications Using Python: Chatbots and Face, Object, and Speech Recognition with TensorFlow and Keras, pp. 31–43 (2018)
22. Gulli, A., Pal, S.: Deep Learning with Keras. Packt Publishing Ltd (2017)
23. Srivastava, N., et al.: Dropout: a simple way to prevent neural networks from overfitting. J. Machine Learn. Res. **15**(1), 1929–1958 (2014)
24. Kouretas, I., Paliouras, V.: Simplified hardware implementation of the softmax activation function. In: 2019 8th International Conference on Modern Circuits and Systems Technologies (MOCAST), pp. 1–4. IEEE (2019)
25. Dalianis, H., Dalianis, H.: Evaluation metrics and evaluation. In: Clinical Text Mining: secondary use of electronic patient records, pp. 45–53 (2018)
26. Meliboev, A., Alikhanov, J., Kim, W.: Performance evaluation of deep learning-based network intrusion detection system across multiple balanced and imbalanced datasets. Electronics. **11**(4), 515 (2022)

# An Analytical Model for Time-Dependent Battery Level Distribution for UAVs

**Ayda Baheri Eslami and Mehmet Akif Yazici**

## 1 Introduction

Drones, or unmanned aerial vehicles (UAVs), have emerged as indispensable elements of industries ranging from logistics to agriculture and military to disaster management, construction, mining, telecommunications, and environmental conservation. Thanks to the development of IoT technologies, UAVs are now commonly equipped with sensor networks and can conduct autonomous data collection, transmission, and analysis. This integration elevates the role of UAVs by allowing them to work alongside ground sensor systems to track environmental changes, follow assets, and react quickly in a real-time manner.

However, the battery life span has always been among the biggest challenges of UAV operations, affecting their mission duration and operational radius. Even though embedding sensor networks would enable UAVs to have optimal performance due to obtaining real-time updates and route adjustments, the battery is the most considerable constraint to overcome. For drones, energy-efficient strategies such as hardware-optimized scheduling and battery-aware charging are crucial for improving the performance of sensor-equipped UAVs in IoT environments, ensuring effective data collection and sufficient energy to complete their missions on time.

This work was supported by the Scientific and Technological Research Council of Türkiye (TUBITAK) under Project EEEAG-120E307. The work of Ayda Baheri Eslami was supported by the Research Fund (BAP) of the Istanbul Technical University under Project MDK-2022-43741.

A. B. Eslami · M. A. Yazici (✉)
Information and Communications Research Group (ICRG), Informatics Institute, Istanbul Technical University, Istanbul, Turkey
e-mail: eslami20@itu.edu.tr; yazicima@itu.edu.tr

 
B. U. Töreyin et al. (eds.), *8th EAI International Conference on Robotic Sensor Networks*, EAI/Springer Innovations in Communication and Computing,
https://doi.org/10.1007/978-3-031-92143-8_13

The stochastic nature of UAVs' battery usage, particularly in dynamic scenarios, and practical tools necessitate versatile models that can forecast battery status and then optimize UAV operations. Various factors should be considered when determining these operations, such as UAV's current battery level, its operating modes (i.e., linear motion, hovering etc.), and battery depletion characteristics in those operating modes. The trajectory of a UAV can be selected to maximize the service time it provides. Due to decreasing battery level, the UAV has to come back to its platform to recharge itself. Its position as well as battery level must be taken into account for deciding exactly when the UAV needs to return.

The literature on battery modeling focuses on either electrical dynamics [1–4] or simple stochastic models based on discrete-time Markov chains [5–8]. Fluid queue model studies, on the other hand, focus on factors such as average battery lifetime [9] and the chance of battery depletion within a particular time frame [10]. However, to the best of our knowledge, no studies have been conducted in order to derive the actual distribution of the battery level within a time span.

In this chapter, we propose an analytical framework to compute the distribution of the battery level after a certain amount of time, when the current battery level and the dynamics of the stochastic behavior are known. In the sequel, we describe the system model we investigate and the methodology in Sect. 2. We present numerical results in Sect. 3. Finally, we conclude in Sect. 4.

## 2 System Model and Methodology

We investigate the battery level dynamics of a UAV, where the current battery level between 0 (empty/depleted battery) and 1 (fully charged battery) is known, and the UAV dynamics are modeled as a continuous-time Markov chain (CTMC). The UAV can have multiple states representing various modes of operation such as linear motion with or without energy harvesting, hovering with or without energy harvesting, and there are different (but constant) rates of battery depletion associated with each mode. Depending on the physical conditions and/or chosen flight speeds and altitudes, multiple states could be defined for the same kind of motion. The exact nature of energy harvesting is irrelevant to the mathematical model, but it can typically be assumed to be solar.

Among the elements of battery consumption are (i) linear movement performed by the UAV, (ii) hovering, (iii) communication, and (iv) data processing and decision-making operations. As a result of these operations occurring together at different rates and intensities, a certain number of states can be defined where the rate of battery depletion/charging varies over time. Generally, a system where the transitions between these states are determined by a CTMC has been examined. The state transition rate matrix of this $N$-state CTMC is defined as

$$Q_s = \begin{bmatrix} q_{1,1} & q_{1,2} & \cdots & q_{1,N} \\ q_{2,1} & q_{2,2} & \cdots & q_{2,N} \\ \vdots & \vdots & \ddots & \vdots \\ q_{N,1} & q_{N,2} & \cdots & q_{N,N} \end{bmatrix}. \tag{1}$$

In this matrix, for $i \neq j$, $q_{i,j} \geq 0$, and the relationship $\sum_{j=1}^{N} q_{i,j} = 0$ holds for all $i \in \{1, \ldots, N\}$. Accordingly, when the CTMC is in state $i$, it remains in this state for an exponentially distributed random duration with a mean of $-1/q_{i,i}$, and when it leaves state $i$, it chooses the next state with probabilities proportional to $q_{i,j}$ values. If the rate of change in the battery level in state $i$ is denoted as $r_i$, the diagonal rate matrix is defined as

$$R_s = \begin{bmatrix} r_1 & & & \\ & r_2 & & \\ & & \ddots & \\ & & & r_N \end{bmatrix}. \tag{2}$$

The UAV battery, starting with a certain level $u$, will be depleted or recharged at a rate determined by the state of a continuous-time Markov chain (CTMC). If $r_i < 0$, the battery is being discharged in state $i$, and if $r_i > 0$, the battery is being charged in state $i$. Figure 1 provides an example of this dynamic. According to this example scenario, the battery is discharged between 0 and 6, and 11 and 16 time units and recharged through energy harvesting between 10 and 11, and 16 and 22 time units. The rate of battery depletion varies between 0 and 4, 4 and 6, and 11 and 15 time units. Similarly, the rate of energy harvesting varies between 6 and 11, and 15 and 22 time units.

The primary objective of this work is to find the probability distribution of the battery level after $H$ time units, given that the UAV's battery level currently is $u$, and $Q$ and $R$ matrices are known. In other words, we propose a method to compute

$$F_b(x; u, H) = \mathrm{P}\{X(H) \leq x \mid X(0) = u\}, \tag{3}$$

where $X(t)$ denotes the battery level (between 0 and 1) at time $t$.

## 2.1 Markov Fluid Queues

Markov fluid queues are characterized by a joint process $\{X(t), M(t); t \geq 0\}$, where $\{X(t); t \geq 0\}$ denotes the buffer content process and $\{M(t); t \geq 0\}$ is a continuous-time Markov chain that controls the net rate (difference between entry and exit rates, or drift) at which the buffer content $X(t)$ changes. The latter process, $\{M(t); t \geq 0\}$,

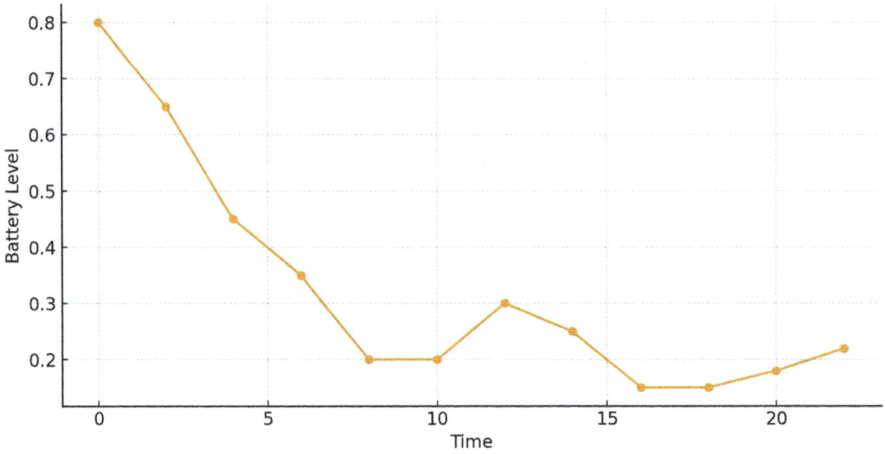


**Fig. 1** A sample scenario for battery consumption dynamics

is often referred to as the background or modulating process for the Markov fluid queue.

Markov fluid queues are categorized based on the characteristics of the underlying Markov process and the fluid flow dynamics, into single-regime, multi-regime, and feedback fluid queues. An extensive overview of Markov fluid queues is represented in [11], which uses the spectral approach to analyze and investigate the behavior of Markov fluid queues. The numerical instability problems regarding the spectral solution are identified, and a more stable alternative solution is described in [12].

The drift of the buffer level is determined by the modulating process as follows:

$$\frac{d}{dt}X(t) = r(M(t)). \tag{4}$$

The cumulative distribution function (cdf), probability density function (pdf), and their steady-state distributions for the buffer level $X(t)$ are defined as follows:

$$F_i(x,t) = \mathrm{P}\{X(t) \le x, M(t) = i\} \tag{5}$$

$$F(x,t) = [F_1(x,t), \ldots, F_N(x,t)], \tag{6}$$

$$F(x) = \lim_{t\to\infty} F(x,t), \tag{7}$$

$$f(x,t) = \frac{d}{dx}F(x,t), \tag{8}$$

$$f(x) = \lim_{t\to\infty} f(x,t). \tag{9}$$

Here, it is assumed that the background process $M(t)$ has $N$ different states. When the CTMC $M(t)$ is represented by the infinitesimal generator matrix $Q$, and the queue change rates corresponding to each state of $M(t)$ from 1 to $N$ are contained in the diagonal matrix $R = \text{diag}[r_1, \ldots, r_N]$, it is known that the steady-state pdf of $X(t)$ satisfies the following system of equations:

$$\frac{d}{dx} f(x)R = f(x)Q. \tag{10}$$

The solution of this system of equations based on spectral expansion that relies on eigenvalue calculations is laid out in [11], whereas the additive decomposition method, which addresses the numerical instabilities of the spectral method, was proposed in [12]. In this latter method, using the Schur decomposition of the matrix $A = QR^{-1}$ along with a similarity transform matrix $Y$, the following system of equations is obtained:

$$\frac{d}{dx} f(x) = f(x)\, Y \begin{bmatrix} 0 & & \\ & A_- & \\ & & A_+ \end{bmatrix} Y^{-1}. \tag{11}$$

In this formulation, the eigenvalues of the matrix $A_-$ are negative, while the eigenvalues of the matrix $A_+$ have positive real parts. The solution to this system of equations is given by

$$f(x) = [a_0 \quad a_- \quad a_+] \begin{bmatrix} 1 & & \\ & e^{A_- x} & \\ & & e^{A_+(B-x)} \end{bmatrix} Y^{-1}. \tag{12}$$

Here, the value $B$ represents the capacity of the queue. The coefficients $a_0$, $a_-$, and $a_+$ can be solved using a set of boundary conditions along with the probability masses at the boundaries 0 and $B$. If the boundary $B$ is infinite (indicating an unlimited queue), the solution will only include exponential components that decay to zero (having negative real parts). In such a scenario, $a_0$ and $a_+$ must be 0. For further information, please refer to [12] and [13].

Systems in which the queue occupancy value $X(t)$ has a feedback effect on the system parameters $Q$ and $R$ are called feedback Markov fluid queue (FMFQ) systems [13]. This feedback often appears as a piecewise constant function in many models. In this scenario, the queue is divided into $K$ regimes, with $Q$ and $R$ matrices being constant within each regime but varying from regime to regime. In summary, if $0 = T_0 < T_1 < \cdots < T_K = B$, the queue is said to be in the $k$-th regime when the occupancy value $X(t)$ is between $T_{k-1}$ and $T_k$. The system parameters valid in this interval are denoted by $Q^{(k)}$ and $R^{(k)}$. Such systems with feedback are called *multi-regime Markov fluid queue* (MRMFQ) systems. The solution of MRMFQ systems primarily involves formulating each regime as an MFQ and then obtaining a common solution for all regimes based on certain boundary conditions

[14]. Instantaneous queue occupancy determines the rate of change of the buffer content and the background process in studies proposed in [15] and [16], which give rise to continuous feedback Markov fluid queues. In [17] the drift rates change according to the buffer content, while models and analytical techniques are used for modeling a network system. Another feedback fluid queues system is suggested in [18] in which net drift rates are continuous functions of buffer content. The work in [19] is presented in a more general way that models feedback fluid queues and the specifications of the background Markov process vary according to the buffer value. In [20] a continuous feedback fluid queue with a background Markov process that has two states and infinite buffer capacity is modeled.

In this chapter, we utilize multi-regime Markov fluid queues, which involve systems where the fluid flow rates and transition dynamics vary across multiple regimes, each governed by distinct parameters. Following the notation and conventions described in [14], the buffer capacity is denoted by $B$, and the buffer content ranges in the closed interval $[0, B]$. $X(t)$ denotes the fluid level in the queue, and $M(t)$ denotes the state of the background process at time $t$. The buffer space is divided into $K$ regimes, defined by the thresholds $0 = T_0 < T_1 < \ldots < T_K = B$. The buffer is said to be in regime $k$, if $T_{k-1} < X(t) < T_k$. The buffer is said to be at threshold $k$ when $X(t) = T_k$. $T_0$ and $T_K$ are the terminal boundary points, whereas the other thresholds correspond to intermediate boundary points. It is assumed that the background process$\{M(t)\ ;\ t \geq 0\}$ has the state space $S = \{1, 2, .., N\}$. When the system is in regime $k$ (at threshold $T_k$), the background process $M(t)$ behaves according to a CTMC with infinitesimal generator $Q^{(k)}(\widetilde{Q}^{(k)})$. We let $R^{(k)}(\widetilde{R}^{(k)})$ to denote the diagonal matrix of drifts in regime $k$ (at threshold $T_k$). The steady-state joint distribution vector is defined by

$$F^{(k)}(y) = \left[F_1^{(k)}(y)\ F_2^{(k)}(y)\ \cdots\ F_N^{(k)}(y)\right], \tag{13}$$

$$F_i^{(k)}(y) = \lim_{t\to\infty} \mathrm{P}\{X(t) \leq y,\ M(t) = i\}, \quad T_{k-1} < y < T_k, \tag{14}$$

and the steady-state joint density vector is defined by

$$f^{(k)}(y) = \left[f_1^{(k)}(y)\ f_2^{(k)}(y)\ \cdots\ f_N^{(k)}(y)\right] = \frac{d}{dy}F^{(k)}(y). \tag{15}$$

Moreover, $c^{(k)},\ k \in \{0, 1, \ldots, K\}$ are defined to be the row vector of the probability mass accumulations at the boundary points:

$$c^{(k)} = \left[c_1^{(k)}\ c_2^{(k)}\ \ldots\ c_N^{(k)}\right], \quad k \in \{0, 1, \ldots, K\}, \tag{16}$$

$$c_i^{(k)} = \lim_{t\to\infty} \mathrm{P}\{X(t) = T_k,\ M(t) = i\}. \tag{17}$$

The steady-state joint density vector $f^{(k)}(y)$ of the feedback Markov fluid queue satisfies the set of differential equations

$$\frac{d}{dy} f^{(k)}(y) R^{(k)} = f^{(k)}(y) Q^{(k)}, \qquad T^{(k-1)} < y < T^{(k)},\ 1 \le k \le K \tag{18}$$

along with the following boundary conditions:

$$c_m^{(0)} = 0, \quad \forall m \in S_+^{(1)} \tag{19}$$

$$c_m^{(k)} = 0, \quad \forall m \in \left(S_+^{(k)} \cap S_+^{(k+1)}\right) \cup \left(S_-^{(k)} \cap S_-^{(k+1)}\right), \quad 1 \le k \le K-1 \tag{20}$$

$$c_m^{(k)} = 0, \quad \forall m \in \left(S_-^{(k)} \cap S_+^{(k+1)}\right) \cap \left(\tilde{S}_+^{(k)} \cup \tilde{S}_-^{(k)}\right), \quad 1 \le k \le K-1 \tag{21}$$

$$c_m^{(K)} = 0, \quad \forall m \in S_-^{(K)} \tag{22}$$

$$f^{(0)}(0^+) R^{(1)} = c^{(0)} \tilde{Q}^{(0)} \tag{23}$$

$$f^{(k+1)}(T_k^+) R^{(k+1)} - f^{(k)}(T_k^-) R^{(k)} = c^{(k)} \tilde{Q}^{(k)}, \quad 1 \le k \le K-1 \tag{24}$$

$$f_m^{(k)}(T_k^-) = 0, \quad \forall m \in S_-^{(k)} \cap \left(\tilde{S}_0^{(k)} \cup \tilde{S}_+^{(k)}\right), \quad 1 \le k \le K-1 \tag{25}$$

$$f_m^{(k+1)}(T_k^+) = 0, \quad \forall m \in \left(\tilde{S}_0^{(k)} \cup \tilde{S}_-^{(k)}\right) \cap S_+^{(k+1)}, \quad 1 \le k \le K-1 \tag{26}$$

$$f^{(K)}(T_K^-) R^{(K)} = -c^{(K)} \tilde{Q}^{(K)} \tag{27}$$

$$\sum_{m=1}^{N} \left( \sum_{k=1}^{K} \int_{T_{k-1}^+}^{T_k^-} f_m^{(k)}(x)\, dx + \sum_{k=0}^{K} c_m^{(k)} \right) = 1, \tag{28}$$

where

$$S_-^{(k)} = \{s \mid r_s^{(k)} < 0\}, \qquad S_0^{(k)} = \{s \mid r_s^{(k)} = 0\}, \qquad S_+^{(k)} = \{s \mid r_s^{(k)} > 0\},$$

$$\tilde{S}_-^{(k)} = \{s \mid \tilde{r}_s^{(k)} < 0\}, \qquad \tilde{S}_0^{(k)} = \{s \mid \tilde{r}_s^{(k)} = 0\}, \qquad \tilde{S}_+^{(k)} = \{s \mid \tilde{r}_s^{(k)} > 0\}.$$

For detailed descriptions of the boundary conditions, we refer the reader to [14].

Methods for obtaining the steady-state distributions of MFQ and MRMFQ models have reached a certain maturity. On the other hand, time-dependent (transient) problems are much more difficult than steady-state analyses. There are some approximation methods in the literature [21]. An important example of time-dependent problems arises in the context of risk theory [22, 23]. In a scenario where a company enters the insurance business with a certain capital, collects a certain amount of premiums from its customers, and meets random-sized claims at random moments, the probability of bankruptcy within a certain period was solved in [24]. Based on this work, which expresses time-dependent distributions in MFQ and MRMFQ models in terms of the steady-state distributions of the same model, [10] modeled the battery occupancy of energy-harvesting IoT devices and optimized the measurement frequency in a scenario where sensors could perform measurements and communication at adaptive rates. In this problem, the battery level parallels the real-time reserve of the insurance company in the bankruptcy problem, the energy harvesting function of the IoT device parallels the premiums collected by the insurance company, and energy consumption parallels the claims payments, which mathematically match exactly. The method proposed for the probability of the queue value hitting zero within a certain period in [24] was generalized by Gürsoy [25] to cover any value of the queue, and a framework for numerically obtaining time-dependent solutions of MFQ and MRMFQ models was established. This framework is Markovian, and Erlangization [26, 27] or concentrated exponential matrix distributions [28] can be used for approximating deterministic quantities.

## *2.2 Markov Fluid Queue Model for Battery Level*

The existing studies in the literature are mostly concerned with calculating the probability that the queue level reaches a certain level within a given period. In contrast, our work has taken a step further by developing a method to obtain the entire probability distribution of the queue level at the end of a certain period. Essentially, starting from the same point as [10, 24, 25], an MRMFQ with an artificial "Reset" state in addition to system states was defined. Using the Erlangization method (using an Erlang-$L$ distribution instead of a deterministic quantity) to represent the deterministic time period, states are defined as $(i, j)$, where $i$ represents the system state and $j$ represents the Erlang phase. When the Erlang-$L$ distribution is used, a total of $NL + 1$ states emerge. Erlangization helps us to accommodate a non-Markovian (that is, deterministic) quantity in a Markovian model. As the number of Erlang phases, $L$, increases, the corresponding Erlang-$L$ distribution approaches a deterministic quantity.

The CTMC will start operating in one of the $N$ system states corresponding to the first Erlang phase. This movement will continue until it exits the last Erlang phase (i.e., until the specified period ends) or until the battery level drops to 0. In both scenarios, a transition to the Reset state occurs. The purpose of the Reset state is to restore the system to its initial state. Therefore, when exiting the last Erlang phase and transitioning to the Reset state, the battery level will be returned to its initial level denoted as $u$. If the battery level is below $u$ when transitioning to the Reset state, the drift rate in the Reset state must be positive to increase the level to $u$. Conversely, if the battery level is above $u$ when transitioning into the Reset state, the drift rate in the Reset state must be negative to decrease the level down to $u$. This shows that the drift rate in the Reset state depends on the queue level, indicating that our model is an MRMFQ model.

The battery level is divided into two regimes in terms of occupancy. The boundary between the two regimes is the initial value of the battery level $u$. When the time period, denoted as $H$, is represented by an Erlang-$L$ distribution, the phase-type distribution parameters of this distribution are written as follows:

$$\boldsymbol{\nu} = [1\,0\,\ldots\,0], \tag{29}$$

$$\mathbf{T} = \begin{bmatrix} -\frac{L}{H} & \frac{L}{H} & 0 & \cdots & 0 \\ 0 & -\frac{L}{H} & \frac{L}{H} & \cdots & 0 \\ 0 & 0 & -\frac{L}{H} & \cdots & 0 \\ \vdots & \vdots & \vdots & \ddots & \vdots \\ 0 & 0 & 0 & \cdots & -\frac{L}{H} \end{bmatrix}, \tag{30}$$

$$\mathbf{T_0} = -\mathbf{T1} = \left[0 \cdots 0\ \tfrac{L}{H}\right]^T. \tag{31}$$

If the state transition rate matrix of the CTMC representing the battery level is denoted by $\mathbf{Q_s}$, the state transition rate matrix $\mathbf{Q}$ of the MRMFQ model for both regimes, the first regime $(0, u)$ and the second regime $(u, 1)$, is expressed as follows:

$$\mathbf{Q} = \begin{bmatrix} \mathbf{I}_L \otimes \mathbf{Q_s} + \mathbf{T} \otimes \mathbf{I}_N & \mathbf{T_0} \otimes \mathbf{1}_N \\ \mathbf{0}_{NL} & 0 \end{bmatrix}. \tag{32}$$

Here, $\mathbf{I}_L$ and $\mathbf{I}_N$ are identity matrices of sizes $L \times L$ and $N \times N$, respectively; $\mathbf{0}$ and $\mathbf{1}$ denote column or row vectors of zeros and ones of the indicated length, and $\otimes$ denotes the Kronecker product. In this structure, the system states $(i, j)$ are ordered first, and the Reset state is ordered last. While the $\mathbf{Q}$ matrix has the same structure in both regimes, since the Reset state will bring the system back to the first Erlang phase at level $u$, it will be

$$\mathbf{Q}(u) = \begin{bmatrix} \mathbf{I}_L \otimes \mathbf{Q_s} + \mathbf{T} \otimes \mathbf{I}_N & \mathbf{T_0} \otimes \mathbf{1}_N \\ \boldsymbol{\nu} \otimes \boldsymbol{\alpha} & -1 \end{bmatrix}. \tag{33}$$

Here, $\alpha$ represents the initial state probability distribution of the system. If the battery level drops to 0, it means the UAV's battery is depleted and can no longer operate. In this case, battery regeneration through energy harvesting in subsequent phases will not be possible. Therefore, the process is restarted by transitioning to the Reset state even if the Erlang phases are not completed. Consequently, at level 0, the state transition rate matrix will be

$$\mathbf{Q}(0) = \begin{bmatrix} -\mathbf{I}_{NL} & \mathbf{1}_{NL} \\ \mathbf{0}_{NL} & 0 \end{bmatrix}. \tag{34}$$

The diagonal drift rate matrices are written as follows for the first regime:

$$\mathbf{R_1} = \begin{bmatrix} \mathbf{I}_L \otimes \mathbf{R_s} & \\ & 1 \end{bmatrix}, \tag{35}$$

and for the second regime:

$$\mathbf{R_2} = \begin{bmatrix} \mathbf{I}_L \otimes \mathbf{R_s} & \\ & -1 \end{bmatrix}. \tag{36}$$

The steady-state distribution solution of this system can be obtained by following the methodology in [14]. As the final step, to obtain the distribution of the battery level at the end of a certain time period, it is necessary to determine the distribution at the moment of transition to the Reset state. The cdf of the queue level at the moment of transition to some state $j$ is given by

$$\begin{aligned} F_b(x, j) &= \lim_{t\to\infty, \Delta t\downarrow 0} \mathrm{P}\{X(t) \le x \mid M(t+\Delta t) = j, M(t) \ne j\} \qquad (37) \\ &= \lim_{t\to\infty, \Delta t\downarrow 0} \frac{\mathrm{P}\{M(t+\Delta t) = j, M(t) \ne j \mid X(t) \le x\}\mathrm{P}\{X(t) \le x\}}{\mathrm{P}\{M(t+\Delta t) = j, M(t) \ne j\}} \\ &= \lim_{t\to\infty, \Delta t\downarrow 0} \frac{\sum_{i\ne j} \mathrm{P}\{M(t+\Delta t)=j, M(t) = i \mid X(t) \le x\}\mathrm{P}\{X(t) \le x\}}{\sum_{i\ne j} \mathrm{P}\{M(t+\Delta t) = j, M(t) = i\}} \\ &= \begin{cases} \dfrac{\sum_{i\ne j} \tilde{Q}^{(0)}_{i,j} c^{(0)}_i + Q^{(1)}_{i,j}(F_i(x) - c^{(0)}_i)}{\sum_{i\ne j} \tilde{Q}^{(0)}_{i,j} c^{(0)}_i + \tilde{Q}^{(1)}_{i,j} c^{(1)}_i + Q^{(1)}_{i,j}(F_i(1) - c^{(0)}_i - c^{(1)}_i)}, & 0 \le x < u, \\ \dfrac{\sum_{i\ne j} \tilde{Q}^{(0)}_{i,j} c^{(0)}_i + \tilde{Q}^{(1)}_{i,j} c^{(1)}_i + Q^{(1)}_{i,j}(F_i(x) - c^{(0)}_i - c^{(1)}_i)}{\sum_{i\ne j} \tilde{Q}^{(0)}_{i,j} c^{(0)}_i + \tilde{Q}^{(1)}_{i,j} c^{(1)}_i + Q^{(1)}_{i,j}(F_i(1) - c^{(0)}_i - c^{(1)}_i)}, & u \le x \le 1, \end{cases} \end{aligned}$$

since $Q^{(1)} = Q^{(2)} = \tilde{Q}^{(2)}$. The distribution of the battery level can now be obtained by substituting the Reset state for $j$.

## 3 Numerical Experimentation

Throughout this section, we consider a system with

$$\mathbf{Q_s} = \begin{bmatrix} -2 & 1.5 & 0.5 \\ 0.5 & -1 & 0.5 \\ 0.5 & 1 & -1.5 \end{bmatrix}, \qquad \mathbf{R_s} = \begin{bmatrix} -0.2 & & \\ & -0.1 & \\ & & 0.05 \end{bmatrix}.$$

Accordingly, the background Markov chain has three states, and the battery discharges at rates of 0.2 and 0.1 units per time unit in the first two states, while it charges at a rate of 0.05 in the third state as the $\mathbf{R_s}$ matrix shows. The first two states could be thought of as states in which the battery is discharged with different rates, corresponding to either different modes of motion such as linear motion and hovering, or different physical conditions such as wind or altitude. The third state can be considered as a state where the battery is actually filling, albeit with a rate of smaller magnitude, due to possibly solar harvesting. More complex models can be solved, but this small system is sufficiently representative. How to obtain the actual values of $\mathbf{Q_S}$ and $\mathbf{R_S}$ is an important question, but it is left out of scope in this study as our focus is on the computation of the battery level distribution when these are known.

We start by demonstrating the numerical accuracy of the model. Two different simulations have been written for this purpose. As the analytical model uses Erlangization to represent the deterministic time horizon, one of the simulations actually simulates the system with an Erlang-distributed time horizon. This simulation is used for the validation of the analytical model. The second simulation is for the real system, where the time horizon is indeed fixed (deterministic). For all the results presented in this section, the corresponding simulation was repeated for $10^4$ iterations.

In Fig. 2, we present the distribution of the battery level after $H$ time units, having started from initial level $u$, for various values of $u$, $H$, and the number of Erlang phases, $L$. The initial state distribution is assumed to be $\alpha = [1\ 0\ 0]$ for Fig. 2, as well as in the remainder of this section, unless specified otherwise. The numerical

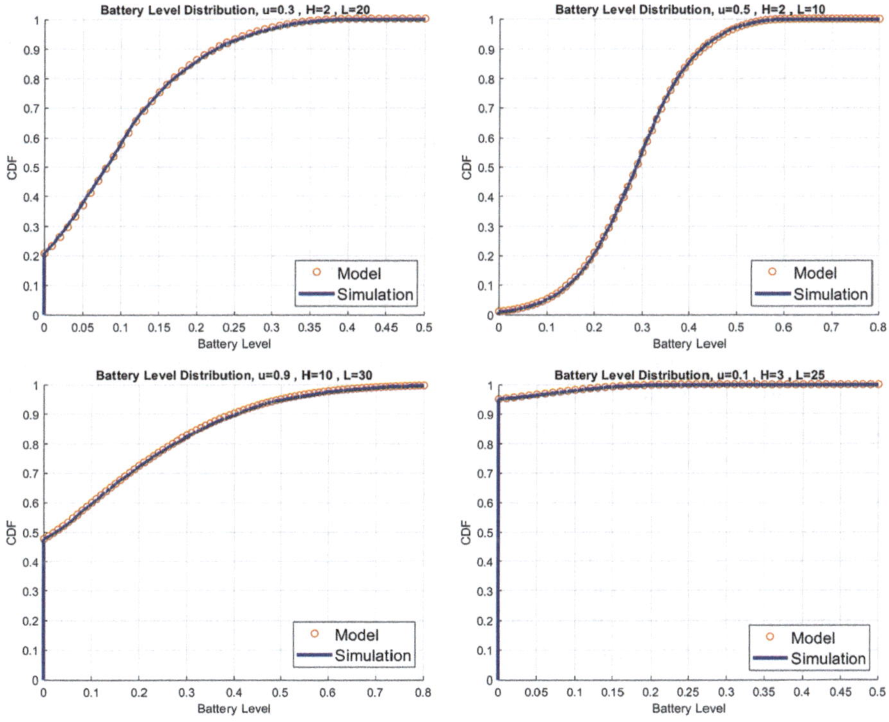


**Fig. 2** Battery level distribution for various scenarios

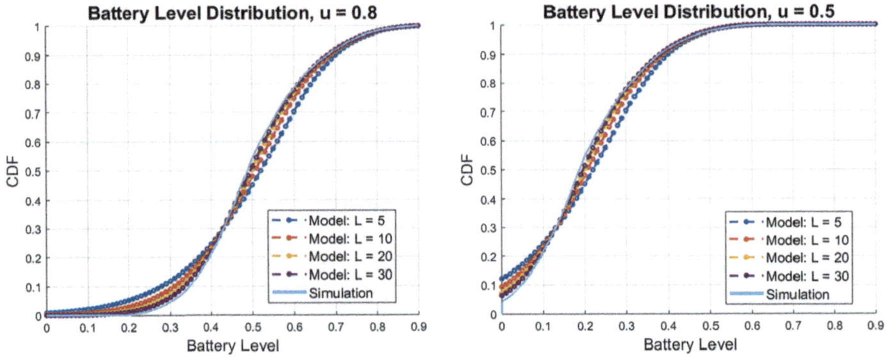


**Fig. 3** The convergence of the distributions obtained from the proposed model to the simulation curve with increasing number of Erlang phases. $H = 3$ for both figures

results obtained from the proposed analytical model are plotted against the results obtained via the simulation (with Erlangization). The plots demonstrate that the analytical model very closely follows the simulation curve, confirming the validity of the proposed model.

Next, we show the validity of Erlangization by plotting in Fig. 3 the distributions obtained from the model with increasing $L$ values, alongside the distribution obtained via the simulation (without Erlangization). Results from two different scenarios are presented with $u = 0.8$ and $u = 0.5$, whereas $H$ is 3 for both scenarios. The number of Erlang phases is increased from 5 to 30, and the convergence of the distribution to the simulation curve as $L$ is increased demonstrates that an Erlang distribution with a sufficient number of phases is useful for representing deterministic quantities in a Markovian setting.

As we have demonstrated the accuracy and the validity of the proposed analytical model, we do not plot simulation curves in the remainder of this section to avoid cluttering of the figures, all in which $L = 25$ Erlang phases are used.

Next, the effect of the initial value of the battery level is examined. Battery level distributions for $u$ taking values in $\{0.1, 0.2, \ldots, 0.9\}$ and $H = 3$ are given in Fig. 4. It is no surprise that with increasing $u$, the cdf curves shift toward right. Moreover, the probability masses at 0 can be considered of particular importance, as we would never want for the UAV to deplete its battery completely. This figure could also be interpreted as showing that under the assumed $\mathbf{Q_s}$ and $\mathbf{R_s}$ matrices, starting to operate the UAV with less than 0.7 battery level has a nonzero risk of losing the UAV, and the figure also quantifies this risk.

In Fig. 5, the distributions of the battery level after various horizon $H$ values ranging from 1 to 10, starting from initial level $u = 0.8$. Since the two dominant states (1 and 2) have negative drift values, the cdf shifts toward left as $H$ increases as the battery has more time to deplete. This analysis is very useful in forecasting the status of the battery level after a range of possible horizon values and could

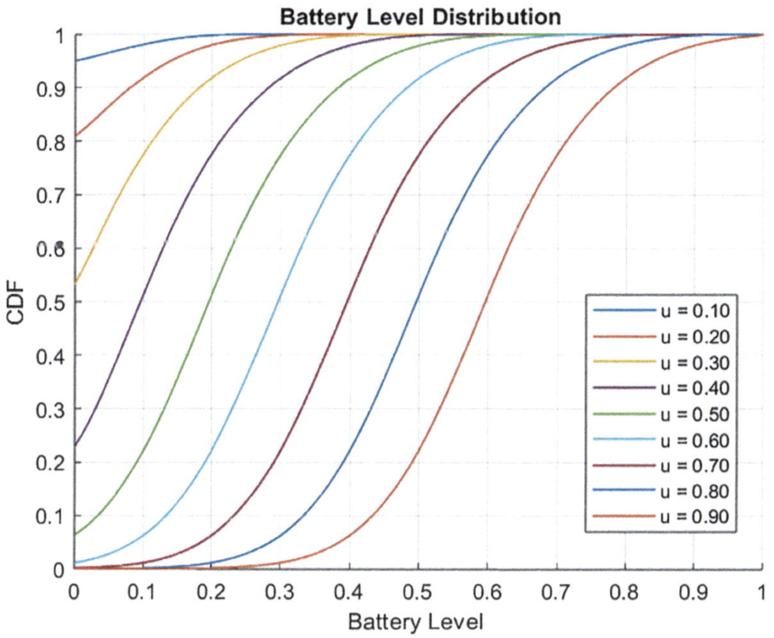


**Fig. 4** Distribution of the battery level at the end of $H = 3$ time units for varying initial levels

be beneficial for trajectory optimization in determining the point of critical battery level below which the UAV is forced to return to its charging station.

Finally, the effect of the initial state distribution on the final distribution is examined. In transient (time-dependent) problems such as the one investigated in this chapter, the outcome inevitably depends on the initial state. Figure 6 shows the battery level distributions after $H = 1$ and $H = 10$ time units for scenarios initiated from different initial states, with initial levels of $u = 0.5$ and $u = 0.9$. Particularly, if $H = 1$, we observe a characteristic difference between the cases starting from 1st and 2nd states that have negative drift rates, and the case starting from the 3rd state that has positive drift rate. This difference is less profound if $H = 10$, as the effect of the initial state will lose its significance with passing of more time. We should also observe that in the scenarios starting from the third state, where energy harvesting occurs and thus the drift rate is positive, the risk of battery depletion is lower after $H = 10$ time units.

As a final note, the analytical solution and the simulations were implemented in Matlab. The results of the model are obtained almost instantaneously, unlike the simulations. Although the system investigated by itself is not considerably big, relatively large $L$ values lead to much larger systems. Hence, computational complexity is not an issue for the proposed model.

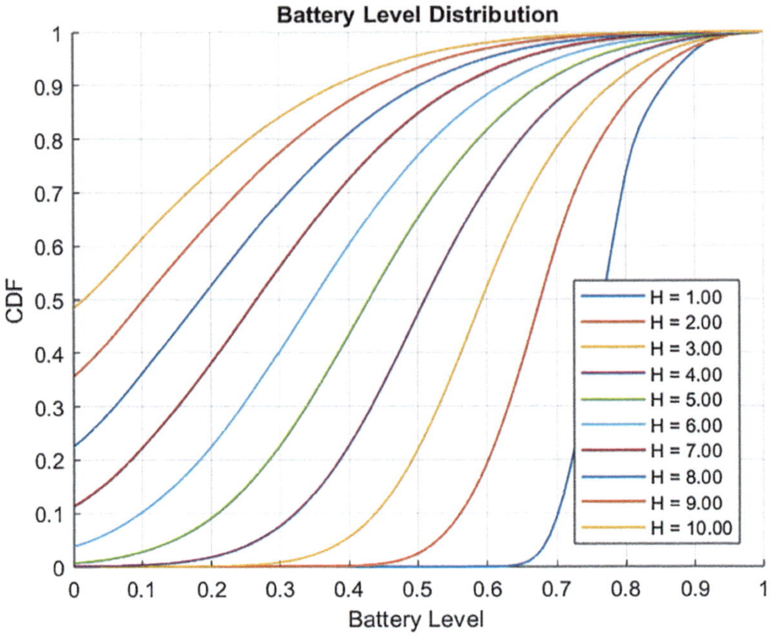


**Fig. 5** Battery level distributions for varying time horizons with initial value $u = 0.8$

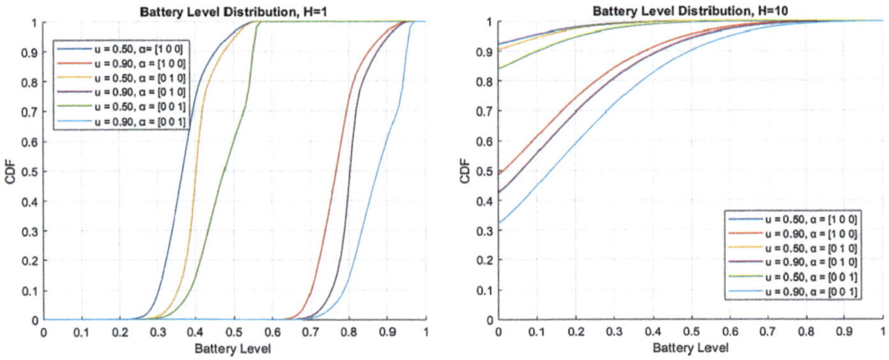


**Fig. 6** Distribution of battery level at the end of $H = 1$ and $H = 10$ time units depending on the initial distribution of the Markov chain and initial level of the battery

## 4 Conclusion

UAVs play a crucial role in many applications including IoT and sensor networks. Battery management and trajectory design are two major issues related to UAV operations. We propose an analytical model based on multi-regime Markov fluid queues for obtaining the time-dependent (transient) distribution of the UAV battery level after a specific time horizon, given a CTMC modeling the system dynamics, corresponding drift rates, and initial battery level. Although there exist studies in the literature that investigate depletion probability for UAV batteries, our study is the first to obtain the entire distribution of the battery level after a time horizon, to the best of the authors' knowledge. Future studies will be conducted on incorporating the proposed model into UAV trajectory design.

## References


1. Ure, N.K., Chowdhary, G., Toksoz, T., How, J.P., Vavrina, M.A., Vian, J.: An automated battery management system to enable persistent missions with multiple aerial vehicles. IEEE/ASME Trans. Mechatron. **20**(1), 275–286 (2014)
2. Jongerden, M.R., Haverkort, B.R.: Which battery model to use? IET Softw. **3**(6), 445–457 (2009)
3. Chen, M., Rincon-Mora, G.A.: Accurate electrical battery model capable of predicting runtime and I-V performance. IEEE Trans. Energy Convers. **21**(2), 504–511 (2006)
4. Handy, M., Timmermann, D.: Simulation of mobile wireless networks with accurate modelling of non-linear battery effects. In: Proceedings of the International Conference on Applied Simulation and Modeling, pp. 532–537 (2003)
5. Chiasserini, C.F., Rao, R.R.: Energy efficient battery management. IEEE J. Select. Areas Commun. **19**(7), 1235–1245 (2001)
6. Chiasserini, C.F., Rao, R.R.: Pulsed battery discharge in communication devices. In: Proceedings of the 5th Annual ACM/IEEE International Conference on Mobile Computing and Networking, pp. 88–95 (1999)
7. Chiasserini, C.F., Rao, R.R.: A model for battery pulsed discharge with recovery effect. In: Proceedings of the 1999 IEEE Wireless Communications and Networking Conference (WCNC) (1999)
8. Chiasserini, C.F., Rao, R.R.: Improving battery performance by using traffic shaping techniques. IEEE J. Select. Areas Commun. **19**(7), 1385–1394 (2001)
9. Jones, G.L., Harrison, P.G., Harder, U., Field, T.: Fluid queue models of battery life. In: Proceedings of the IEEE 19th Annual International Symposium on Modelling, Analysis, and Simulation of Computer and Telecommunication Systems, pp. 278–285 (2011)
10. Tunc, C., Akar, N.: Markov fluid queue model of an energy harvesting IoT device with adaptive sensing. Perform. Eval. **111**, 1–16 (2017)
11. Kulkarni, V.G.: Fluid models for single buffer systems. In: Dshalalow, J.H. (ed.) Frontiers in Queuing: Models and Applications in Science and Engineering, pp. 321–338. CRC Press, Boca Raton (1997)
12. Akar, N., Sohraby, K.: Infinite- and finite-buffer Markov fluid queues: a unified analysis. J. Appl. Probab. **41**(2), 557–569 (2004)
13. Mandjes, M., Mitra, D., Scheinhardt, W.: Models of network access using feedback fluid queues. Queueing Syst. **44**(4), 365–398 (2003)
14. Kankaya, H.E., Akar, N.: Solving multi-regime feedback fluid queues. Stoch. Model. **24**(3), 425–450 (2008)
15. Adan, J., van Doorn, E.A., Resing, J., Scheinhardt, W.R.: Analysis of a single-server queue interacting with a fluid reservoir. Queueing Syst. **29**(2), 313–336 (1998)
16. Scheinhardt, W.R.W.: Markov-modulated and feedback fluid queues. Ph.D. dissertation, University of Twente, The Netherlands (2000)
17. Elwalid, A.I., Mitra, D.: Statistical multiplexing with loss priorities in rate-based congestion control of high-speed networks. IEEE Trans. Commun. **42**(11), 2989–3002 (1994)

18. Kella, O., Stadje, W.: Exact results for a fluid model with state-dependent flow rates. Probab. Eng. Informat. Sci. **16**(4), 389–402 (2002)
19. Scheinhardt, W., Van Foreest, N., Mandjes, M.: Continuous feedback fluid queues. Oper. Res. Lett. **33**(6), 551–559 (2005)
20. Boxma, O., Kaspi, H., Kella, O., Perry, D.: On/off storage systems with state-dependent input, output, and switching rates. Probab. Eng. Informat. Sci. **19**(1), 1–14 (2005)
21. Dendievel, S., Latouche, G.: Approximations for time-dependent distributions in Markovian fluid models. Methodol. Comput. Appl. Probab. **19**(1), 285–309 (2017)
22. Asmussen, S.: Risk theory in a Markovian environment. Scand. Actuar. J. **1989**(2), 69–100 (1989)
23. Asmussen, S., Albrecher, H.: Ruin Probabilities, vol. 14. World Scientific, Singapore (2010)
24. Yazici, M.A., Akar, N.: The finite/infinite horizon ruin problem with multi-threshold premiums: a Markov fluid queue approach. Ann. Oper. Res. **252**(1), 85–99 (2017)
25. Gürsoy, Ö.: Numerical methods for the transient analysis of multi-regime Markov fluid queues. Ph.D. dissertation, Bilkent University, Turkey (2019)
26. Ramaswami, V., Woolford, D.G., Stanford, D.A.: The Erlangization method for Markovian fluid flows. Ann. Oper. Res. **160**, 215–225 (2008)
27. Asmussen, S., Avram, F., Usabel, M.: Erlangian approximations for finite-horizon ruin probabilities. ASTIN Bull. J. IAA **32**(2), 267–281 (2002)
28. Horváth, A., Sáfár, O., Telek, M., Zámbá, B.: Concentrated matrix exponential distributions. In: European Workshop on Performance Engineering, pp. 18–31. Springer, Berlin (2016)

# Semantic Search on Turkish E-Commerce Data: Benchmarking Language Models

**Selami Ugur Civelek and Duygu Cakir**

## 1 Introduction

Rapid development of artificial intelligence, particularly in machine learning and natural language processing (NLP), has revolutionized various domains of computer technology, including search systems [1]. Search engines serve as the primary entry point for users interacting with web resources, especially in e-commerce platforms where efficient product discovery is crucial [2]. Traditional keyword-based search and matching techniques, while foundational, often fail to capture the minor relationships between terms and the contextual patterns inherent in user queries [3, 4]. This limitation becomes particularly pronounced in e-commerce settings, where users' search intents can be complex and varied [5].

To address these challenges, modern approaches leverage semantic-based methods, utilizing state-of-the-art language models to enhance query understanding and retrieval accuracy [6, 7]. These methods employ vectorial representations known as embeddings, which can capture the semantic content of text data more effectively than traditional approaches [8].

This chapter focuses on evaluating the performance of various embedding models in extracting semantic content from a Turkish e-commerce product dataset. Specifically, we compare the retrieval performances of embeddings extracted using

S. U. Civelek
Bahcesehir University, Istanbul, Turkey

Obase, Istanbul, Turkey
e-mail: selamiugur.civelek@bahcesehir.edu.tr; ugur.civelek@obase.com

D. Cakir (✉)
Bahcesehir University, Istanbul, Turkey
e-mail: duygu.cakir@bau.edu.tr



B. U. Töreyin et al. (eds.), *8th EAI International Conference on Robotic Sensor Networks*, EAI/Springer Innovations in Communication and Computing,
https://doi.org/10.1007/978-3-031-92143-8_14

Turkish BERT [9], LLaMA 3.1-8B [10], and Voyage AI's Multilingual-2 models when incorporated into a search system.

Our research aims to contribute to the growing body of work on semantic search in diverse linguistic contexts, with a particular focus on the Turkish language. By benchmarking these models, we seek to provide insight into their effectiveness in capturing and utilizing semantic relationships in e-commerce search scenarios. The rest of this chapter is organized as follows: Sect. 2 reviews related work in the field of semantic search and language models. Section 3 describes our proposed method, including the dataset, preprocessing steps, and evaluation metrics. Section 4 presents the results of our comparative analysis, and Sect. 5 concludes the paper with a discussion of our findings and potential future directions.

## 2 Related Work

The domain of search systems has undergone significant transformation with the evolution of natural language processing (NLP) and machine learning (ML). Traditional keyword-based search systems have long been the cornerstone of information retrieval (IR), where search engines match user queries against indexed data using simple term matching techniques. However, these systems have struggled to understand the semantic meaning behind user queries, leading to the development of more sophisticated methods. This section provides an overview of key developments relevant to our study.

### *2.1 Traditional Keyword-Based Search Systems*

Traditional search methods are highly dependent on matching exact terms from user queries with terms in the indexed documents. This approach, often referred to as keyword-based search, operates on the principles of Boolean logic, where documents are retrieved based on the presence or absence of specific keywords. Key techniques like the term frequency-inverse document frequency (TF-IDF), introduced by Jones [11], and the vector space model (VSM), developed by Salton, Wong and Yang [12], have been fundamental in this approach. These methods weigh the importance of each keyword and represent documents and queries as vectors in a multidimensional space to enable more detailed matching.

Despite their foundational importance, keyword-based search systems have several limitations, particularly in capturing the semantic meaning of queries. The reliance on exact keyword matching often results in the retrieval of irrelevant documents, especially when queries involve synonyms, morphological variations, or require an understanding of the broader context. Advanced probabilistic models, such as the BM25 algorithm introduced by Robertson et al. [13], have attempted to mitigate some of these issues by incorporating probabilistic elements to improve the accuracy of the retrieval. Nevertheless, the limitations of these traditional methods have driven the search for more sophisticated approaches that can better understand the semantics of user queries.

## 2.2 Modern Approaches: Word Embeddings and Semantic Representations

The limitations of keyword-based search led to the development of semantic search methods, which focus on understanding the meaning behind the words in a query. One of the significant advances in this area has been the introduction of word embeddings, which represent words as dense vectors in a continuous space, capturing both syntactic and semantic similarities between them.

The introduction of Word2Vec by Mikolov et al. [6] marked a pivotal moment in the field, as it used shallow neural networks to learn word representations based on their context. This method was particularly effective in capturing semantic relationships between words. Another notable advancement was the GloVe model introduced by Pennington, Socher, and Manning [7], which combined the advantages of global matrix factorization and local context window methods to produce more robust word representations.

These embeddings provided the foundation for modern NLP models, enabling a more sophisticated and contextually aware analysis of language data. As a result, semantic search, which utilizes these embeddings to better understand and interpret user queries, has become a more viable alternative to traditional keyword-based methods.

## 2.3 Advanced Language Models

The advent of deep learning and transformer-based architectures further revolutionized semantic search. Models such as BERT (Bidirectional Encoder Representations from Transformers) introduced by Devlin et al. [1] have been instrumental in advancing the capabilities of semantic search systems. BERT, with its bidirectional attention mechanism, allows for a deeper understanding of context within user queries, significantly improving the accuracy of search results.

Subsequent models, such as RoBERTa, ALBERT, and GPT-3, have built on BERT's architecture, refining the approach to achieve even better performance across various NLP tasks, including semantic search. These models have been particularly effective in extracting embeddings that can capture complex linguistic relationships, thus enabling more accurate and context-aware search results.

### 2.4 Semantic Search in the Turkish Language

Adapting these advanced models to languages with complex morphologies, such as Turkish, presents unique challenges. Turkish BERT models, such as BERTurk [9], have been specifically trained in Turkish corpora to better capture the details of the language. These models have shown improved performance on tasks such as sentiment analysis and named entity recognition compared to multilingual models. The adaptation of these models to the Turkish language has been crucial in enhancing the effectiveness of semantic search systems in Turkish, allowing more accurate interpretation of user queries in a linguistically rich environment.

## 3 Proposed Method

This chapter aims to evaluate the effectiveness of advanced language models in performing semantic search on Turkish e-commerce data. The proposed method involves a comparative analysis of three distinct language models: Turkish BERT, LLaMA 3.1-8B, and Voyage AI's Multilingual-2. These models are employed to extract semantic embeddings from product descriptions and titles, which are then used to assess their performance in retrieving relevant search results. The following sections outline the methodology in detail.

### 3.1 Research Design

The research design involves several key steps: dataset preparation, preprocessing, embedding extraction, and performance evaluation. The dataset comprises product information from a Turkish industrial goods seller, including fields such as product title, description, and category. The study focuses on evaluating the models' ability to generate embeddings that capture the semantic content of these texts and their effectiveness in a search system.

### 3.2 Dataset and Preprocessing

The dataset used in this study consists of approximately 10,000 products from a Turkish e-commerce platform. The primary fields of interest are the product titles and descriptions (Fig. 1), which are used to generate the embeddings. The preprocessing step involves normalizing the text to lowercase, removing extra whitespaces, and eliminating special characters such as HTML tags. The processed text is then concatenated, with the product title and description separated by a special token (///), to form the final input for the models.

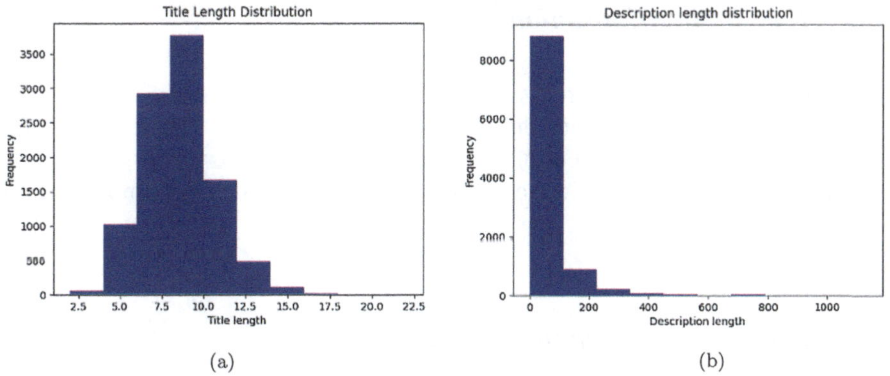


**Fig. 1** Word count distribution for the title and description fields. (**a**) Title length distribution. (**b**) Description length distribution

## *3.3 Embedding Models*

Three models are used to extract semantic embeddings:

### 3.3.1 Turkish BERT

A variant of BERT specifically pretrained on Turkish text, producing 768-dimensional embeddings. This model is well-suited for capturing the details of the Turkish language and is expected to perform strongly in this context.

### 3.3.2 LLaMA 3.1-8B

A large-scale language model with eight billion parameters, known for its advanced language understanding capabilities. It produces 4096-dimensional embeddings, making it a powerful candidate for capturing complex semantic relationships.

### 3.3.3 Voyage AI's Multilingual-2

A model designed to handle multiple languages, producing 1024-dimensional embeddings. Its multilingual capabilities allow it to capture cross-lingual semantic relationships effectively, which could be advantageous in the diverse linguistic landscape of Turkish e-commerce data.

### 3.4 Embedding Extraction

The embedding extraction process involves feeding the preprocessed product texts into each model to generate dense vector representations. These embeddings represent the semantic content of the product titles and descriptions. The process is repeated for all products in the dataset, resulting in a matrix of embeddings for each model.

### 3.5 Search System Implementation

To evaluate the performance of the embeddings in a search system, a k-nearest neighbors (k-NN) algorithm is employed. The k-NN algorithm identifies the top-k most similar product embeddings to a given query embedding based on cosine similarity. For this study, the value of $k$ is set to 10, meaning that the search system retrieves the 10 most relevant products for each query (Table 1).

Cosine similarity, defined as the cosine of the angle between two vectors in a multidimensional space, is used to measure the similarity between the query embedding and the product embeddings. This metric is well-suited for high-dimensional spaces and provides a robust measure of semantic similarity. The employed cosine similarity is given in Eq. 1.

$$\text{cosine similarity}(\mathbf{v}_1, \mathbf{v}_2) = \frac{\mathbf{v}_1 \cdot \mathbf{v}_2}{\|\mathbf{v}_1\| \|\mathbf{v}_2\|} \tag{1}$$

Where:

- $\mathbf{v}_1$ and $\mathbf{v}_2$ are the vectors being compared.
- $\mathbf{v}_1 \cdot \mathbf{v}_2$ is the dot product of $\mathbf{v}_1$ and $\mathbf{v}_2$.
- $\|\mathbf{v}_1\|$ is the magnitude (or length) of vector $\mathbf{v}_1$.
- $\|\mathbf{v}_2\|$ is the magnitude (or length) of vector $\mathbf{v}_2$.

**Table 1** Turkish queries with English translations

| Query # | Turkish text | English translation |
|---|---|---|
| 1 | türk kahve makinası kırmızı | Red Turkish coffee machine |
| 2 | meyve sıkacağı | Fruit juicer |
| 3 | 12li yemek çatalı | 12-piece dining fork set |
| 4 | ekmek servis için ünite | Unit for serving bread |
| 5 | çekirdek kahve | Coffee beans |
| 6 | beyaz kahve değirmeni | White coffee grinder |
| 7 | elektrikli tost makinesi | Electric toaster |
| 8 | yuvarlak ve beyaz fanus tepsi | Round and white cloche tray |
| 9 | paslanmaz elektrikli fırın | Stainless steel electric oven |
| 10 | hızlı pişirmeli wok tava | Fast-cooking wok pan |

### 3.6 Performance Metrics

The performance of each model is assessed using standard information retrieval metrics:

**Precision@K (Prec@K):** Evaluates the number of relevant documents at or above the rank $K$ divided by the $K$ itself and is given by Eq. 2.

$$\text{Prec@K}(q, D) = \frac{|Ret(D, K) \cap Rel(D, q)|}{K} \tag{2}$$

where $Ret(D, K)$ is the set of documents at or above the rank $K$ in total retrieved documents $D$, and $Rel(D, q)$ is the set of all documents that are relevant for query $q$. This metric evaluates how relevant the documents retrieved by the search system are based on the rank $K$.

**Recall@K (Rec@K):** Evaluates how well the system succeeded overall in retrieving relevant documents and is given by Eq. 3.

$$\text{Rec@K}(q, D) = \frac{|Ret(D, K) \cap Rel(D, q)|}{|Rel(D, q)|} \tag{3}$$

where $Ret(D, K)$ is the set of documents at or above the rank $K$ in total retrieved documents $D$, and $Rel(D, q)$ is the set of all documents that are relevant for query $q$.

**F1@K:** is the harmonic mean of Prec@K and Rec@K. It balances the two metrics and is defined Eq. 4.

$$\text{F1@K}(q, D) = 2 \times \frac{\text{Prec@K}(q, D) \times \text{Rec@K}(q, D)}{\text{Prec@K}(q, D) + \text{Rec@K}(q, D)} \tag{4}$$

**Average Precision (AvgPrec):** Average precision is the sum of precisions computed not only at a static K value but for a set of i values where each i is the sorted rank of each relevant document retrieved by the system. It gives the finest-grained distinctions of the previous metrics and is sensitive to rank, precision, and recall.

## 4 Results

In this section, we present the results of the experiments conducted to evaluate the performance of the Turkish BERT, LLaMA 3.1-8B, and Voyage AI's Multilingual-2 models. The performance of each model is assessed based on the information retrieval metrics described in the previous section: Precision@K, Recall@K, F1@K, and average precision (AvgPrec). The value of $k$ is set to 5 for all metrics.

**Table 2** Comparison of model performance

| Model | Prec@5 | Rec@5 | F1@5 | AvgPrec |
|---|---|---|---|---|
| Turkish BERT | 0.40 | 0.43 | 0.40 | 0.52 |
| LLaMA 3.1-8B | 0.32 | 0.53 | 0.39 | 0.43 |
| Voyage AI's Multilingual-2 | 0.82 | 0.57 | 0.65 | 0.89 |

**Table 3** Results for Turkish BERT

| Query # | Prec@5 | Rec@5 | F1@5 | AvgPrec |
|---|---|---|---|---|
| 1 | 0.40 | 0.67 | 0.50 | 0.36 |
| 2 | 1.00 | 0.50 | 0.67 | 1.00 |
| 3 | 0.60 | 0.50 | 0.55 | 0.86 |
| 4 | 0.00 | 0.00 | 0.00 | 0.14 |
| 5 | 0.40 | 0.50 | 0.44 | 0.48 |
| 6 | 0.40 | 0.67 | 0.50 | 0.47 |
| 7 | 0.00 | 0.00 | 0.00 | 0.19 |
| 8 | 0.40 | 0.67 | 0.50 | 0.67 |
| 9 | 0.80 | 0.80 | 0.80 | 0.91 |
| 10 | 0.00 | 0.00 | 0.00 | 0.10 |
| Average | 0.40 | 0.43 | 0.40 | 0.52 |

Table 2 presents a comparison of the performance metrics for the three models: Turkish BERT, LLaMA 3.1-8B, and Voyage AI's Multilingual-2. The performance metrics include Precision@5, Recall@5, F1@5, and Average Precision. The results show that Voyage AI's Multilingual-2 model generally outperforms the other two models.

The Turkish BERT model demonstrates moderate performance with an average Precision@5 of 0.40, a Recall@5 of 0.43, an F1@5 of 0.40, and an average precision of 0.52. The LLaMA 3.1-8B model shows variability in its performance, with an average Precision@5 of 0.32, a Recall@5 of 0.53, an F1@5 of 0.39, and an average precision of 0.43. Voyage AI's Multilingual-2 model outperforms the other two models, achieving an average Precision@5 of 0.82, a Recall@5 of 0.57, an F1@5 of 0.65, and an average precision of 0.89. Detailed results on 10 different queries and 3 different models can be found on Tables 3, 4, and 5.

### *4.1 Summary of Results*

The results demonstrate that the Voyage AI's Multilingual-2 model significantly outperforms Turkish BERT and LLaMA 3.1-8B in terms of Precision@5, F1@5, and average precision. The superior performance of the Voyage AI model indicates its robustness and effectiveness in handling semantic search tasks within the context of Turkish e-commerce data. This model's multilingual design allows it to capture detailed semantic relationships, making it a valuable tool for improving search accuracy in linguistically diverse datasets.

**Table 4** Results for LLaMA 3.1-8B

| Query # | Prec@5 | Rec@5 | F1@5 | AvgPrec |
|---|---|---|---|---|
| 1 | 0.40 | 0.50 | 0.44 | 0.37 |
| 2 | 0.40 | 0.67 | 0.50 | 0.83 |
| 3 | 0.40 | 0.67 | 0.50 | 0.38 |
| 4 | 0.20 | 0.50 | 0.29 | 0.20 |
| 5 | 0.00 | 0.00 | 0.00 | 0.17 |
| 6 | 0.80 | 0.80 | 0.80 | 0.81 |
| 7 | 0.20 | 0.50 | 0.29 | 0.20 |
| 8 | 0.20 | 0.50 | 0.29 | 0.29 |
| 9 | 0.40 | 0.67 | 0.50 | 0.83 |
| 10 | 0.20 | 0.50 | 0.29 | 0.20 |
| Average | 0.32 | 0.53 | 0.39 | 0.43 |

**Table 5** Results for Voyage AI's Multilingual-2

| Query # | Prec@5 | Rec@5 | F1@5 | AvgPrec |
|---|---|---|---|---|
| 1 | 0.40 | 0.67 | 0.50 | 0.57 |
| 2 | 1.00 | 0.50 | 0.67 | 1.00 |
| 3 | 1.00 | 0.50 | 0.67 | 1.00 |
| 4 | 1.00 | 0.71 | 0.83 | 1.00 |
| 5 | 0.80 | 0.57 | 0.67 | 0.91 |
| 6 | 1.00 | 0.63 | 0.77 | 0.94 |
| 7 | 1.00 | 0.50 | 0.67 | 1.00 |
| 8 | 0.80 | 0.57 | 0.67 | 0.84 |
| 9 | 0.60 | 0.43 | 0.50 | 0.80 |
| 10 | 0.60 | 0.60 | 0.60 | 0.80 |
| Average | 0.82 | 0.57 | 0.65 | 0.89 |

## 5 Conclusion and Future Work

The advancements in natural language processing (NLP) and machine learning (ML) have significantly impacted the field of information retrieval, particularly in enhancing the capabilities of search systems. This study investigated the performance of three advanced language models—Turkish BERT, LLaMA 3.1-8B, and Voyage AI's Multilingual-2—in conducting semantic search on Turkish e-commerce data. The models were evaluated based on their ability to extract semantic embeddings and improve the relevance of search results.

The experimental results demonstrated that Voyage AI's Multilingual-2 model outperformed the other models in terms of Precision@5, F1@5, and average precision. This model's superior performance can be attributed to its multilingual capabilities, which allow it to capture minor semantic relationships within the dataset. Turkish BERT, while specifically trained on Turkish text, showed moderate performance, indicating its effectiveness in Turkish-specific tasks but highlighting the advantages of models designed for broader linguistic diversity. LLaMA 3.1-8B, despite being a large-scale and powerful model, exhibited variability in its performance, suggesting that its architecture might be less suited to the specific details of the Turkish language and the e-commerce context.

The findings of this study underscore the importance of selecting the appropriate language model based on the specific linguistic and contextual requirements of the task. For applications involving Turkish e-commerce data, the Voyage AI's Multilingual-2 model presents a robust and effective solution for semantic search, offering significant improvements in the relevance and accuracy of search results.

Future research could explore the integration of other advanced language models, including those optimized for specific tasks such as sentence embeddings or cross-lingual information retrieval. Additionally, the incorporation of contrastive learning techniques and multimodal embeddings could further enhance the quality of semantic search in diverse and complex datasets. The scalability of these models in real-time applications also warrants further investigation, particularly in resource-constrained environments.

Longitudinal studies that monitor the performance of embedding models over time as user behavior and underlying data evolve would provide valuable insights. Regular updates and reevaluations could ensure that these models continue to meet the dynamic needs of users and keep pace with technological advancements in the field.

## References

1. Devlin, J., Chang, M.-W., Lee, K., Toutanova, K.: BERT: Pre-training of deep bidirectional transformers for language understanding. In: Proceedings of the 2019 Conference of the North American Chapter of the Association for Computational Linguistics: Human Language Technologies, vol. 1, pp. 4171–4186 (long and short papers) (2019)
2. Karpukhin, V., Oguz, B., Min, S., Lewis, P., Wu, L., Edunov, S., Chen, D., Yih, W.-t.: Dense passage retrieval for open-domain question answering. In: Proceedings of the 43rd International ACM SIGIR Conference on Research and Development in Information Retrieval, pp. 675–684 (2020)
3. Salton, G., Wong, A., Yang, C.: A vector space model for automatic indexing. Commun. ACM **18**(11), 613–620 (1975)
4. Robertson, S.E., Sparck Jones, K.: Relevance weighting of search terms. J. Am. Soc. Inf. Sci. **27**(3), 129–146 (1976)
5. Lewis, P., Liu, Y., Goyal, N., Stiennon, N., et al.: Retrieval-augmented generation for knowledge-intensive NLP tasks. In: Proceedings of the 43rd International ACM SIGIR Conference on Research and Development in Information Retrieval, pp. 964–973 (2020)
6. Mikolov, T., Sutskever, I., Chen, K., Corrado, G.S., Dean, J.: Distributed representations of words and phrases and their compositionality. In: Advances in Neural Information Processing Systems, pp. 3111–3119 (2013)
7. Pennington, J., Socher, R., Manning, C.D.: Glove: global vectors for word representation. In: Proceedings of the 2014 Conference on Empirical Methods in Natural Language Processing (EMNLP), pp. 1532–1543 (2014)
8. Peters, M.E., Neumann, M., Iyyer, M., Gardner, M., Clark, C., Lee, K., Zettlemoyer, L.: Deep contextualized word representations. In: Proceedings of the 2018 Conference of the North American Chapter of the Association for Computational Linguistics: Human Language Technologies, vol. 1 (Long Papers), pp. 2227–2237 (2018)

9. Schweter, S.: BERTurk–BERT models for Turkish. Preprint. arXiv:2009.10753 (2020)
10. Brown, T.B., Mann, B., Ryder, N., Subbiah, M., Kaplan, J., Dhariwal, P., Neelakantan, A., Shyam, P., Sastry, G., Askell, A., et al.: Language models are few-shot learners. Adv. Neural Inf. Proces. Syst. **33**, 1877–1901 (2020)
11. Sparck Jones, K.: A statistical interpretation of term specificity and its application in retrieval. J. Document. **28**(1), 11–21 (1972)
12. Salton, G., Wong, A., Yang, C.: A vector space model for automatic indexing. Commun. ACM **18**(11), 613–620 (1975)
13. Robertson, S.E., Walker, S., Jones, S., Hancock-Beaulieu, M., Gatford, M.: Okapi at TREC-3. In: Proceedings of the Third Text REtrieval Conference (TREC-3), pp. 109–126 (1995)

# Application of an Asymmetric Ciphering Computational Process Based on the Complexity of Multiplications with Parameters and Finite Group Logarithm Problem to the Electronic Digital Authentication Protocol

**Akhmadaliev Shakhobiddin, Abdumannon Jumakulov, Muzaffar Botirov, and Rustam Toxirov**

## 1 Introduction

Major Cybersecurity Threats in Robotics

1. Malware Attacks on Robot Software: Hackers can deploy malicious software to alter a robot's behavior or disable it entirely.
2. Data Interception and Manipulation: Robots often use sensors to collect data. Attacks on communication networks can result in data tampering, posing significant risks, especially in critical systems.
3. Unauthorized Remote Control: Cybercriminals gaining control of robots can lead to sabotage or physical damage.
4. Cloud System Vulnerabilities: Many robots rely on cloud platforms for data storage or processing. Security flaws in these platforms can create significant risks.
5. Privacy Violations Robots, especially in healthcare or domestic settings, collect sensitive user data. A breach can lead to serious privacy concerns.

The application of an asymmetric ciphering computational process based on the complexity of the multiplications with parameters and the finite group logarithm problem (DLP) to the Electronic Digital authentication Protocol (EDSP) offers a high level of security for digital communications. This approach leverages the difficulty of solving certain number-theoretic problems in order to secure the integrity and authenticity of electronic messages.

A. Shakhobiddin (✉) · A. Jumakulov · M. Botirov · R. Toxirov
Kokand University, Kokand, Uzbekistan
e-mail: shaxmadaliyev@kokanduni.uz; x.jumaqulov@kokanduni.uz; mbotirov@kokanduni.uz; r.tohirov@kokanduni.uz

 
B. U. Töreyin et al. (eds.), *8th EAI International Conference on Robotic Sensor Networks*, EAI/Springer Innovations in Communication and Computing,
https://doi.org/10.1007/978-3-031-92143-8_15

1. Asymmetric Ciphering and Public-Key Cryptography:

   Asymmetric ciphering involves the use of two distinct keys: a public key (known to everyone) and a private key (kept secret). The core advantage of asymmetric ciphering is that anyone can encrypt a message using the public key, but only the holder of the corresponding private key can decrypt it. This principle is fundamental to many cryptographic protocols, including digital authentications.
2. Complexity of multiplications with parameters and DLP:

   The security of many asymmetric ciphering schemes (including RSA and elliptic curve cryptography) relies on the difficulty of solving certain mathematical problems. In this case, two problems are particularly relevant:

   Multiplications with parameters: This problem refers to the difficulty of solving a multiplication with parameters problem where large numbers or matrices are involved. The concept of multiplications with parameters can be seen as a generalization of modular arithmetic in ciphering schemes.

   Finite group logarithm problem (DLP): The DLP is the problem of finding the exponent $x$ given $g^x \bmod p$ for a known base $g$ and modulus $p$, where $p$ is typically a large prime number. The DLP is considered computationally infeasible for large values of $p$, which is the foundation of many cryptographic systems, such as the Digital Signature Algorithm (DSA), ElGamal, and cryptography using elliptic curves.

   The DLP has been widely adopted in many ciphering systems due to its assumed intractability with current computational resources.
3. Digital authentication Protocol (EDSP):

   Digital authentications are used to verify the authenticity and integrity of digital messages. The process typically involves the following steps:

   Message Hashing: The message to be signed is first hashed using a cryptographic hash function (like SHA-256), which produces a fixed-size digest (hash value) of the message.

   Signing: The sender then signs the hash of the message using their private key. This ensures that only the sender (who possesses the private key) can produce the signature for that specific message. In the case of an asymmetric ciphering system based on the DLP, the signing process involves performing operations that rely on the difficulty of solving the finite group logarithm problem.

   Verification: The recipient of the signed message can verify the authenticity of the signature by checking it with the sender's public key. If the signature is valid, this proves that the message has not been tampered with and that it was indeed signed by the holder of the private key. The verification process often uses the public key to decrypt the signature and check that it matches the hash of the original message.
4. Application to Digital Authentications:

   Signature Generation: In an asymmetric ciphering system based on the DLP (e.g., the Digital authentication Computational process), the private key is used to sign the hash of the message. The signing process involves generating a signature using a computational process (like DSA) that uses the difficulty of solving the DLP as a core part of its security. The private key allows the sender to generate a unique signature for the hash of the message.

Signature Verification: Verification of the signature by the recipient uses the public key of the sender. The recipient calculates the hash of the message and compares it to the signature generated by the sender using the public key. If the signature corresponds correctly to the message's hash and can be verified using the public key, the recipient can be confident that the message has not been altered and that it was signed by the correct private key holder.

5. Advantages of Using Multiplications with parameters and DLP in EDSP:

   Security: The use of multiplications with parameters and the DLP-based asymmetric ciphering ensures that the digital authentication system is secure, as the DLP is a hard problem to solve in polynomial time, even with large computational resources. This makes digital authentications resistant to forgery.

   Efficiency: While the DLP-based systems can be computationally intensive, modern computational processes like the Elliptic Curve Digital Signature Algorithm (ECDSA) provide efficient and secure alternatives based on the DLP.

   Non-repudiation: The sender cannot deny having signed the message, as the signature can only be generated using their private key. This property is crucial for legal and contractual contexts.

6. Possible Computational processes based on DLP:

   Several computational processes and protocols make use of the DLP and related number-theoretic problems to implement digital authentications:

   Digital authentication Computational process (DSA): DSA is widely used in various cryptographic standards and relies on the DLP for security.

   ElGamal Signature Scheme: Based on the DLP, the ElGamal signature scheme is another example of how the DLP can be applied to digital authentications.

   Elliptic Curve Digital authentication Computational process (ECDSA): ECDSA is based on elliptic curve cryptography, which is a variant of DLP, and is more efficient in terms of key size compared to traditional DSA.

   According to traditional practices, signing a document grants it official status and legally guarantees that the person who signed it is the author (responsible party) of the document.

   The development of information and communication technologies, as well as computer networks, has created the foundation for implementing electronic document interchange. With the introduction of electronic documents, it has also become necessary to develop reliable methods for identifying the author of a document and verifying its legitimacy. In practice, disputes sometimes arise when individuals challenge the authenticity of a document, claiming that the signature or its contents are forged.

   These issues are resolved through the use of an electronic digital authentication (EDS).

The following problems are resolved by digital authentication [1–3]:

- gives a way to confirm the document's integrity using its hash value,
- enables the ability to authenticate the electronic document's source (assert its validity),
- guarantees that the signature cannot be revoked.

## 2 Related Works

There are now three categories of different approaches (schemes) for putting digital authentications into practice:

1. Symmetric ciphering-based digital authentication systems (schemes).
2. Public-key ciphering-based systems (schemes).
3. Systems (schemes) based on computational processes specifically created for creating and validating digital authentications.

In this article, the use of asymmetric ciphering computational processes based on parametric matrix multiplication [4] is analyzed for solving cryptographic tasks such as:

- Ensuring the confidentiality of electronic information.
- Implementing an electronic digital authentication alongside a given hash function (HF), enabling the verification of data integrity and ensuring its authentication. Practical application methods and protocols are developed for these solutions.

There are two methods for implementing an electronic digital authentication (EDS) protocol technique that uses an asymmetric public key ciphering computational process:

1. Direct Signature Method: In this approach, the document or its hash is directly encrypted using the private key of the signer, and the recipient verifies the signature by decrypting it with the corresponding public key.
2. Hash-Based Signature Method: In this method, a hash value of the document is first computed and then encrypted using the signer's private key. The recipient protocol technique that uses an asymmetric public key the hash with the public key and comparing it to the computed hash of the received document.

In the first method, to ensure the EDS (Electronic Digital authentication) of an electronic document, the encrypted representation of that document using the sender's private key can be used. In this case, the length of the EDS will be the same as the length of the document being signed. To create such a signed document, the data is encrypted with the signer's private key. Then, by utilizing the signer's (owner's) public key to decrypt the encrypted data, any user of the information-communication network can confirm that the signature is accurate.

## 3 Proposed Method

The proposed method for enhancing the security and efficiency of the Electronic Digital authentication Protocol (EDSP) combines asymmetric ciphering computational processes based on the complexity of multiplications with parameters and the finite group logarithm problem (DLP). This method aims to leverage the inherent computational difficulty of the DLP to ensure the authenticity, integrity, and non-repudiation of digital messages within a secure communication framework.

The first method protocol scheme will be reviewed. Each user i and user j in the information-communication network should have their respective pairs of public $(k_i^o, k_i^M)$ and $\left(k_j^o, k_j^M\right)$ private keys based on asymmetric ciphering computational processes (for instance: El-Gamal, RSA).

If the electronic document's originator $M$– the data—is not private, then

1. $i$—the digital authentication is created by the user. $P$ in the form that follows:

$$M \parallel E_{k_i^M} = M',$$

where $M'$-digitally signed information -is an electronic document.
$j$ through a network of open communication and information.

Having received the information, user $j$performs the following:

1. $P = E_{k_i^M}(M)$ is distinct from the information.
2. The creator of the signature's public key is used to decrypt the digital authentication $k_i^o$:

$$D_{k_i^o}(P) = D_{k_i^o}\left(E_{k_i^M}\right) = M.$$

3. The electronic document is deemed legitimate if the information gleaned from decoding the digital authentication matches the signed information; if not, it is not.

When the data that needs to be signed is private:

1. In this form, user I generates a digital authentication:

$$E_{k_j^o}\left[M \parallel E_{k_i^M}\right] = M',$$

where $M'$ an electronic document containing signed information is used.
via an open channel of the communication and information network.
User $j$ does the following after receiving the signed secret information:
using its private key to carry out this decryption $k_j^M$:

$$D_{k_j^M}\left(M'\right) = M \parallel E_{k_i^M} = M'.$$

2. $M'$ decryption yields information that is used to generate a digital authentication $L = E_{k_i^M}(M)$.
3. Uses the author of the digital authentication's public key to perform decryption $k_i^o$:

$$D_{k_i^o}(L) = D_{k_i^o}\left(E_{k_i^M}\right) = M$$

4. The electronic document is deemed legitimate if the information gleaned from decoding the digital authentication matches the signed information it is not
5. The second method looks at the protocol scheme. It is regarded as digital authentication since the information is first signed using the signer's secret key, and the significance of the outcome of mapping the data into a digital block (verify the sum *HY*, H-function). $H(M) = h$ the procedure of EDS formation, transmission, obtaining, and confirming digitally signed information is considered.

If the electronic organizer for documents *M*- if the information is not private, it is without:

1. Using this form, user *i* completes the digital authentication *L*:

$$M \parallel L = M',$$

where $M'$- information that has been digitally signed is an electronic document.

*j* through a network of open communication and information.

After obtaining the data, user j does the following:

1. Acknowledged $M'$- a digital authentication $L = E_{k_i^M}(HY)$is distinct from the information.
2. $k_i^o$the public key of the signature's creator is used to decrypt digital authentication:

$$D_{k_i^o}(L) = D_{k_i^o}\left(E_{k_i^M}(HY)\right) = M.$$

3. The electronic document is deemed legitimate if the information gleaned from decoding the digital authentication matches the signed information it is not.

When the data that needs to be digitally signed is private:

1. The following is how user *i* generates a digital authentication *L*:

$$E_{k_j^o}\left[M \parallel E_{k_i^M}(HY)\right] = E_{k_j^o}[M \parallel L] = M',$$

where $M'$- information that has been signed is an electronic document.

$j$ through a communication and information network channel that is open. The j-user does the following after receiving the signed secret information: $k_j^M$ using its private key to carry out this decryption:

$$D_{k_j^M}\left(M'\right) = D_{k_j^M}\left(E_{k_j^o}\left[M \parallel E_{k_i^M}(HY)\right]\right) = M \parallel E_{k_i^M}(HY) = M.$$

2. The data $M'$ gathered via decryption is used to establish a digital authentication $P = E_{k_i^M}(HY)$.

   Uses the author of the digital authentication's public key to perform decryption $k_i^o$:

$$D_{k_i^o}(P) = D_{k_i^o}\left(E_{k_i^M}(HY)\right) = HY$$

   If the data acquired by decryption of digital authentication is identical to the $HY$ acquired through the decryption of digital authentication, the electronic document is deemed authentic it is not.

The drawback of the *EDS* there is only one potential digital authentication that corresponds to the provided public key ciphering computational procedure $M$-information. Due to the fact that the checksum $HY(M)$ according to the specified $M$-the value of information is always equal to $HY$ the computation of the hash-function value is based on a keyless computational procedure and consistently yields the same expression as the outcome of decryption and ciphering $H(M) = H$is separated. Such a circumstances make it challenging to apply such *EDS* computational processes.

Particularly designed (made) the aforementioned flaws are absent from the computational procedures used in the derivation of eds calculations and their verification. Since the hash-function value of the data to be signed is utilized in these computations, the eds is formed using a parameter selected by the signer in addition to the signer's secret key.

Using a parameter selected by the signer, practical application methods and protocols will be developed for the implementation of electronic digital authentication, verification of the integrity of information, and solutions for ensuring its authentication. The symmetric ciphering computational process [4] below was developed based on the difficulty of solving the problem of parameter multiplication of matrices and discrete logarithmization in a finite field along with the given HF.

$R_{il}^t = g^{x_{il}^t} \bmod p R_{il}^t$ $m \times n$ -matrix are calculated, where $x_{il}^t$- unknowns (which could be bytes), $i = 1, \ldots, m$; $l = 1, \ldots, n$. Then $(p, g, A^t{}_{m \times n}\ , R^t{}_{m \times n}\ )$ is the quaternion's public key, $x_{il}^t$additionally, the unknowns are $t$- designated as private key components by the user.

J-user t- to user cryptosystem $M_{n \times m}$- encrypts public data and transmits it with digital authentication based on the hash value in the manner described below:

1. The value of its hash $M = M_{n \times m} H(M) = H(M_{n \times m}) = H_{n \times m} = H$ based on the computational procedure of hash-functions of certain open data.

   $R$—this hash value's numeric signature, j, is created by the user's $x_{il}^{j}$ private key along with a number he selects $k_1$:

   (a) $R^{jk_1} = R^{jk_1}{}_{m \times n} = g^{x_{il}^{j} + k_1} \pmod p$ matrix components are computed
   (b) Ciphering $A_{n \times m}^{j}$is performed in the form ® $H = A_{n \times m}^{j} + H_{n \times m} + A_{n \times m}^{j} R_{m \times n}^{jk_1} H_{n \times m} \pmod p = L_{n \times m} = L$ and a digital authentication is formed.

2. Combining the data to be signed (M) with its digital authentication (L) yields a $M \parallel L = M^{'}$ - lengthy E-document.
3. $j$ - the elements of the matrix are computed using the public key of the matrix by selecting at random a number k that is only known to the user. $R = R^{tk}{}_{m \times n} = \left(R_{il}^{t}\right)^{k} \bmod p = g^{kx_{il}^{t}} \pmod p)\ t$ user $.R^{t}{}_{m \times n}$,
4. Encrypting in the form $A_{n \times m}^{t} ® M_{n \times m}^{\prime} + A_{n \times m}^{t} + M_{n \times m}^{\prime} + A_{n \times m}^{t} R_{n \times m}^{tk} M_{n \times m}^{\prime} \pmod p = w_{n \times m}$ and $C_{n \times m} = \left(w_{n \times m}; d_t = g^{k} \bmod p; d_p = g^{k_1} \bmod p\right)$ sending a triplet as ciphering data.

The cipher $C_{n \times m} = (w_{n \times m}; d_t; d_p)$ has received the data; the user decrypts it as follows:

1. Making use of a hidden key $d_t{}^{x_{il}^{t}} \bmod p = g^{kx_{il}^{t}} \bmod p = D_{il}^{tk}$ known $x_{il}^{t}$ solely to the user, the figures are computed and $D_{mn}$the matrix is produced.
2. The matrix is open $A_{nm}^{t} = \left(I_{nn} + A_{nm}^{t} D_{mn}^{tk}\right)^{-1} \left(-A_{nm}^{t}\right) \pmod p$; this is the key's parametric inverse $\left(A_{nm}^{t}\right)^{-1}$.
3. $R = D_{n \times m}^{tk}$; this value replacement procedure is used to carry out decryption:

$$
\begin{aligned}
&\left(A_{nm}^{t}\right)^{\backslash -1} ® w_{nm} = \left(I_{nn} + A_{nm}^{t} D_{mn}^{tk}\right)^{-1} \left(-A_{nm}^{t}\right) ® \Big(A_{nm}^{t} + M^{/}{}_{nm} + \\
&+ A_{nm}^{t} R_{mn}^{tk} M^{/}{}_{nm}\Big) \pmod p = \left(I_{nn} + A_{nm}^{t} R_{mn}^{tk}\right)^{-1} \left(-A_{nm}^{t}\right) + \\
&\left(A_{nm}^{t} + M^{/}{}_{nm} + A_{nm}^{t} R_{mn}^{tk} M^{/}{}_{nm}\right) + \left(I_{nn} + A_{nm}^{t} R_{mn}^{tk}\right)^{-1} \left(-A_{nm}^{t}\right) R_{mn}^{tk} + \\
&\left(A_{nm}^{t} + M^{/}{}_{nm} + A_{nm}^{t} R_{mn}^{tk} M^{/}{}_{nm}\right) + \left(I_{nn} + A_{nm}^{t} R_{mn}^{tk}\right)^{-1} \left(-A_{nm}^{t}\right) R_{mn}^{tk} + \\
&\left(A_{nm}^{t} + M^{/}{}_{nm} + A_{nm}^{t} R_{mn}^{tk} M^{/}{}_{nm}\right) \pmod p = \left(I_{nn} + A_{nm}^{t} R_{mn}^{tk}\right)^{-1} \left(-A_{nm}^{t}\right) \\
&\left(I_{mm} + R_{mn}^{tk} A_{nm}^{t}\right) + A_{nm}^{t} + \left(I_{nn} + A_{nm}^{t} R_{mn}^{tk}\right) M_{nm}^{\prime} + \left(I_{nn} + A_{nm}^{t} R_{mn}^{tk}\right)^{-1} \\
&\left(-A_{nm}^{t}\right) R_{mn}^{tk} \left(I_{nn} + A_{nm}^{t} R_{mn}^{tk}\right) M^{/}{}_{nm} \pmod p
\end{aligned}
$$

Even if they are swapped in terms involving matrix products, the equality remains unchanged if all of the diagonal elements of the matrices in this expression of the last equality are non-zero and all other components are zero (such matrices have the feature of commutativity). For these matrices, this equality is true:

$$\begin{aligned}
&\left(A^t_{nm}\right)^{\backslash-1} ®w_{nm} = \left(I_{nn} + A^t_{nm} R^{tk}_{mn}\right)^{-1} \left(-A^t_{nm}\right)\\
&\left(I_{mm} + R^{tk}_{mn} A^t_{nm}\right) + A^t_{nm} + \left(I_{nn} + A^t_{nm} R^{tk}_{mn}\right) M^{/}{}_{nm} +\\
&\left(I_{nn} + A^t_{nm} R^{tk}_{mn}\right)^{-1} \left(-A^t_{nm}\right) R^{tk}_{mn} \left(I_{nn} + A^t_{nm} R^{tk}_{mn}\right) M^{/}{}_{nm} \pmod p =\\
&\left(-A^t_{nm}\right) \left(I_{nn} + A^t_{nm} R^{tk}_{mn}\right)^{-1} \left(I_{mm} + R^{tk}_{mn} A^t_{nm}\right) + A^t_{nm} +\\
&\left(I_{nn} + A^t_{nm} R^{tk}_{mn}\right) M^{/}{}_{nm} + \left(-A^t_{nm}\right) R^{tk}_{mn} \left(I_{nn} + A^t_{nm} R^{tk}_{mn}\right)^{-1}\\
&\left(I_{nn} + A^t_{nm} R^{tk}_{mn}\right) M_{nm} \pmod p = -A^t_{nm} + A^t_{nm} +\\
&M^{/}{}_{nm} + A^t_{nm} R^{tk}_{mn} M^{/}{}_{nm} - A^t_{nm} R^{tk}_{mn} M^{/}{}_{nm} \pmod p = M^{/}{}_{nm}
\end{aligned}$$

In general, if the matrices participating in these equality expressions are chosen so that they have the property of commutativity, the decoding process mentioned above can be easily implemented. Here

$$M^{/}{}_{nm} = M_{nm} \parallel L_{nm}$$

and

$$L_{nm} = A^j_{nm} + H_{nm} + A^j_{nm} R^{jk_1}_{mn} H_{nm} \pmod p = A^j_{nm} ® H = L$$

that the process of determining the correctness of an electronic digital authentication continues as follows:

4. $j$- is calculated $R^j_{mn}$by multiplying $R^{jk_1}_{il} = g^{x^j_{il}+k_1} \bmod p = D^{jk_1}_{il}$by the public key of the user $d_p = g^{k_1} \bmod p$, $D^{jk_1}_{n\times m}$the -matrix is generated.
5. Open is the matrix $A^j_{nm} = \left(I_{nn} + A^j_{nm} D^{jk_1}_{mn}\right)^{-1} \left(-A^{jk_1}_{nm}\right) \pmod p$; this is the key's parametric inverse $\left(A^j_{nm}\right)^{\backslash-1}$.
6. $R = D^{jk_1}_{n\times m}$; this value replacement procedure is used to carry out decryption:

$$\begin{aligned}
&\left(A^j_{nm}\right)^{\backslash-1} ®w_{nm} = \left(I_{nn} + A^j_{nm} D^{jk_1}_{mn}\right)^{-1} \left(-A^j_{nm}\right) ®\\
&\left(A^j_{nm} + H_{nm} + A^j_{nm} R^{jk_1}_{mn} H_{nm}\right) \pmod p =\\
&\left(I_{nn} + A^t_{nm} R^{jk_1}_{mn}\right)^{-1} \left(-A^t_{nm}\right) + \left(A^t_{nm} + H_{nm} + A^t_{nm} R^{jk_1}_{mn} H_{nm}\right) +\\
&\left(I_{nn} + A^j_{nm} R^{jk_1}_{mn}\right)^{-1} \left(-A^j_{nm}\right) R^{jk_1}_{mn} \left(A^j_{nm} + H_{nm} + A^j_{nm} R^{jk_1}_{mn} H_{nm}\right) \pmod p =\\
&\left(I_{nn} + A^j_{nm} R^{jk_1}_{mn}\right)^{-1} \left(-A^j_{nm}\right) \left(I_{mm} + R^{jk_1}_{mn} A^j_{nm}\right) + A^j_{nm} +\\
&+ \left(I_{nn} + A^t_{nm} R^t_{mn}\right) H_{nm} + \left(I_{nn} + A^j_{nm} R^{jk_1}_{mn}\right)^{-1} \left(-A^j_{nm}\right) R^{jk_1}_{mn}\\
&\left(I_{nn} + A^j_{nm} R^{jk_1}_{mn}\right) H_{nm} \pmod p
\end{aligned}$$

Given that all other elements of the matrices in the expression of this last equality are zero (such matrices have the property of commutativity) and that only the diagonal elements are not all zero, we can approach this equality as follows by using the fact that the equality remains unchanged even if the positions of the diagonal elements are altered in the terms involving matrix products:

$$\left(A_{nm}^{j}\right)^{\backslash-1}®w_{nm} = \left(I_{nn} + A_{nm}^{j}R_{mn}^{jk_1}\right)^{-1}\left(-A_{nm}^{j}\right)$$
$$\left(I_{mm} + R_{mn}^{jk_1}A_{nm}^{j}\right) + A_{nm}^{j} + \left(I_{nn} + A_{nm}^{j}R_{mn}^{jk_1}\right)H_{nm} +$$
$$\left(I_{nn} + A_{nm}^{j}R_{mn}^{jk_1}\right)^{-1}\left(-A_{nm}^{j}\right)R_{mn}^{jk_1}\left(I_{nn} + A_{nm}^{t}R_{mn}^{t}\right)H_{nm} =$$
$$\left(-A_{nm}^{j}\right)\left(I_{nn} + A_{nm}^{j}R_{mn}^{jk_1}\right)^{-1} + \left(I_{nn} + A_{nm}^{j}R_{mn}^{jk_1}\right)H_{nm} +$$
$$\left(-A_{nm}^{j}\right)R_{mn}^{jk_1}\left(I_{nn} + A_{nm}^{jt}R_{mn}^{jk_1}\right)^{-1}\left(I_{nn} + A_{nm}^{j}R_{mn}^{jk_1}\right)H_{nm} \pmod{p} =$$
$$-A_{nm}^{j} + A_{nm}^{j} + H_{nm} + A_{nm}^{j}R_{mn}^{jk_1}H_{nm}H_{nm} + A_{nm}^{j}R_{nm}^{jk_1}H_{nm}$$
$$-A_{nm}^{t}R_{mn}^{t}H_{nm} \pmod{p} = H_{nm}$$

7. $t$- the section of the enlarged document $M_1$coming from the deciphering of the $M_1{}^{/}$user's encrypted data is $C_{n\times m}$ hashed $M_1 \parallel P_1 : H(M_1) = H(M_{1nm}) = H_1$
8. $H_1 = H(M_{1nm}) = H_{nm} = H$ if so, the legitimacy of its source is determined by the authenticity of its electronic digital authentication, and the electronic document is regarded as complete (authentic).

## 4 Conclusion

Using a computational method for asymmetric ciphering based on the complexity of multiplications with parameters and the finite group logarithm problem in the Electronic Digital authentication Protocol provides a robust foundation for secure digital communications. The security of these systems is underpinned by the difficulty of the DLP, ensuring that digital authentications are both secure and resistant to attacks. By combining these mathematical challenges with cryptographic computational processes, electronic signatures can offer strong guarantees of authenticity, integrity, and non-repudiation, making them suitable for a wide range of applications, from securing online transactions to ensuring the validity of digital contracts.

Along with the signer's private key, a parameter selected by the signer was also employed in the suggested EDS computing procedure. Consequently, this EDS computational procedure shares the same characteristics as EDS computational procedures that rely on the development and validation of a custom (made) digital authentication computation.

With advancements in technologies like Artificial Intelligence (AI) and the Internet of Things (IoT), ensuring cybersecurity in robotics will only grow more complex. Collaboration between engineers, cybersecurity experts, and policymakers will be essential for developing robust standards and regulations.

## References

1. Алферов, А.: П., Зубов А. Ю., Кузьмин А. С., Черемушкин А. В. Основы криптографии: Учебное пособие, 2-е изд. –М.: Гелиос АРВ, 2002.-480 с
2. Шнайер, Б.: Прикладная криптография. Протоколы, алгоритмы, исходные тексты на языке Си. –М.: издательство ТРИУМФ, 2003–816 с
3. Акбаров, Д.Е.: Ахборот хавфсизлигини таъминлашнинг криптографик усуллари ва уларнинг қлланилиши—Тошкент, "збекистон маркаси", 2009–434 бет
4. Khasanov, P.F., Axmadaliev, A.D.E., Sh, S.: Methods for creating new asymmetric computational processs based on existing computational complexities using parametric algebra operations. Infokommunikatsii. **1**(9), 31–35 (2009)

# Part II
# Marine Robotics Workshop

# Sensitivity Analysis of Hydrodynamic Coefficients in Ship Manoeuvring: A Case Study Using VLCC Turning Simulations

**Suat Adnar, Ali Doğrul, and Ömer Kemal Kınacı**

## 1 Introduction

A successful ship design should perform a good hydrodynamic performance and for the hydrodynamic performance, it is necessary to consider three main issues: ship resistance and propulsion, seakeeping, and manoeuvrability. It is essential to conduct the ship design process by making a priori estimations for ship performance parameters prior to the construction of the ship. It is possible to make estimates with a rough approach, which provide an idea of the likely outcome, as well as estimates with high accuracy. However, the increase in the accuracy rate leads to cost and time problems in the method to be used. Manoeuvring performances can be estimated by comparing similar vessels, experimental model tests, computational methods, numerical simulations using system identification method, or empirical methods. The comparison of similar ships based on full scale tests and experimental model tests are both costly methods. On the other hand, computational methods require a tough preparation process. In addition to these, empirical methods and numerical simulations represent the most cost and process effective means of achieving results in a timely manner. Empirical methods are inadequate for the purpose of making accurate predictions. At this point, numerical simulations using

S. Adnar (✉) · A. Doğrul
National Defence University, Turkish Naval Academy, Istanbul, Turkey
e-mail: sadnar@dho.edu.tr

Ö. K. Kınacı
Faculty of Naval Architecture and Ocean Engineering, Istanbul Technical University, Istanbul, Turkey

Marine Cybernetics Advanced Vehicle Technologies, ITU Ariteknokent, Istanbul, Turkey

Marine Robotics Laboratory, Istanbul Technical University, Istanbul, Turkey



B. U. Töreyin et al. (eds.), *8th EAI International Conference on Robotic Sensor Networks*, EAI/Springer Innovations in Communication and Computing,
https://doi.org/10.1007/978-3-031-92143-8_16

mathematical model defined by system identification methods are more efficient in terms of process, cost, and accuracy. The system identification method also requires hydrodynamic coefficients. The hydrodynamic coefficients are used in a mathematical model (MMG Model, Abkowitz Model, etc.) to make predictions. Obtaining these coefficients requires model experiments and CFD analyses. Once these coefficients have been obtained, it is possible to make simple and rapid predictions of the manoeuvring performance of ships of similar forms.

On the other hand, the model must ensure the precision of its results. Sensitivity analysis gives idea about the responses of the hydrodynamic coefficients in the mathematical model at specific deviations. In this manner, some studies were presented in the literature.

The sensitivity analysis of hydrodynamic coefficients was first presented by Hwang [1] to identify how sensitive the ship's response is to the change in each hydrodynamic coefficient. Kose et al. [2] studied a modular mathematics model that belongs to the MMG (Manoeuvring Modelling Group) family. The study revealed that the linear manoeuvring derivatives had the greatest influence on the parameters, while the added mass coefficients to the equations were the least influential. Rhee et al. [3] pointed to a problem with sensitivity in the sense of Hwang's original definition. Hydrodynamic coefficients are estimated using simulated data of turning and zigzag tests, the estimation performances are analysed by coefficient sensitivity, and a new experimental scheme is proposed to increase the estimation efficiency. Wang et al. [4] carried out a sensitivity study using a mathematical model of a container vessel that has 4DOF (Degree Of Freedom). It was shown first-order coefficients have the most effect. The sensitivity of the hydrodynamic derivatives in the equations of motion of a container ship (S175) was studied by Shenoi et al. [5]. The sensitivity analysis is carried out by 10% deviations. It was concluded that $N'_r$ has a high effect on all standard manoeuvring parameters. Acceleration derivatives have a small effect on the turning circle test indices, with the exception of $Y'_{\dot{v}}$, $Y'_{\dot{r}}$ which has an average to large effect on steady turning radius and tactical diameter. $Y'_v$ and $X'_{uu}$ have average to high impact on all the standard turning circle test indices. Kim et al. [6] have carried out a sensitivity study for a container ship. A series of simulations were performed in deep and shallow water. $N'_r$, $N'_v$, and $N'_\delta$ are the most influence on turning manoeuvrings for both deep and shallow water. The effect of $N'_\delta$ on tactic diameter in deep water was relatively low, while its effect on advance remained high. The values of $Y'_r$, $Y'_v$, and $Y'_\delta$ are significantly increased in shallow water condition for turning manoeuvring. Pires da Silva et al. [7] have conducted a sensitivity analysis on a 3DOF half-modular mathematical model comprising 19 coefficients with Euclidean metric. A typical of a frigate or destroyer-class surface-displacement ship was conducted to the sensitivity study. The results show that the first-order hydrodynamic coefficients have most impact on turning manoeuvre indices.

Within this literature review and to the best of the authors' knowledge, the studies in the literature were made using MMG, Abkowitz, and similar mathematical models. The present study, however, focuses on the sensitivity analysis using the mathematical model reported by ABS (American Bureau of Shipping). The sensitivity analysis was made for a VLCC (Very Large Crude Carrier) type ship form and the mathematical model was created using MATLAB platform. The turning circle manoeuvres are simulated by ±%10 deviations of hydrodynamic coefficients.

## 2 Theoretical Background

The mathematical model reported by the ABS was actually first proposed by MARAD (US Maritime Administration). This model, developed as a result of a series of towing tank experiments based on large and full-formed ship models called MARAD Systematic Series, is therefore suitable for manoeuvring predictions on large and full-formed ships. It has been finalized with some modifications by ABS. The following equations express the components of ABS model [8].

### 2.1 *Equations of Motions*

#### 2.1.1 Surge Motion

The forces acting in the surge direction are expressed as follows:

$$\mathrm{m}\left[\dot{u} - vr - x_{cg}r^2\right] = X + X_{Rd} \tag{1}$$

In this context, non-linear terms are incorporated by opening X, which represents the forces acting on the hull, to the Taylor series.

$$X = 0.5\rho L^3 X'_{\dot{u}}\dot{u} + 0.5\rho L^4 X'_{rr}r^2 + 0.5\rho L^2 X'_{vv}v^2 + 0.5\rho L^3 X'_{vr}vr \tag{2}$$

The term with $X'_{\dot{u}}\dot{u}$ is defined as an inertial quantity that reflects the changes in the inertial properties of a body which moving in a fluid compared with a body moving in a vacuum. The terms $X'_{rr}r^2$, $X'_{vv}v^2$, and $X'_{vr}vr$ represent additional resistance resulting from the turning motion. In one respect, this may be considered a drag component due to lift force if a ship hull is considered to be a lifting surface.

A change in the axial velocity ($u$) may exert a considerable impact on the additional resistance caused by the turning motion (drag component due to the lift force). It should be calculated with:

$$X_\eta = 0.5\rho L^2 X'_{vv\eta}v^2\left(\eta - 1\right) \tag{3}$$

$X'_{vv\eta}$= The derivative due to the model not being at the self-propulsion point.

The $\eta$ is the self-propulsion point correction, described in the reference [8].

$X_{Rd}$ is the rudder force and the flow around the rudder may vary depending on whether the vessel is in self-propulsion or not, as the flow around the rudder depends on the propeller. This is accounted for by the following formula:

$$X_{Rd} = 0.5\rho L^2 u^2 \left( X'_{\delta\delta}\delta_R^2 + X'_{\delta\delta\eta\eta}\delta_R^2\eta^2 \right) \tag{4}$$

$X'_{\delta\delta}$= coefficient of the rudder drag force measured at the self-propulsion point.
$X'_{\delta\delta\eta\eta}$= correction if a vessel is not in the self-propulsion point.
In addition, the polynomial fit consisting of the coefficients $a_i$, $b_i$, and $c_i$ in eq. (5) is described in the reference [8].
Finally, the complete surge equation is:

$$m\left[\dot{u} - vr - x_{cg}r^2\right] = 0.5\rho L^3 X'_{\dot{u}}\dot{u} + 0.5\rho L^2 u^2\left[a_i + b_i\eta + c_i\eta^2\right]$$
$$+0.5\rho L^4 X'_{rr}r^2 + 0.5\rho L^2 X'_{vv}v^2 + 0.5\rho L^3 X'_{vr}vr + 0.5\rho L^2 X'_{vv\eta}v^2(\eta - 1)$$
$$+0.5\rho L^2 u^2\left(X'_{\delta\delta}\delta_R^2 + X'_{\delta\delta\eta\eta}\delta_R^2\eta^2\right) \tag{5}$$

### 2.1.2 Sway Motion

The forces acting in the sway direction are expressed as follows:

$$m\left[\dot{v} - ur - x_{cg}\dot{r}\right] = Y + Y_{Rd} \tag{6}$$

In this context, non-linear terms are incorporated by opening $Y$, which represents the forces acting on the hull, to the Taylor series.

$$Y = 0.5\rho L^3 Y'_{\dot{v}}\dot{v} + 0.5\rho L^4 Y'_{\dot{r}}\dot{r} + 0.5\rho L^2 u Y'_v v + 0.5\rho L^3 u Y'_r r + +0.5\rho L^2 Y'_{v|v|}v|v|$$
$$+0.5\rho L^3 Y'_{v|r|}v|r| \tag{7}$$

The first two terms are expressions of the inertial forces in the fluid and represent non-dimensional added mass. The terms $0.5\rho L^2 Y'_{v|v|}v|v|$ and $0.5\rho L^3 Y'_{v|r|}v|r|$ are a non-linear part of the force of lift on the hull of the vessel.

The loss of speed in turn results in a vessel being unable to maintain self-propulsion. Therefore, a correction is given as follows:

$$Y_\eta = \left(0.5\rho L^2 u Y'_{v\eta}v + 0.5\rho L^3 u Y'_{r\eta}r + 0.5\rho L^2 Y'_{v|v|\eta}v|v|\right).(\eta - 1) \tag{8}$$

Flow around the propeller is not symmetric, which results in the creation of a lateral force component by any propeller. This phenomenon is particularly pronounced in the case of a vessel has a single-screw. This force will be modelled as follows:

$$Y_* = 0.5\rho L^2 u^2 Y'_* + 0.5\rho L^2 u^2 Y'_{*\eta}(\eta - 1) \tag{9}$$

The first term of eq. (14) represents the lateral force of the propeller at the point of self-propulsion. The second term serves to implement a correction if the vessel is not in the self-propulsion point. The coefficients $Y'_*$ and $Y'_{*\eta}$ represent non-dimensional measures of the aforementioned effect. The rudder force comprises a non-linear term and a correction for a vessel which is not in its self-propulsion point.

$$Y_{Rd} = 0.5\rho L^2 u^2 Y'_\delta \delta_R + 0.5\rho L^3 u Y'_{\delta|r|}\delta_R |r| + 0.5\rho L^2 u^2 Y'_{\delta\eta}\delta_R(\eta - 1) \tag{10}$$

$Y'_\delta$= rudder force coefficient.

$Y'_{\delta|r|}$= rudder force coefficient taking into account for yaw rate influence on rudder.

$Y'_{\delta\eta}$= the coefficient for correction if a vessel is not in the self-propulsion point.

Finally, the complete sway equation is:

$$\begin{aligned}
m\left[\dot{v} - ur - x_{cg}\dot{r}\right] &= 0.5\rho L^3 Y'_{\dot{v}}\dot{v} + 0.5\rho L^4 Y'_{\dot{r}}\dot{r} + 0.5\rho L^2 u Y'_v v + 0.5\rho L^3 u Y'_r r \\
&+ 0.5\rho L^2 Y'_{v|v|} v|v| + 0.5\rho L^3 Y'_{v|r|} v|r| + \Big(0.5\rho L^2 u Y'_{v\eta} v + 0.5\rho L^3 u Y'_{r\eta} r \\
&+ 0.5\rho L^2 Y'_{v|v|\eta} v|v|\Big).(\eta - 1) + 0.5\rho L^2 u^2 Y'_\delta \delta_R + 0.5\rho L^3 u Y'_{\delta|r|}\delta_R |r| \\
&+ 0.5\rho L^2 u^2 Y'_{\delta\eta}\delta_R(\eta - 1) + 0.5\rho L^2 u^2 Y'_* + 0.5\rho L^2 u^2 Y'_{*\eta}(\eta - 1)
\end{aligned} \tag{11}$$

### 2.1.3 Yaw Motion

The moments acting in the yaw motion are expressed as follows:

$$I_z \dot{r} + m x_{cg}[\dot{v} + ur] = N + N_{Rd} \tag{12}$$

The expanded hull force moments, $N$ in form of Taylor series, contain non-linear terms.

$$\begin{aligned}
N = {} & 0.5\rho L^5 N_{\dot{r}}\dot{r} + 0.5\rho L^4 N'_{\dot{v}}\dot{v} + 0.5\rho L^4 u N'_r r + 0.5\rho L^3 u N'_v v + \\
& 0.5\rho L^5 N'_{r|r|} r|r| + 0.5\rho L^3 N'_{v|v|} v|v| + 0.5\rho L^4 N'_{v|r|} v|r|
\end{aligned} \tag{13}$$

A correction is required to account for the speed loss during turning manoeuvre and the leaving from the self-propulsion point.

$$N_\eta = \Big(0.5\rho L^4 u N'_{r\eta} r + 0.5\rho L^3 u N'_{v\eta} v + 0.5\rho L^3 Y'_{v|v|\eta} v|v|\Big).(\eta - 1) \tag{14}$$

The lateral force generated by the asymmetric flow around the propeller, as expressed in eq. (9), results in a moment.

$$N_* = 0.5\rho L^3 u^2 N'_* + 0.5\rho L^3 u^2 N'_{*\eta} (\eta - 1) \tag{15}$$

Similar to eq. (10), for the moment of rudder force, the expression is:

$$N_{Rd} = 0.5\rho L^3 u^2 N'_\delta \delta_R + 0.5\rho L^4 u N'_{\delta|r|} \delta_R |r| + 0.5\rho L^3 u^2 N'_{\delta\eta} \delta_R (\eta - 1) \tag{16}$$

Finally, the complete yaw equation is:

$$\begin{aligned} I_z \dot{r} + m x_{cg} [\dot{v} + ur] &= 0.5\rho L^5 N_{\dot{r}} \dot{r} + 0.5\rho L^4 N'_{\dot{v}} \dot{v} + 0.5\rho L^4 u N'_r r \\ &+ 0.5\rho L^3 u N'_v v + 0.5\rho L^5 N'_{r|r|} r |r| + 0.5\rho L^3 N'_{v|v|} v |v| + 0.5\rho L^4 N'_{v|r|} v |r| \\ &+ \left(0.5\rho L^4 u N'_{r\eta} r + 0.5\rho L^3 u N'_{v\eta} v + 0.5\rho L^3 N'_{v|v|\eta} v |v|\right) . (\eta - 1) \\ &+ 0.5\rho L^3 u^2 N'_\delta \delta_R + 0.5\rho L^4 u N'_{\delta|r|} \delta_R |r| \\ &+ 0.5\rho L^3 u^2 N'_{\delta\eta} \delta_R (\eta - 1) + 0.5\rho L^3 u^2 N'_* + 0.5\rho L^3 u^2 N'_{*\eta} (\eta - 1) \end{aligned} \tag{17}$$

# 3 Numerical Study

## 3.1 General Data

In this study, it is aimed to examine the effect of hydrodynamic coefficients on the turning manoeuvre characteristics of a large and fully formed ship. In order to achieve this objective, the effect of varying hydrodynamic coefficients on turning manoeuvre indices (advance, transfer, tactical diameter, steady turning circle) was investigated in accordance with the ABS mathematical model [8], with 10% deviations applied to coefficients on both port and starboard sides at two distinct speeds. The hydrodynamic coefficients were sourced from the ABS manoeuvrability guide [8]. The hydrodynamic coefficients were derived from the data obtained from the MARAD systematic series of full-form ship model experimental tests [9]. The main particulars of the "sample ship" from the MARAD Systematic series forms are presented in Table 1.

## 3.2 Validation of the Code

For the validation purpose, the numerical simulation was carried out with some assumptions given below:

- The command speed (only surge direction) is assumed to be 15 knots.

**Table 1** Sample ship particulars [8]

| Specification | Value |
|---|---|
| Length between perpendiculars, $L$, m | 349.8 |
| Breadth moulded, $B$, m | 58.3 |
| Draft in full load, $T$, m | 19.4 |
| Block coefficient | 0.875 |
| Displacement, metric tons | 355,600 |
| Longitudinal position of centre of gravity from amidships, $x_{cg}$, m | 0 |
| Speed, corresponding to 85% of engine output, $V_s$, knots | 15 |

**Table 2** Comparison of the results

| Parameter | This study | ABS | Relative Diff. (%) |
|---|---|---|---|
| Time for 90 degrees course change, seconds | 169 | 173 | 2.31 |
| Time for 180 degrees course change, seconds | 406 | 409 | 0.73 |
| Tactical diameter in ship lengths | 2.77 | 2.79 | 0.71 |
| Advance in ship lengths | 2.57 | 2.77 | 7.22 |
| Transfer in ship lengths | 1.02 | – | – |
| Diameter of steady turn in ship lengths | 2.73 | 2.73 | 0.00 |
| Velocity of steady turn, knots | 5.33 | 5.33 | 0.00 |

- Sway velocity (v) and yaw rate (r) are zero.
- Before the rudder execution, the ship reaches the approach speed of 15 knots and maintains this for a period of 60 s.
- The rudder would make a direct port or starboard hardover. The time taken for the rudder to reach an angle of 35 degrees ($\delta_R$) was neglected.
- The simulation was continued for 60 min.

The numerical code was validated with the ABS numerical results at the command speed of 15 knots. The results of the present study and ABS for the turning circle manoeuvre are shown in Table 2. The two studies differ by a maximum of 7.22% in relative terms, with some values identical. It is presumed that these differences were caused by the course change of ship to the port side before the rudder execution. It is obvious that the values will be closer if the course change of the ship before the rudder execution is ignored. This is demonstrated by the fact that the diameter of the steady turning and the speed after entering this course are perfectly aligned. Consequently, the results of this study will be consistent, as the initial conditions differ only slightly.

### *3.3 Sensitivity Analysis*

In this section, the results of the sensitivity analyses in terms of advance, tactical diameter, steady turning circle, and transfer were presented. The numerical simulations were conducted with the assumptions as described briefly in the validation section. In addition to the command speed of 15 knots corresponding to 0.132 Froude number, 3 knots speed corresponding to 0.026 Froude number was also investigated in the sensitivity analyses. The changes are shown in relative values. In the following results, 15 knot speed results will be referred to as "high speed" investigation and 3 knot speed results will be referred to as "low speed" investigation. The effects of hydrodynamic coefficients were divided into four different groups:

0 < Negligible effect ≤ %0.25 < Low effect ≤ %1 < Average effect ≤3% < High effect.

Numerical simulations were performed on both port and starboard sides. Due to the lateral forces and moments caused by the propeller, it is observed that the manoeuvres on the port side naturally take place in a slightly narrower radius. In other words, the manoeuvre index values differ. However, it was observed that the responses of the hydrodynamic coefficients to 10% changes were almost similar. These data can be found in the related study of the first author [10]. In this study, only evaluations on the starboard manoeuvre are presented.

Considering the speed difference at 10% deviations of the hydrodynamic coefficients, it is seen that the rates of change in tactical diameter and permanent rotational diameter do not differ much. However, the speed difference on the advance and transfer indices completed relatively early in the manoeuvre is noticeable, and these effects show a common behaviour. It is observed that the changes in the coefficients that have a positive effect on the turning circle manoeuvre (the ship makes a narrower manoeuvre) are better with increasing speed and give better results, while the positive effect decreases with lowering speed. As for the coefficients affecting the turning circle manoeuvre in a negative way (the ship makes a wider manoeuvre), it is observed that this negative effect increases with higher speed and the negative effect decreases with lower speed. More detailed observations can be found in the first author's thesis [10].

Since the changes show the same behaviour in the sensitivity analyses performed at starboard/port and 3/15 knot speeds and since the most obvious changes are obtained in the manoeuvre simulation performed in the starboard direction, the starboard direction graphs are presented. Therefore, the results were evaluated on the basis of the starboard manoeuvre.

### 3.3.1 Effects on Advance

The first-order derivatives ($N_v$, $N_r$, and $N_\delta$) of the yaw moment have the highest effect on the advance manoeuvring index (see Fig. 1) and effects are in the range of 3.5% to 4.5%

While the yaw moment derivative $N_{vr}$ has an average effect, $N_{vv}$, $N_{\dot{r}}$, and $N_{\delta\eta}$ have low effect. For derivatives of sway forces $Y_{vv}$, $Y_{vr}$, $Y_v$, and $Y_r$ have a low effect on some manoeuvring indices and negligible effect on others. The lateral

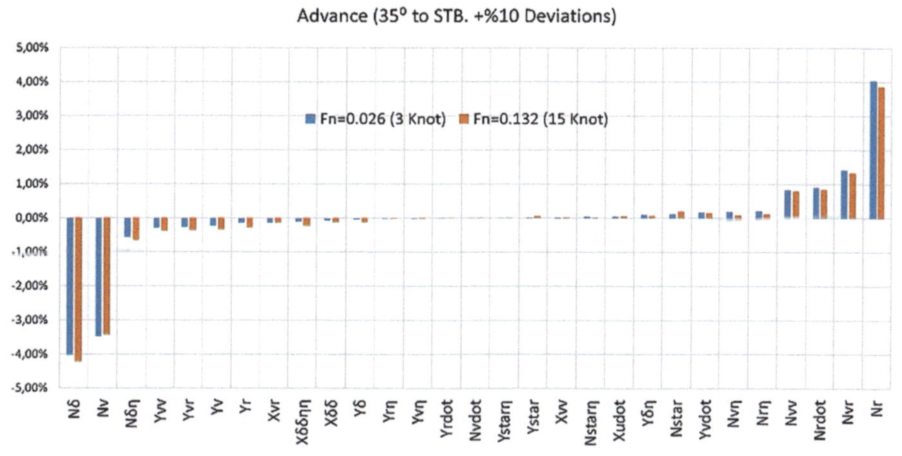


**Fig. 1** Effects of 10% increase of hydrodynamic coefficients on the advance distance in starboard manoeuvring

force derivatives ($Y_*$, $Y_{*\eta}$) and moment derivatives ($N_*$, $N_{*\eta}$) of the propeller have low and negligible effect.

The added mass moment coefficient $N_{\dot{r}}$ has average and low effect for different manoeuvres on advance.

For $X_{\dot{u}}$, $X_{vv}$, $X_{\delta\delta}$, $X_{\delta\delta\eta\eta}$, $Y_{\dot{v}}$, $Y_{v\eta}$, $Y_{\dot{r}}$,$Y_{r\eta}$, $Y_\delta$, $Y_*$, $Y_{*\eta}$, $N_{\dot{v}}$, and $N_{*\eta}$, the effects of these hydrodynamic derivatives are negligible.

### 3.3.2 Effects on Tactical Diameter

While the effects of $N_v$ and $N_r$, which are first-order derivatives of yaw, on the tactical diameter are high as for the advance index, the effect of $N_\delta$ is weaker on the tactical diameter (see Fig. 2). In addition, the $N_v$ effect is approximately 2% greater than the $N_r$ effect. For $N_{vr}$, $N_{vv}$, $N_{\delta\eta}$, $N_{r\eta}$, and $N_{v\eta}$, the effects of these hydrodynamic derivatives are average.

For $X_{\dot{u}}$, $X_{vv}$, $X_{\delta\delta}$, $X_{\delta\delta\eta\eta}$, $Y_{v\eta}$, $Y_{\dot{r}}$,,$Y_{r\eta}$, $Y_\delta$, $Y_*$, $Y_{*\eta}$, $N_{\dot{v}}$, $N_*$, and $N_{*\eta}$, the effects of these hydrodynamic derivatives are negligible.

### 3.3.3 Effects on Steady Turning Circle

While the effects of $N_v$ and $N_r$ on the steady turning diameter index are still significantly high, the effect of $N_\delta$ decreases from high to average (see Fig. 3).

$N_{\delta\eta}$ has more effect on the steady turning diameter than any other manoeuvring indices. This effect is around 5.5%.

$N_{vr}$, $N_{vv}$, $N_\delta$, $N_{r\eta}$, and $N_{v\eta}$ derivatives have average effect.

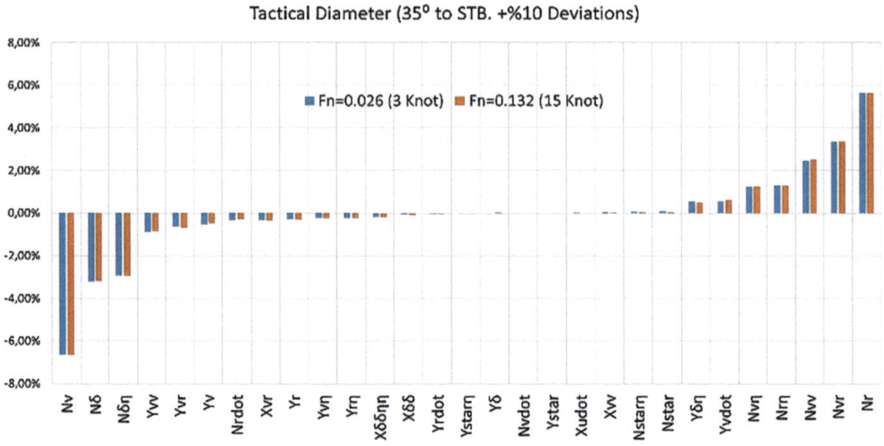


**Fig. 2** Effects of 10% increase of hydrodynamic coefficients on the tactical diameter in starboard manoeuvring

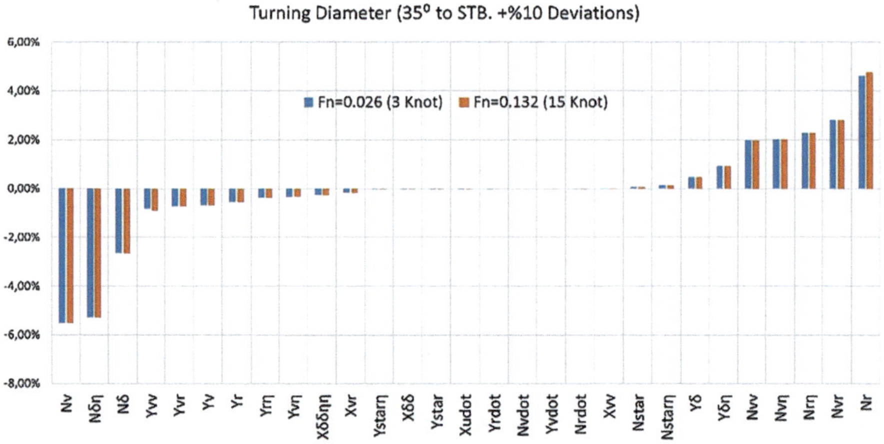


**Fig. 3** Effects of 10% increase of hydrodynamic coefficients on the turning diameter in starboard manoeuvring

$Y_{v\eta}$, $Y_{r\eta}$, and $Y_{\delta\eta}$ derivatives which contain self-propulsion point correction have low effect.

For $X_{\dot{u}}$, $X_{vv}$, $X_{vr}$, $X_{\delta\delta}$, $Y_{\dot{v}}$, $Y_{\dot{r}}$, $Y_*$, $Y_{*\eta}$, $N_{\dot{v}}$, $N_{\dot{r}}$, $N_*$, and $N_{*\eta}$, the effects of these hydrodynamic derivatives are negligible.

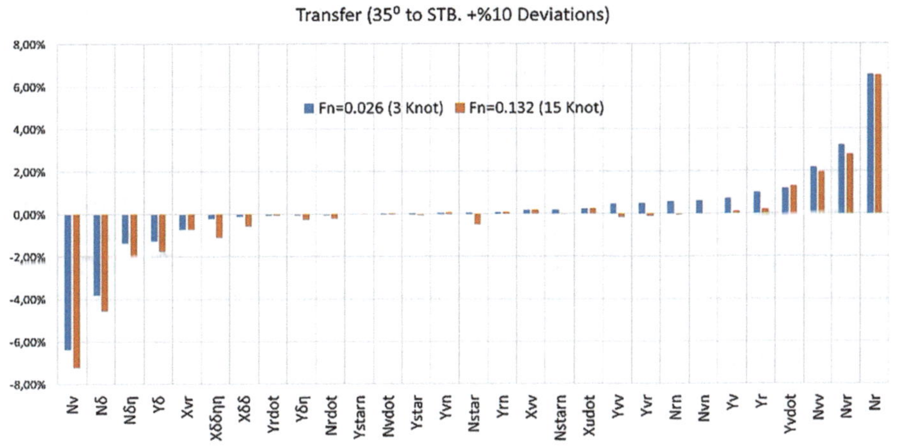


**Fig. 4** Effects of 10% increase of hydrodynamic coefficients on the transfer distance in starboard manoeuvring

### 3.3.4 Effects of Transfer

In terms of transfer distance, first-order derivatives of yaw moment have the greatest effect. $Y_\delta$, $Y_{\dot v}$, and $Y_r$, which are derivatives of the sway force, have the most effect on the transfer distance among the manoeuvring indices (see Fig. 4). On the other hand, it has low effect on the advance, tactical diameter and steady turning diameter indices. Most of the hydrodynamic coefficients have a low effect.

$X_{\dot u}$, $X_{vv}$, $Y_{\dot r}$, $Y_{v\eta}$, $Y_{r\eta}$ $Y_*$, $Y_{*\eta}$, $N_{\dot v}$, and $N_{*\eta}$ coefficients have a negligible effect.

## 4 Conclusion

In this study, it was investigated that the impact of ±10% deviations in the hydrodynamic coefficients on the turning circle manoeuvring indices of the ABS model. It was investigated for both port and starboard sides. The mathematical model and the required equations were coded in MATLAB platform as a m-file. The simulations were carried out for each hydrodynamic coefficient deviations. The results observed on manoeuvring indices; transfer, advance, tactical diameter and steady turning diameter. The following results were obtained in this study.

The sensitivity analysis investigation in the port direction and the necessary explanation for the different speed investigations are explained in Subsection 3.3.

The first-order derivatives of the yaw moment ($N_v$, $N_r$ and $N_\delta$) mostly exert the most significant influence on the manoeuvring indices. It is seen that the effects of $N_{vv}$, $N_{vr}$, and $N_{\delta\eta}$ on tactical diameter, $N_{vr}$and $N_{\delta\eta}$ on steady turning circle, and $N_{vr}$

on transfer reach high degrees. Especially $N_{\delta\eta}$ has a very high effect on the steady turning diameter.

The effects of surge and sway forces are quite low compared to yaw. $Y_\delta$, $Y_{\dot{v}}$, and $Y_r$, which are derivatives of the sway force, have the most influence on the transfer index among the manoeuvring indices, but this effect remains at low levels.

The terms representing moments and forces of inertia have a very low effect.

Most of the terms that include self-propulsion point correction have a low effect.

The derivatives representing the propeller-induced lateral force and moment exert a positive influence on the manoeuvres towards the port, due to the propeller rotation direction. However, it is affected at low and negligible levels from the deviations of the coefficients.

It is concluded that $X_{\dot{u}}$, $X_{vv}$, $Y_{\dot{r}}$, $Y_*$, $Y_{*\eta}$, $N_{\dot{v}}$, and $N_{*\eta}$ hydrodynamic derivatives have negligible effect in all turning manoeuvre simulations.

**Acknowledgements** This study is derived from the master thesis of the first author [10].

## References

1. Hwang, W.: "Application of system identification to ship maneuvering." Thesis, Massachusetts Institute of Technology. (1980). Accessed 5 October 2024. [Online]. https://dspace.mit.edu/handle/1721.1/15820
2. Kose, K., Misiag, W. A.:"A Systematic procedure for predicting manoeuvring performance." (1993). Accessed 8 March 2024. [Online]. https://trid.trb.org/View/444577
3. Rhee, K.-P., Kim, K.: A New Sea Trial Method for Estimating Hydrodynamic Derivatives. J. Ship Ocean Technol. **3**(3), 25–44 (1999)
4. Wang, X.-G., Zou, Z.-J., Xu, F., Ren, R.-Y.: Sensitivity analysis and parametric identification for ship manoeuvring in 4 degrees of freedom. J. Mar. Sci. Technol. **19**(4), 394–405 (2014). https://doi.org/10.1007/s00773-014-0277-y
5. Shenoi, R.R., Krishnankutty, P., Selvam, R.P.: Sensitivity study of hydrodynamic derivative variations on the maneuverability prediction of a container ship. In: Presented at the ASME 2015 34th International Conference on Ocean, Offshore and Arctic Engineering, American Society of Mechanical Engineers Digital Collection (2015). https://doi.org/10.1115/OMAE2015-41490
6. Kim, D., Kim, S.-H., Kim, S.-J., Paik, K.-J.: A study on the sensitivity analysis of the hydrodynamic derivatives On The Maneuverability Of KVLCC2 in shallow water. Brodogr. Int. J. Nav. Archit. Ocean Eng. Res. Dev. **68**(4), 1–22 (2017). https://doi.org/10.21278/brod68401
7. Pires da Silva, P., Sutulo, S., Guedes Soares, C.: Sensitivity Analysis of Ship Manoeuvring Mathematical Models. J Mar Sci Eng. **11**(2), 2 (2023). https://doi.org/10.3390/jmse11020416
8. Palfy, J.: "Guide for Vessel Maneuverability 2017," (2006)
9. Roseman, D. P.: The MARAD Systematic Series of Full-Form Ship Models
10. Adnar, S.: The sensitivity analysis of manoeuvring derivatives on a VLCC with turning circle simulation tests. Istanbul Technical University, Istanbul (2024) Accessed 17 October 2024. [Online]. https://tez.yok.gov.tr/UlusalTezMerkezi/tezSorguSonucYeni.jsp

# Maneuvering Model for Ship Dynamic Positioning

**Omer Kemal Kinaci and Burak Gunguder**

## 1 Introduction

Dynamic positioning systems (DPS) enable ships to automatically maintain their position and heading using their propellers, thrusters, and/or rudders. Ships with DPS are equipped with sensors that can measure the ship's accelerations, velocities, and positions. The data from these sensors are used to calculate the required response from the ship using its actuators. The ship, then, can either maintain its position or follow a prescribed trajectory.

A DPS for a conventional surface ship typically needs to control the vessel's motions in three degrees of freedom (3-DOF). These motions—surge, sway, and yaw—occur in the horizontal plane. In this chapter, we present a simplified mathematical model to simulate DPS operations, which typically occur at low ship speeds. To achieve this, we utilize a modular structure to set up a new DPS model, implementing the MMG mathematical model [5] for modeling the hull. The interactions between the hull, thrusters, rudders, and propellers are disregarded.

O. K. Kinaci (✉)
Faculty of Naval Architecture and Ocean Engineering, Istanbul Technical University, Istanbul, Turkey

Marine Robotics Laboratory, Istanbul Technical University, Istanbul, Turkey

Marine Cybernetics Advanced Vehicle Technologies Ltd., ITU ARI Teknokent, Istanbul, Turkey
e-mail: kinacio@itu.edu.tr

B. Gunguder
Faculty of Naval Architecture and Ocean Engineering, Istanbul Technical University, Istanbul, Turkey

Marine Robotics Laboratory, Istanbul Technical University, Istanbul, Turkey
e-mail: gunguder@itu.edu.tr



B. U. Töreyin et al. (eds.), *8th EAI International Conference on Robotic Sensor Networks*, EAI/Springer Innovations in Communication and Computing,
https://doi.org/10.1007/978-3-031-92143-8_17

Wave effects are accounted for by solving the empirical relations proposed by [1]. After a brief introduction in this section, we present the details of the ship in Sect. 2. The actuators, their positions, and their hydrodynamic properties are also given in the same section. Section 3 is dedicated to our mathematical model developed for low-ship-speed DPS. The details of the hull and actuator modeling along with the wave effects are presented in this section. We present the results using our model in Sect. 4 and conclude the study in Sect. 5.

## 2 The Details of the Model Ship

The ship studied in this chapter is DTC containership at 1/80 model scale whose geometric properties are given in Table 1. The original ship only has one rudder and one propeller. However, in this chapter, we equipped the model with a total of eight actuators: two propellers, two rudders, and four thrusters (consisting of two-stern and two-bow thrusters). The origin of the ship's fixed coordinate system (CS) is located amidships. The origins of the other coordinate systems are indicated by dots on the actuators. Figure 1 provides an illustrative image of the ship's actuators. The names of the actuators, their types, and symbols are given in Table 2. In the ship-fixed CS, the z-axis points downward.

The positions of the actuators are defined according to the amidships of the ship-fixed CS, i.e., $T_{S2}(x, y, z)$ denotes the $x$, $y$, and $z$ distances of the front stern thruster from (0, 0, 0). We also make use of the numbering given in Table 2 to simplify the notations. The rudder CS is defined at $\frac{1}{4}$ chord length from their leading edges. The propeller and thruster CS are defined at the centers of their disks.

The rudder has a maximum turning angle of $\delta_{\max} = 35°$ and a saturation rate of $\dot{\delta}_{\max} = 2.25°/\text{s}$. The form of the propellers is selected from the Wageningen B-Series database [2]. They have five blades, $Z = 5$, a pitch ratio of $P/D = 1$, and an

**Table 1** Geometric properties of the full-scale ship

| | $L_{PP}$ | $B_{WL}$ | $T$ | $\nabla$ | $C_B$ | $S_W$ | $x_G$/L |
|---|---|---|---|---|---|---|---|
| Units | $m$ | $m$ | $m$ | $m^3$ | – | $m^2$ | – |
| Full-scale ship | 355 | 51 | 14.5 | $1.735 \times 10^5$ | 0.661 | 22032 | – 0.01 |

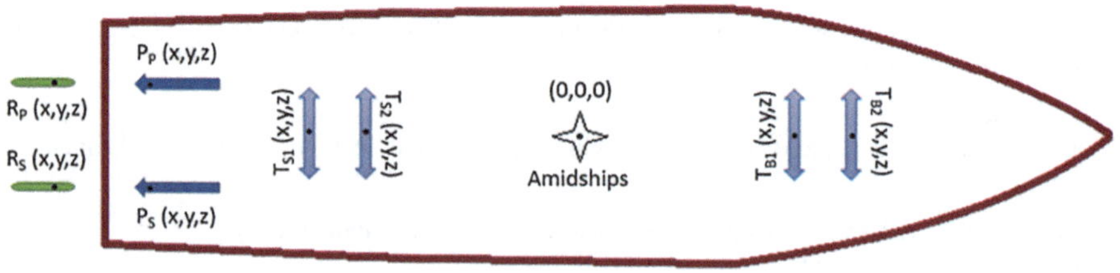


**Fig. 1** The actuators and their positions along the ship

**Table 2** Names, symbols, and properties of the actuators

| No. | Type | Symbol | Name | Properties |
|---|---|---|---|---|
| 1 | Rudder | $R_P$ | Port rudder | Profile area: $A_R = 47.5\,\text{m}^2$ |
| | | | | Position: $x_R = -0.5L$ |
| | | | | Position: $y_R = -0.4B/L$ |
| 2 | | $R_S$ | Starboard rudder | Profile area: $A_R = 47.5\,\text{m}^2$ |
| | | | | Position: $x_R = -0.5L$ |
| | | | | Position: $y_R = 0.4B/L$ |
| 3 | Propeller | $P_P$ | Port propeller | Diameter: $D_P = 8.9\,\text{m}$ |
| | | | | Position: $x_P = -0.48L$ |
| | | | | Position: $y_P = -0.4B/L$ |
| 4 | | $P_S$ | Starboard propeller | Diameter: $D_P = 8.9\,\text{m}$ |
| | | | | Position: $x_P = -0.48L$ |
| | | | | Position: $y_P = 0.4B/L$ |
| 5 | Thruster | $T_{S1}$ | Rear stern thruster | Diameter: $D_P = 4.8\,\text{m}$ |
| | | | | Position: $x_P = -0.35L$ |
| 6 | | $T_{S2}$ | Front stern thruster | Diameter: $D_P = 4.8\,\text{m}$ |
| | | | | Position: $x_P = -0.3L$ |
| 7 | | $T_{B1}$ | Rear bow thruster | Diameter: $D_P = 4.8\,\text{m}$ |
| | | | | Position: $x_P = 0.3L$ |
| 8 | | $T_{B2}$ | Front bow thruster | Diameter: $D_P = 4.8\,\text{m}$ |
| | | | | Position: $x_P = 0.35L$ |

expanded area ratio of $A_E/A_0 = 0.8$. Using our in-house code [3], we digitalized the database and defined the propellers and thrusters by a second-order polynomial, which reads as $K_T = k_0 + k_1 J + k_2 J^2$. Here, the parameters are calculated by the in-house code as $k_{0,P} = 0.484$, $k_{1,P} = -0.330$ for the propellers, and $k_{0,T} = 0.513$, $k_{1,T} = -0.419$, and $k_{2,T} = -0.098$ for the thrusters.

The motion of the ship model is represented by nonlinear hydrodynamic coefficients defined by the three-degree-of-freedom MMG maneuvering mathematical model [5]. In the MMG model, $R_0$ represents the straight-ahead forward motion, where $u > 0$. However, in this chapter, the ship can also move astern. In this case, this hydrodynamic coefficient will get a negative sign. The coefficients to be used in the mathematical model are used to calculate the ship's response, which is comprehensively explained in the following section.

## 3 Mathematical Model

In this chapter, we utilize the modular mathematical approach of the MMG maneuvering model [5] to solve the equations of motion in the horizontal plane. This approach makes use of the superposition principle to add forces and moments acting on each component of the ship. The ship motions are reduced to solely include the motions in the horizontal plane, leading to three-degree-of-freedom in this model. The equations of motion to be solved are given below:

$$\begin{aligned}(m+m_x)\dot{u}-(m+m_y)vr-mx_Gr^2&=X_H+X_P+X_R+X_W\\(m+m_y)\dot{v}+(m+m_x)ur+mx_G\dot{r}&=Y_H+Y_T+Y_R+Y_W\\(I_z+J_z+mx_G^2)\dot{r}+mx_G(\dot{v}+ur)&=N_H+N_P+N_T+N_R+N_W.\end{aligned}\tag{1}$$

Here, the subscripts $H$, $P$, $T$, $R$, and $W$ refer to the hull, propeller, thruster, rudder, and wave, respectively. The left-hand sides of these equations contain the masses, including the added masses (which are assumed to be constant), and time-varying velocities and accelerations. The right-hand sides, on the other hand, include the forces and moments acting on (or created by) the ship and its actuators. The terms of the right-hand sides have different modeling aspects, which are given in the following subsections. A similar mathematical model was used without thruster and wave modeling in [6].

## 3.1 Ship Modeling

The ship-fixed CS is chosen in the traditional manner, with the $z$-axis pointing downward and the $y$-axis pointing to the starboard side. The origin is located amidships, as depicted in Fig. 1. Consequently, the mathematical model includes the center of gravity as a parameter. Any changes in the center of gravity during the design stage can be easily incorporated into the model. The ship is represented by hydrodynamic coefficients, given in [4], in accordance with the MMG maneuvering model. The details of the model can be found in [5], in which they calculate the forces and moments acting on the body by

$$\begin{aligned}X_H&=-R_0(u)+X_{vv}v^2+X_{vr}vr+X_{rr}r^2+X_{vvvv}v^4\\Y_H&=Y_vv+Y_rr+Y_{vvv}v^3+Y_{vvr}v^2r+Y_{vrr}vr^2+Y_{rrr}r^3\\N_H&=N_vv+N_rr+N_{vvv}v^3+N_{vvr}v^2r+N_{vrr}vr^2+N_{rrr}r^3.\end{aligned}\tag{2}$$

The hull forces are expressed by the constant hydrodynamic coefficients at the right-hand sides of Eq. 2, except the term related to the surge motion, $R_0$. In the original MMG model, the ship is only moving forward. In this chapter, we have modified the model so that the ship can also move astern. Therefore, $R_0$ is defined as a variable that can change depending on the ship's direction, and has two different values, as positive for $u > 0$ and negative for $u < 0$.

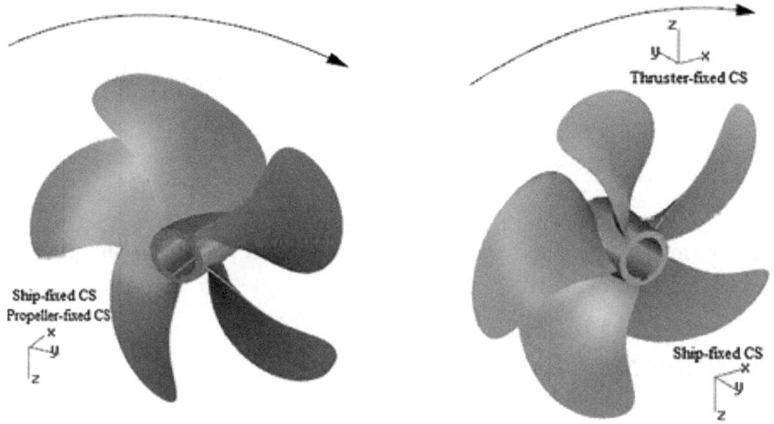


**Fig. 2** Propeller-fixed (left) and thruster-fixed (right) coordinate systems

## *3.2 Propeller Modeling*

The propeller-fixed CS is similar to the ship-fixed CS, where the $z$-axis points downward. The origin is defined at the center of the propeller disk. Considering the direction of the $x$-axis, propeller rotation is accepted to be positive in the clockwise direction. The propeller-fixed CS is shown in Fig. 2 (left).

In this model, we assume that the propeller only produces surge force $X_P$. The side force generated by the propeller is ignored; therefore, $Y_P = 0$. The surge force $X_P$ is equal to the propeller thrust $T$, which we express by

$$X_P = K_T \rho n_P |n_P| {D_P}^4. \tag{3}$$

Here, the second propeller rotation rate is given in an absolute value sign because we want to keep the direction of rotation. The surge force creates a yaw moment on the ship. This moment is found by multiplying $X_P$ by the transversal distance of the propeller from amidships $y_P$:

$$N_P = y_P \cdot X_P. \tag{4}$$

Here, $K_T$ is the thrust coefficient, $n_P$ is the propeller rotation rate, and $D_P$ is the propeller diameter. The thrust coefficient $K_T$ can be found from the open-water propeller performance graph, in which it is defined as a function of the advance coefficient $J_P$. It reads as

$$K_T = k_0 + k_1 J_P + k_2 J_P |J_P|. \tag{5}$$

The advance coefficient $J_P$ is a nondimensional number that normally accounts for the interactions between the ship and the propeller in self-propulsion calculations. However, in this model, we assume that the propeller receives uniform flow. In this case, we may define $J_P$ by

$$J_P = \frac{u}{n_P D_P}. \tag{6}$$

In this equation, $u$ is the ship's velocity due to the uniform flow assumption.

### 3.3 Thruster Modeling

The thruster model is very similar to the propeller model. The origin is defined at the center of the thruster disk. Thrusters generate thrust along the $y$-axis, and considering its direction, thruster rotation is accepted to be positive in the counter-clockwise direction. The thruster-fixed CS is shown in Fig. 2 (right).

In this model, we assume that the thruster only produces sway force $Y_T$. The surge force generated by the thruster is ignored; therefore, $X_T = 0$. The sway force $Y_T$ is equal to the thruster thrust $T$, which we express by

$$Y_T = K_T \rho n_T |n_T| {D_T}^4. \tag{7}$$

Here, the second thruster rotation rate is given in an absolute value sign because we want to keep the direction of rotation. The surge force creates a yaw moment on the ship. This moment is found by multiplying $Y_T$ by the longitudinal distance of the thruster from amidships, $x_T$:

$$N_T = x_T \cdot Y_T. \tag{8}$$

Here, $K_T$ is the thrust coefficient, $n_T$ is the thruster rotation rate, and $D_T$ is the thruster diameter. The thrust coefficient $K_T$ can be found from the open-water thruster performance graph, in which it is defined as a function of the advance coefficient $J_T$. It reads as

$$K_T = k_0 + k_1 J_T + k_2 J_T |J_T|. \tag{9}$$

The advance coefficient $J_T$ is a nondimensional number, and we define $J_T$ by

$$J_T = \frac{v}{n_T D_T}. \tag{10}$$

In this equation, $v$ is the ship's sway velocity.

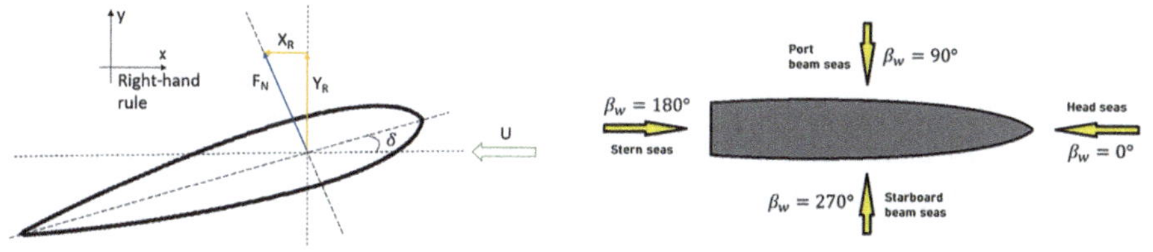


**Fig. 3** Rudder-fixed coordinate system (left). Wave directions (right)

## 3.4 Rudder Modeling

The rudder-fixed CS is the reverse of the ship-fixed CS where the $z$-axis points upward from the rudder rod. When the right-hand rule applies, the rudder angle $\delta$ rotates positively in the counter-clockwise direction. In the modular mathematical model, we are interested in finding the rudder forces $X_R$ and $Y_R$, and moment $N_R$. Here, we have to remember that we want to calculate these forces and moments in the ship-fixed coordinate system. When the rudder is rotated at an angle of $\delta$ while receiving the flow from the $-x$ direction, it creates a normal force, depicted as $F_N$ in Fig. 3. The components of this normal force are the rudder forces $X_R$ and $Y_R$, which we are interested in.

Due to small propeller rotations in the dynamic positioning system, we ignore the interactions between the propeller and the rudder. In this case, we may write

$$X_R = -F_N \sin(\delta) \tag{11}$$

$$Y_R = -F_N \cos(\delta). \tag{12}$$

$Y_R$ is negative because in the ship-fixed coordinate system, the $y$-axis is positive in the starboard direction. The rudder will additionally create a yaw moment due to its distance from the amidships. Considering that the longitudinal and transversal positions of the rudder are denoted by $x_R$ and $y_R$, respectively, the rudder yaw moment becomes

$$N_R = -(x_R \cdot Y_R + y_R \cdot X_R). \tag{13}$$

Here, $N_R$ is negative because the positive rotation is defined in the clockwise direction in the ship-fixed coordinate system. To be able to calculate $X_R$, $Y_R$, and $N_R$, we need to find the rudder normal force, $F_N$. If we assume that the rudder has a slender form and the rotation angles are small, we may utilize the flat-plate approximation, which gives the lift coefficient as $C_L = 2\pi \cdot \delta$. The rudder normal force $F_N$ can be calculated using the Morison equation:

$$F_N = \frac{1}{2}\rho A_R U^2 C_L = \rho A_R U^2 \pi \delta. \tag{14}$$

Here, $A_R$ is the movable profile area of the rudder.

### 3.5 Wave Modeling

In a seaway, external disturbances can include waves, winds, and currents. In this chapter, we focus exclusively on modeling waves, using the empirical relationships established by [1] for irregular seas, while neglecting other disturbances. Waves exert two types of forces on ships. The first type, first-order oscillatory forces, fluctuates with the wave encounter frequency and varies over time. The second type, second-order drift forces, is constant. In this chapter, we focus solely on the wave drift forces to model irregular seas as given by the relations in [1]:

These equations are given for zero ship speed. Considering that DPS works at low ship speeds, this is considered to be a fair assumption. The equations are modified for infinite wave period and water depth, as compared to the original version. They are added to the right-hand sides of the maneuvering equations, as given in Eq. 1. Interactions between the waves and the ship are neglected. In the model, $\beta_w = 0°$ refers to the head seas, while $\beta_w = 180°$ refers to the stern seas. $\beta_w = 90°$ and $\beta_w = 270°$ are the port and starboard side beam seas, respectively. These are shown schematically in Fig. 3.

## 4 Results

This section is devoted to the results obtained from the simulation tool. The different models in the tool are tested one by one to investigate its consistency. First, the actuators (propellers, rudders, and thrusters) are tested. Then the ship's response is simulated in irregular seas.

### 4.1 Test Results of the Propeller Model

In this part, we will check if the propeller model works correctly. It should be noted here that this is not a validation, as we have made several assumptions that might alter the model's accuracy. However, the model should be consistent. The simulation tool is tested under the following conditions:

- Rudders are at their neutral positions, $\delta_p = \delta_s = 0$.
- Thrusters are not working.

Different cases are tested for the propellers in the simulation tool. Conditions of the propellers and the obtained results are given below:

- When both propellers are given $n_{Pp} = n_{Ps} = 0.6$ rps, the ship is moving straight ahead at a speed of $U = 5.77$ knots.

- When the port propeller is not working $n_{Pp} = 0$ and the starboard propeller is given $n_{Ps} = 0.6$ rps, the ship is moving port, the steady turning diameter is STD = 4.29, and the ship speed is $U = 3.13$ knots.
- When the starboard propeller is not working $n_{Ps} = 0$ and the port propeller is given $n_{Pp} = 0.6$ rps, the ship is moving starboard, the steady turning diameter is STD = 4.29, and the ship speed is $U = 3.13$ knots.

The obtained results seem plausible due to the following:

- The ship moves straight ahead when both propellers are given the same rotation rate. This is expected as there is nothing to break the symmetry.
- The ship turns to the side at which the propeller is not working. The static state of the case dictates the ship to move to the side of the working actuator.
- The steady turning diameters are equal on both sides. This is expected due to the symmetry in the model.

We have tested forward motion with positive propeller rotation rate. When the propeller works reverse, the ship moves astern. The consistency of the model is additionally tested with negative propeller rotation rates:

- When both propellers are given $n_{Pp} = n_{Ps} = -0.6$ rps, the ship is moving straight astern at a speed of $U = -5.77$ knots.

The simulation does not distinguish the difference between the forward and astern motions and assumes that the ship resistances to both motions are equal. The tool also assumes similar propeller performance in reversed rotation (when moving astern). Therefore, it returns the same ship speed as in the first case, with the difference of moving backward (the negative sign implies this).

## 4.2 *Test Results of the Thruster Model*

In this part, we will check if the thruster model works correctly. It should be noted here that this is not a validation; however, the model should be consistent. The simulation tool is tested in the following conditions:

- Rudders are at their neutral positions, $\delta_p = \delta_s = 0$.
- Propellers are not working.
- The initial surge speed is $U = 1$ knot.

Different cases are tested for the thrusters in the simulation tool. Conditions of the thrusters and the obtained results are given below:

- When stern thrusters are given $n_{Ts1} = n_{Ts2} = 0.6$ rps and the bow thrusters are not working $n_{Tb1} = n_{Tb2} = 0$, the ship is turning starboard and forward. The advance is $Ad = 1.34$, transfer $Tr = -0.11$, steady turning diameter STD = 1.72, and ship speed $U = 0.48$ knots.

- When the rear stern thruster is given $n_{Ts1} = 0.6$ rps and the other thrusters are not working $n_{Ts2} = n_{Tb1} = n_{Tb2} = 0$, the ship is turning starboard and forward. The advance is $Ad = 1.57$, transfer $Tr = -0.27$, steady turning diameter STD = 1.55, and ship speed $U = 0.32$ knots.
- When the front stern thruster is given $n_{Ts2} = 0.6$ rps and the other thrusters are not working $n_{Ts1} = n_{Tb1} = n_{Tb2} = 0$, the ship is turning starboard and forward. The advance is $Ad = 1.81$, transfer $Tr = -0.31$, steady turning diameter STD = 1.89, and ship speed $U = 0.36$ knots.
- When bow thrusters are given $n_{Tb1} = n_{Tb2} = 0.6$ rps and the stern thrusters are not working $n_{Ts1} = n_{Ts2} = 0$, the ship is turning port and astern. The advance is $Ad = 1.21$, transfer $Tr = 2.29$, steady turning diameter STD = 14.17, and ship speed $U = 1.03$ knots.
- When the rear bow thruster is given $n_{Tb1} = 0.6$ rps and the other thrusters are not working $n_{Tb2} = n_{Ts1} = n_{Ts2} = 0$, the ship is turning port and astern. The advance is $Ad = 0.92$, transfer $Tr = 6.05$, steady turning diameter STD = 21.19, and ship speed $U = 0.70$ knots.
- When the front bow thruster is given $n_{Tb2} = 0.6$ rps and the other thrusters are not working $n_{Tb1} = n_{Ts1} = n_{Ts2} = 0$, the ship is turning port and forward. The maneuvering indices are transient in the first two turnings and become time independent at the third.

The obtained results seem logical due to the following:

- The ship turns starboard when stern thrusters have a positive rotation rate, at which the thruster generates a thrust in the port direction. Stern thrusters are behind amidships, creating a moment to turn the ship to starboard.
- The ship moves forward when only stern thrusters are working and moves astern when only front thrusters are working. This is because stern thrusters are positioned at the back and the front thrusters in front of amidships.
- The ship turns faster when both aft or both front thrusters are working (compared to solo aft or solo front thruster).
- The ship turns faster when the farthest thruster from amidships is working.
- The headings of the ships are consistent with the motion they are conducting.

The thruster model may further be tested as there may be many created scenarios; however, the results of the tests seem consistent.

## 4.3 Test Results of the Rudder Model

In this section, we check if the rudder model works correctly and expect consistent results. Here, our aim is not validating the model. We test the simulation tool under the following conditions:

- Both propellers are rotating at $n_{Pp} = n_{Ps} = 0.6$ rps.
- Thrusters are not working.

Then we tried different cases for the rudder. Conditions of the rudders and the obtained results are outlined briefly:

- When both rudders are given $\delta_p = \delta_s = 35^\circ$, the steady turning diameter is STD = 2.75, and the ship is turning starboard with $U = 2.88$ knots.
- When the port rudder is at its neutral position $\delta_p = 0$ and the starboard rudder is given $\delta_s = 35^\circ$, the steady turning diameter is STD = 3.24, and the ship is turning starboard at a velocity of $U = 3.29$ knots.
- When the starboard rudder is at its neutral position $\delta_s = 0$ and the port rudder is given $\delta_p = 35^\circ$, the steady turning diameter is STD = 3.41, and the ship is turning starboard at a velocity of $U = 3.34$ knots.

The results seem consistent due to the following:

- The ship turns starboard when the rudder angle is positive.
- The steady turning diameter increases when one of the rudders is neutral.
- The steady turning diameter is lower in case 2 compared to case 3. This is expected because the inner rudder has a higher impact in turning.

## 4.4 Test Results of the Wave Model

The wave model is checked via turning self-propulsion simulations. We have used [1] given to represent ship motions in irregular waves. The simulation tool is tested to understand the wave model's consistency under the following conditions:

- Both propellers are rotating at $n_{Pp} = n_{Ps} = 0.6$ rps.
- Thrusters are not working.
- Rudders are at their neutral positions, $\delta_p = \delta_s = 0$.
- Significant wave height is taken as $H_{1/3} = \frac{T}{6}$.

The wave angle and direction are changed, and the ship's response is noted down below under different cases:

- When the wave angle is $\beta_w = 0^\circ$ for the head seas condition, the ship is moving straight ahead at a speed of $U = 5.48$ knots.
- When the wave angle is $\beta_w = 180^\circ$ for the stern seas condition, the ship is moving straight ahead at a speed of $U = 6.06$ knots.

The results are plausible because the ship is sailing at a speed of $U = 5.77$ knots in the calm water case (see Sect. 4.1), slowing down in head seas and accelerating in stern seas. These are parallel to our intuitions.

## 5 Conclusion

In this chapter, we present a simplified maneuvering model as a potential foundation for a feedforward control algorithm in the dynamic positioning of ships. The mathematical model is modular in structure, separately accounting for the hull, propellers, rudders, and thrusters. External disturbances are limited to wave effects, with empirical relations used to calculate the forces and moments acting on the vessel in irregular seas. Interactions between the ship, its actuators, and the environment are disregarded. Each actuator is tested individually to ensure consistency in the results, which proved satisfactory. Additionally, the model successfully captures wave drift due to external disturbances. The results are promising and demonstrate that the model is a robust tool for simulating the small motions of ships, making it suitable for developing control algorithms for dynamic positioning.

**Acknowledgments** This study was supported by Turkish Petroleum—Offshore Technology Center.

## References

1. Ankudinov, V.: Ship manoeuvring simulation model including regimes of slow speeds and large drift angles. In: Report for the First International Maritime Simulation Symposium (Munich) (1985)
2. Oosterveld, M.W.C., van Oossanen, P.: Further computer-analyzed data of the Wageningen B-screw series. Int. Shipbuild. Prog. **22**(251), 251–262 (1975)
3. Ozturk, D., Delen, C., Belhenniche, S.E., Kinaci, O.K.: The effect of propeller pitch on ship propulsion. Trans. Maritime Sci. **11**(01), 133–155 (2022)
4. Sukas, O.F., Kinaci, O.K., Bal, S.: Asymmetric ship maneuvering due to twisted rudder using system-based and direct CFD approaches. Appl. Ocean Res. **108**, 102529 (2021)
5. Yasukawa, H., Yoshimura, Y.: Introduction of MMG standard method for ship maneuvering predictions. J. Marine Sci. Technol. **20**, 37–52 (2015)
6. Yigit, Y., Nguyen, L.D., Ozdem, M., Kinaci, O.K., Hoang, T., Canberk, B., Duong, T.Q. Twinport: 5g drone-assisted data collection with digital twin for smart seaports. Sci. Rep. **13**(1), 12310 (2023)

# Fundamental Study on Forecasting Surface Ship Maneuvering Characteristics for Autonomous Docking Operations

Hasan Timurlek, Mahmutcan Esenkalan, and Taner Cosgun

## 1 Introduction

The development of autopilot systems and docking mechanisms for autonomous marine vehicles is a rapidly evolving field for unmanned marine technology. These systems are crucial for extending the operational capabilities of autonomous underwater vehicles (AUVs) and autonomous surface vehicles (ASVs) by enabling battery recharging, data transfer, and maintenance without the need for frequent surface support.

Autonomous navigation and docking systems have been a subject of significant research and development over the past few decades to explore various approaches to improve efficiency, accuracy, and safety. For instance, artificial neural networks (ANNs) have been employed to develop efficient automatic ship docking systems that can handle environmental disturbances [1]. This approach demonstrates the potential of machine learning techniques in simulating complex berthing scenarios. A system based on Linear Quadratic Regulator (LQR) and Covariance Matrix Adaptation Evolution Strategy (CMA-ES) has been proposed to address challenges such as path planning, optimal control, and adaptive berthing strategies [2]. Such hybrid approaches aim to create more robust and adaptable simulation models. Liu et al. applied optimization techniques to develop optimal berthing trajectories based on ship dynamics, with the goal of minimizing energy consumption [3]. The development of autonomous docking systems for marine vehicles is a multifaceted challenge that involves various aspects such as guidance, control, sensor integration, and mechanical design. Recent advancements, including neural-network-based guidance [4], integrated guidance and control strategies [5], and manipulator-

H. Timurlek · M. Esenkalan · T. Cosgun (✉)
Yildiz Technical University, Istanbul, Turkey
e-mail: tcosgun@yildiz.edu.tr

 
B. U. Töreyin et al. (eds.), *8th EAI International Conference on Robotic Sensor Networks*, EAI/Springer Innovations in Communication and Computing,
https://doi.org/10.1007/978-3-031-92143-8_18

assisted docking [6], demonstrate the ongoing efforts to improve the reliability and efficiency of these systems.

Numerical modeling plays a crucial role in the development and optimization of autonomous docking systems for marine vehicles. It allows researchers to simulate complex scenarios, test control algorithms, and evaluate system performance under various conditions. Martinsen et al. [7] proposes a method using numerical optimal control to frame the docking task as a dynamic positioning problem with spatial constraints for collision avoidance [7]. Yazdani et al. [8] introduced a real- time quasi-optimal trajectory generator based on the inverse dynamics in the virtual domain (IDVD) method, which considers the vehicle's hydrodynamic model and optimizes for mission time and energy expenditure [8]. Some researchers have explored novel methods for offline black-box modeling of ship maneuvering using random maneuvers and advanced machine learning techniques [9]. These approaches aim to improve the accuracy and generalization ability of ship maneuvering models used in simulations. Skulstad et al. [10] proposed a hybrid approach for the motion prediction for ship docking by integrating a supervised machine learning model of the ship into the ship dynamic model. Moreno and Tannuri [11] developed a methodology to reduce large sets of environmental conditions to smaller, representative sets for ship maneuvering simulations. This approach highlights the importance of efficient simulation techniques in capturing the complexity of real-world berthing scenarios.

Docking maneuvers, in particular, are complex operations that require precise control to ensure safe and efficient berthing, and they are becoming increasingly relevant in the context of autonomous shipping technologies. Over the years, various modeling approaches have been developed to simulate ship motion and maneuvering, including empirical methods, potential flow theory, and Computational Fluid Dynamics (CFD) simulations. CFD, in particular, has gained popularity for its ability to capture the intricate details of fluid-structure interactions and provide high-fidelity results. However, CFD simulations are often time-consuming and computationally expensive, which limits their practical application in real-time scenarios such as maneuvering forecasting during autonomous docking operations. Accurate forecasting a ship's velocity and position during its route is indeed crucial for the docking of autonomous or unmanned marine vehicles. It enables precise maneuvering, helps avoid collisions, and ensures smooth berthing. As demonstrated in Andrade et al. [12], advanced technologies like video analytics and computer vision are being developed to enhance vessel velocity estimation and docking processes, further underlining the importance of accurate forecasting in autonomous marine operations. On the other hand, forecasting methods balance accuracy with computational efficiency to be used in real-time autopilots or providing data to train machine learning algorithms.

Building on this body of work, the present study proposes a system-based approach to modeling ship maneuvering, specifically focusing on fundamental motions during ship docking, which are straight-ahead motion, crash-ahead motion, turning and another harbor maneuver, kicking turn at slow speed (accelerating turning test). By introducing an in-house developed simulation code, we propose a practical approach that predicts ship hydrodynamic performance under different conditions by utilizing ship resistance test data and four-quadrant propeller open-water performance map. The proposed model is validated using experimental and numerical data from the literature, ensuring its accuracy and reliability. This approach not only offers a viable solution for engineers engaged in ship design and docking simulations but also contributes to the broader efforts of integrating high-fidelity maneuvering models into autonomous vessel systems, ultimately enhancing their reliability and performance.

## 2 Methodology

The method proposed in the study is based on estimating the exact position and speed of the ship during its operation by initially defining two performance values of the ship. Using the results of the ship's resistance tests and four-quadrant propeller performance curves, the performance of the ship under different motion conditions can be determined. The methodology used consists of two main steps: the surge module and the maneuvering module. The surge module can simulate all possible scenarios of the ship's surge motion during its straight movement. The maneuvering module, on the other hand, estimates the maneuvering performance by considering the rudder interactions of the ship.

The Duisburg Test Case (DTC) ship model was used as a test case in the study [13]. The ship was scaled at a ratio of 1/80 for the calculations. The main dimensions and geometry of the ship are given in Table 1. The represented methodology utilized in the present code requires the propeller open-water performance map and the ship's towed resistance data. The resistance and four-quadrant propeller hydrodynamics values used in this study are the values calculated in an earlier CFD study [14]. Figure 1 shows the resistance and four-quadrant propeller curves for DTC hull.

### *2.1 The Surge Module*

The surge module, which constitutes the first stage of the method used in the study, utilizes the mathematical framework explained in the previous study [15]. Using the

**Table 1** Dimensions of the DTC container ship

| Parameter | Dimension |
|---|---|
| Scale | 1/80 |
| Length, L | 4.4375 m |
| Breadth, B | 0.6375 m |
| Draught, T | 0.1813 m |
| Displacement, $\nabla$ | 0.3388 $m^3$ |

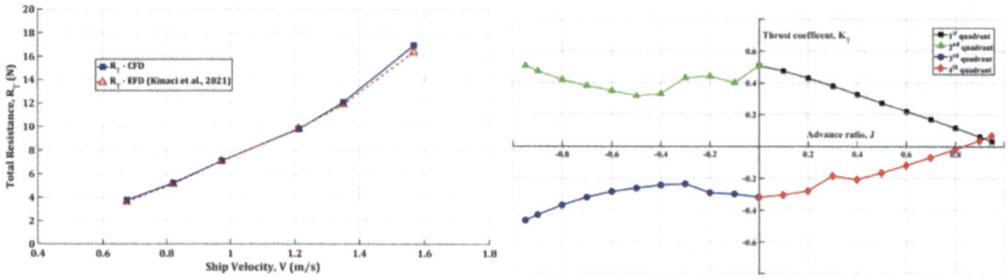


**Fig. 1** Towed resistance (left) and four-quadrant propeller performance curves (right) for DTC hull and propeller

ship resistance and open-water propeller test results as input, the method is capable of calculating all possible scenarios of surge motion: the straight- ahead, crash-ahead, crash-back, and astern motion cases. This section explains the theoretical background of surge module of the code.

A code is developed to calculate the ship's surge velocity during its straight-ahead course. The flow diagram of the code is given in Fig. 2. The simulation initiates from the user-provided initial speed and position. The time is calculated at each iteration using the time step size Δt, which is given as an input. The equation used to determine the velocity received by the propeller $V_A$ is:

$$V_A = V \cdot (1 - w) \tag{1}$$

where $w$ is the wake fraction, $V$ is the ship's speed. The advance coefficient is calculated by

$$J = \frac{V_A}{nD} \tag{2}$$

In this equation, $n$ is the propeller rotation rate and $D$ is the propeller diameter. Following this step, the primary segment proceeds to call the function 1, which encompasses propeller performance characteristics and the ship resistance.

After necessary parameters are extracted from the function 1, the primary segment calculates the thrust coefficient utilizing the equation

$$K_T = K_0 + K_1 \cdot \mathrm{J} + K_2 \cdot J^2 \tag{3}$$

and the thrust by

$$T = K_T \rho n^2 D^4 \tag{4}$$

The coefficients of ship resistance, extracted from function 1, is used to calculate the total resistance by the following equation:

$$R_T = r_1 \cdot V + r_2 \cdot V^2 \tag{5}$$

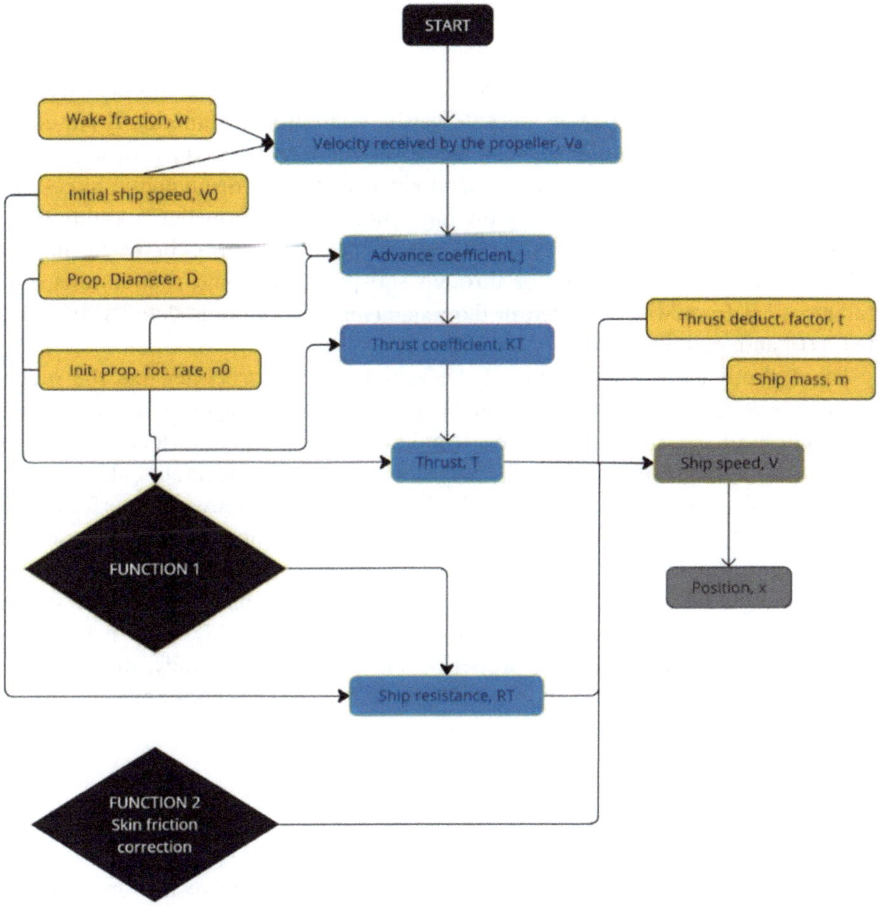


**Fig. 2** Flow diagram of the code for surge motion

If necessary, the code also calculates the skin friction correction arises from differences in Reynolds numbers between model and full-scale ships according to ITTC recommendations. After all the above is completed, the ship velocity can be defined in the form:

$$V_i = V_{i-1} + \frac{\Delta t}{m + m_x} \cdot \left( T \;+\; F_D \;+\; \frac{R_T}{1-t} \right) \tag{6}$$

Here, $t$ is the thrust deduction factor. In the present study, the propeller-hull interaction parameters ($w$ and $t$) were assumed to be constant in all calculations. Finally, the position of the ship x is calculated by

$$x_i = x_{i-1} + V_i \cdot \Delta t \tag{7}$$

## 2.2 The Maneuvering Module

The second phase of the method employed in this study involves the maneuvering module. Here, the ship's 3-DOF (surge, sway, yaw) motion is numerically simulated using a maneuvering module based on the MMG model proposed by Yasukawa and Yoshimura (2015). In essence, the model divides the total hydrodynamic forces into distinct components contributed by the hull, rudders, and propellers. Additionally, the interaction effects between the hull and rudders, as well as between the hull and propellers, are accounted for through self-propulsion and rudder tests. The mathematical framework utilized in the maneuvering module is detailed in Sukas et al. [16] and Yigit et al. [17].

The hydrodynamic coefficients of the vessel and the propeller and the rudder parameters were taken from Sukas et al. [18]. The maneuvering derivatives represent the variation of hydrodynamic forces or moments with respect to local velocity or acceleration and are of very small magnitudes. Therefore, accurate calculation of these derivatives is essential to precisely determine the maneuvering performance. Otherwise, the maneuvering performance of the vessel cannot be correctly assessed. The maneuvering derivatives used in this study are presented in Table 2.

It should also be noted that the mathematical model is based on Yasukawa and Yoshimura (2015), but the maneuvering model in their study does not include asymmetries associated with the propeller. Therefore, in this study, the effects of propeller asymmetry on lateral force and turning moment are considered.

# 3 Results

Numerical results obtained are presented in this section. The results are analyzed based on the fundamental maneuvers used during docking. The maneuvers examined include forward motion (straight-ahead), stopping (in other words, the forward rotation of the propeller) while moving astern (crash-ahead), turning motion with the rudder at a specific angle (turning circle), and an accelerating turning test, which is a maneuver used in harbor and is included among the ITTC maneuvering tests.

**Table 2** Hull derivatives in the study

| Parameter | Value | Parameter | Value | Parameter | Value |
|---|---|---|---|---|---|
| Xvv | −0.0732 | Yv | −0.1847 | Nv | −0.0961 |
| Xvvvv | 0.5248 | Yvvv | −1.5742 | Nvvv | −0.2019 |
| Xrr | −0.0119 | Yr | 0.0390 | Nr | −0.0360 |
| Xvr | −0.0348 | Yrrr | −0.0417 | Nrrr | −0.0431 |
| | | Yvrr | −0.7432 | Nvrr | −0.2219 |
| | | Yvvr | −1.052 | Nvvr | −0.7689 |

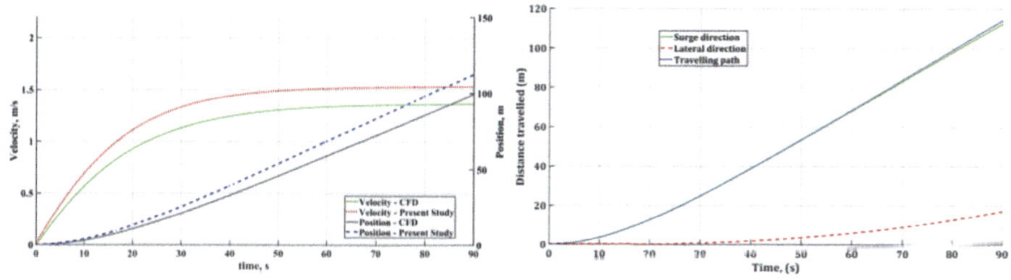


**Fig. 3** Numerical predictions in straight-ahead condition

The results for the straight-ahead motion are shown in Fig. 3. In the test conducted for this case, the initial speed of the ship was V = 0 m/s, and the propeller rotational speed was n = 20 rps. The results show a satisfactory agreement with the reference [14] used for comparison in terms of both speed and the linear distance covered by the ship. However, there is a slight difference in the final speed reached between the numerical predictions (1.52 m/s in the present study and 1.36 m/s in the reference). This difference in the constant final speeds achieved by the ship also leads to a slight variation in the total linear distance traveled.

Distance covered by the ship in the longitudinal and transverse directions during the straight-ahead maneuver can be seen in Fig. 3 (right). Although the ship moves primarily in the surge direction, the lateral force generated by the rudder and propeller causes a small amount of movement in the –y direction. This is a factor that should be considered in applications like docking, where the final position and path of the ship are crucial.

When the ship has a negative speed and the propeller rotates in the positive direction, creating positive thrust that first stops the ship and then forces it to move forward, this is called the crash-ahead maneuver. The results for the crash-ahead maneuver are shown in Fig. 4. In this case, the ship's initial speed was V = −0.6 m/s, and the propeller speed was n = 20 rps. Here, the stopping of a ship moving astern was simulated, so the calculations were terminated when the ship's speed reached zero. The present calculation results are very close to the reference values used for comparison. In the current simulation, it took 8.0 s for the speed of the ship moving astern to reach zero, while in the reference study, this value was 8.4 s. Due to the close match in speed values, the distances covered by the ship while moving astern are also very similar.

To analyze the turning motion of the ship at a specific rudder angle, the turning circle test was used. Here, the initial speed of the ship was V = 1.532 m/s, the propeller rotational speed was n = 20 rps, and the rudder angle was $\delta = 35^0$. In Fig. 4 (right), the trajectory results are presented in comparison with system-based calculations and free- running CFD results conducted by Sukas et al. [18]. It is observed that the ship follows a trajectory similar to that in the reference studies during its starboard turning motion. The maneuvering indices of the present turning test results compared to the literature are given in Table 3. The results show a good

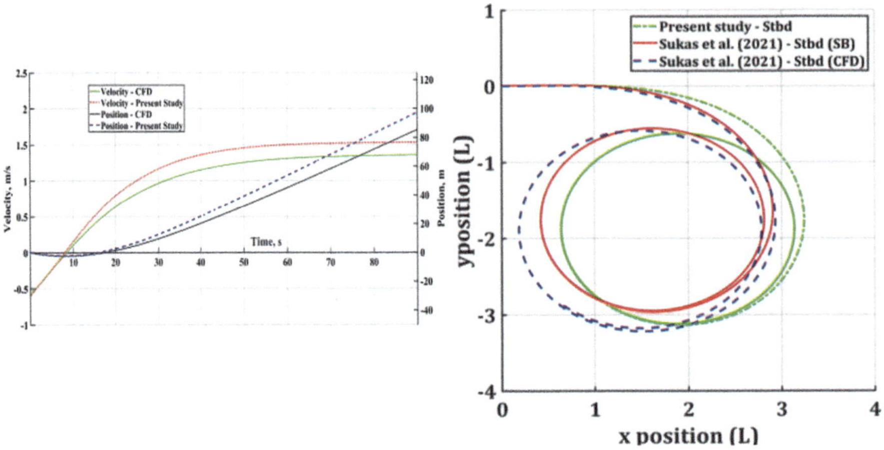


**Fig. 4** Numerical predictions for the ship at negative speed (left), turning circle test (right)

**Table 3** Maneuvering indices in turning test

| | Ad vance | Trans fer | Tactical Diameter | Steady Turning Diameter | Steady Yaw Rate | Steady Turn ing Speed |
|---|---|---|---|---|---|---|
| Kinaci and Ozturk [19] | 2.71 | 0.91 | 2.67 | 2.61 | 0.36 | 0.44 |
| Sukas et al. [18], SB | 2.80 | 1.09 | 2.87 | 2.39 | 0.36 | 0.39 |
| Sukas et al. [18] CFD | 2.83 | 1.21 | 3.12 | 2.60 | 0.34 | 0.43 |
| Present study | 3.47 | 1.18 | 3.03 | 2.49 | 0.31 | 0.39 |

level of agreement with the reference values for the turning diameters, yaw rate, and turning speed values. However, it is noted that the advance and transfer values of the ship in the present calculations are slightly higher, indicating that the ship traveled a bit further before establishing a steady turning motion.

The ability of the ship to make a "kicking" turn at slow speed, which is used in harbor maneuvers, is assessed with the accelerating turning test [20]. In this test, the ship and the propeller are initially stopped. Then, the motion is started at half-ahead ship speed and $\delta = 35°$. The turn continues until a 180-degree change of heading is completed. Here, the initial speed of the ship is $V = 0$ m/s, and the propeller speed is $n = 10.55$ rps. The trajectory of the ship during the accelerating turning test can be seen in Fig. 5. The ship continued its movement until it reached a 180-degree heading angle, at which point the simulation was terminated.

The heading angle and velocity of the ship during the accelerating turning test are shown in Fig. 6. The essential information that the ITTC requires to be obtained from the heading angle test is presented in Table 4. All the required values from this

**Fig. 5** Trajectory of the ship during accelerating turning

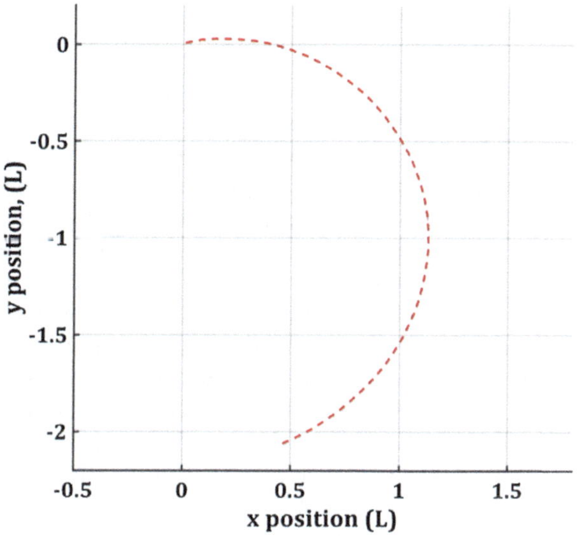


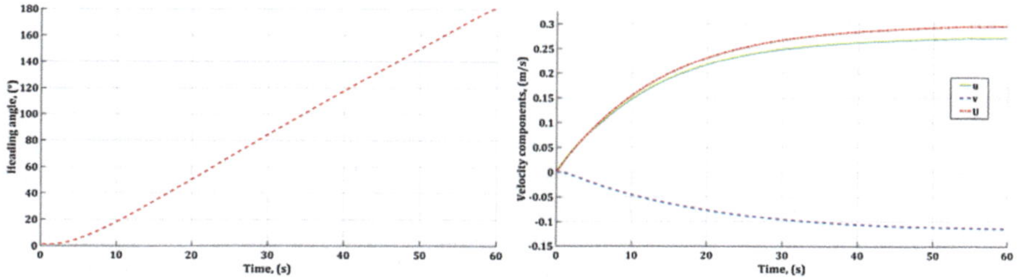


**Fig. 6** Heading angle (left) and velocity of the ship during accelerating turning (right)

**Table 4** Essential information obtained from accelerating turning test

| Parameter | Value |
|---|---|
| Advance | 1.056 L |
| Transfer | 0.612 L |
| Tactical diameter | 2.098 L |
| Final speed | 0.296 m/s |
| Time at $\psi = 90^{\circ}$ | 32.10 s |
| Time at $\psi = 180^{\circ}$ | 59.8 s |

test were predicted within reasonable ranges using the present method. However, as far as is known, there is no data in the literature for the DTC ship containing the results of this test, so no comparison could be made. When examining the heading angle results, it is observed that the ship quickly transitioned to steady turning. The time to change heading by 90 and 180 degrees was obtained as 32.10 s and 59.80 s, respectively. Additionally, the tactical diameter during the turn was found to be 2.098 ship lengths.

## 4 Conclusions

In this study, a method that provides quick results for predicting ship maneuvers is proposed. The method consists of surge and maneuvering modules and essentially uses the ship's resistance curve and open-water propeller test data as inputs to forecast the ship's speed and position under different conditions. The fundamental maneuvers performed by ships during docking were tested in this study, including straight-ahead motion, stopping of a ship in reverse speed (crash-ahead motion), turning, and another harbor maneuver: kicking turn at slow speed (accelerating turning test). The obtained results were compared with the existing data in the literature, and it was observed that the method provided results with a high degree of accuracy. The study concluded that the proposed method could calculate ship maneuvers quickly and reliably. Therefore, its use in systems that need to forecast ship performance in real time or to generate data for training machine learning algorithms managing autonomous ship movement/docking is deemed possible. In future work, it is planned to enhance and integrate the current method into the autonomously managed docking operation of an unmanned ship model.

## References

1. Shuai, Y., et al.: An efficient neural-network based approach to automatic ship docking. Ocean Eng. **191**, 106514 (2019). https://doi.org/10.1016/j.oceaneng.2019.106514
2. Chen, G., Yang, S., Yin, J.: Ship autonomous berthing simulation based on covariance matrix adaptation evolution strategy. J Mar Sci Eng. **11**(7), 1400 (2023). https://doi.org/10.3390/jmse11071400
3. Liu, X., Ren, J., Wang, H.: Optimal Energy Trajectory Planning and Control for Automatic Ship Berthing, vol. 35, pp. 1420–1425. IEEE (2020). https://doi.org/10.23919/ccc50068.2020.9188805
4. Wong, J., Fenton, R., Nejat, G., Benhabib, B.: Neural-Network-Based Docking of Autonomous Vehicles, pp. 1574–1579 (2006). https://doi.org/10.1109/itsc.2006.1707448
5. Batista, P., Silvestre, C., Oliveira, P.: A Two-Step Control Strategy for Docking of Autonomous Underwater Vehicles (2012). https://doi.org/10.1109/acc.2012.6314817
6. Chen, J.L., et al.: A novel autonomous docking method of unmanned marine vehicle based on manipulator. Appl. Mech. Mater. **364**, 370–374 (2013). https://doi.org/10.4028/www.scientific.net/AMM.364.370
7. Martinsen, A.B., Lekkas, A.M., Gros, S.: Autonomous docking using direct optimal control. IFAC-PapersOnLine. **52**(21), 97–102 (2019). https://doi.org/10.1016/j.ifacol.2019.12.290
8. Yazdani, A.M., Tang, Y., Mahmoudzadeh, S., Sammut, K., Yakimenko, O.A., Lammas, A.: Time and Energy Efficient Trajectory Generator for Autonomous Underwater Vehicle Docking Operations. IEEE (2016). https://doi.org/10.1109/oceans.2016.7761021
9. Zhang, Y.-Y., Wang, Z.-H., Zou, Z.-J.: Black-box modeling of ship maneuvering motion based on multi-output nu-support vector regression with random excitation signal. Ocean Eng. **257**, 111279 (2022). https://doi.org/10.1016/j.oceaneng.2022.111279
10. Skulstad, R., Li, G., Fossen, T., Vik, B., Zhang, H.: A hybrid approach to motion prediction for ship docking—integration of a neural network model into the ship dynamic model. IEEE Trans. Instrum. Meas. **1**, 1 (2020). https://doi.org/10.1109/TIM.2020.3018568

11. Moreno, F.M., Tannuri, E.A.: Clustering Applied to Large Sets of Environmental Conditions for Selecting Typical Scenarios for Ship Maneuvering Real-Time Simulations (2021). https://doi.org/10.1115/omae2021-62875
12. De Andrade, J.V.R., Cruz, F., Da Silva Filho, N.M., Fernandes, B.J.T., Izídio, A.R.L.C.: Vessel velocity estimation and docking analysis: a computer vision approach. Algorithms. **16**(7), 326 (2023). https://doi.org/10.3390/a16070326
13. el Moctar, O., Shigunov, V., Zorn, T.: Duisburg test case: post-Panamax container ship for benchmarking. Ship Technol. Res. **59**, 50–64 (2012). https://doi.org/10.1179/str.2012.59.3.004
14. Esenkalan, M.: 2024. A practical approach to compute ships' surge motion utilizing four-quadrant propeller hydrodynamic performance. MSc. Thesis. Yildiz Technical University
15. Cosgun, T., et al.: Four-quadrant propeller hydrodynamic performance mapping for improving ship motion predictions. Brodogradnja. **75**(3), 1 (2024)
16. Sukas, O.F., Kinaci, O.K., Bal, S.: Theoretical background and application of MANSIM for ship maneuvering simulations. Ocean Eng. **192**, 106239 (2019). https://doi.org/10.1016/j.oceaneng.2019.106239
17. Yigit, Y., Nguyen, L.D., Ozdem, M., Kinaci, O.K., Hoang, T., Canberk, B., Duong, T.Q.: Twinport: 5g drone-assisted data collection with digital twin for smart seaports. Sci. Rep. **13**(1), 12310 (2023)
18. Sukas, O.F., Kinaci, O.K., Bal, S.: Asymmetric ship maneuvering due to twisted rudder using system-based and direct CFD approaches. Appl. Ocean Res. **108**, 102529 (2021)
19. Kinaci, O.K., Ozturk, D.: Straight-ahead self-propulsion and turning maneuvers of DTC container ship with direct CFD simulations. Ocean Eng. **244**, 110381 (2022). https://doi.org/10.1016/j.oceaneng.2021.110381
20. ITTC. 7.5-04-02-01 Recommended Procedures and Guidelines: Full Scale Maneuvering Trials (2017)

# Prediction of Course-Keeping Control Coefficients of DARPA Suboff

**Duygu Ünlü and Cihad Delen**

## Nomenclature

| Term | Description | Units | Term | Description | Units |
|---|---|---|---|---|---|
| $A_E/A_0$ | Blade area ratio | – | AFF-8 | Autonomous free-floating target-8 | – |
| $\alpha$ | Proportional gain constant | – | $\beta$ | Integral gain constant | – |
| DARPA | Defense Advanced Research Projects Agency | – | $\delta$ | Rudder angle | Degrees |
| $D_{MAX}$ | Maximum diameter | m | ITTC | International Towing Tank Conference | – |
| $K$ | Nomoto gain constant | – | $K_d$ | Derivative gain | – |
| $K_i$ | Integral gain | – | $K_p$ | Proportional gain | – |
| $K_Q$ | Torque coefficient | – | $K_T$ | Thrust coefficient | – |
| $L_{OA}$ | Length overall | m | $L_{PP}$ | Length between perpendiculars | m |
| $n$ | Propeller rotation rate | rps | P/D | Pitch ratio | – |
| PI | Proportional-integral control | – | PID | Proportional-integral-derivative control | – |
| R | Resistance | N | $r$ | Yaw rate | rad/s |
| $\rho$ | Density | kg/m$^3$ | $S$ | Wetted surface area | m$^2$ |
| T | Thrust | N | $\lambda$ | Scale ratio | – |
| $\psi$ | Heading angle | Degrees | $\zeta$ | Damping ratio | – |
| $\omega_n$ | Natural frequency | Hertz | $z$ | Number of blades | – |

D. Ünlü (✉) · C. Delen
Department of Naval Architecture and Marine Engineering, Istanbul Technical University, Istanbul, Turkey

Marine Robotics Laboratory, Faculty of Naval Architecture and Ocean Engineering, Istanbul Technical University, Istanbul, Turkey
e-mail: unlu22@itu.edu.tr; cdelen@itu.edu.tr



B. U. Töreyin et al. (eds.), *8th EAI International Conference on Robotic Sensor Networks*, EAI/Springer Innovations in Communication and Computing,
https://doi.org/10.1007/978-3-031-92143-8_19

# 1 Introduction

The field of PID control in submarines, particularly concerning maneuverability and course-keeping strategies like zig-zag maneuvers, has garnered considerable scholarly interest in recent years. Various studies have contributed to this vital area through both experimental and numerical methodologies, revealing important insights into the challenges and advancements in this field. Moradi [23] laid the groundwork by addressing the complexities of multivariable system design for submarine depth control. This theme is echoed in Pan et al. [25] who investigated hydrodynamic coefficients of the Suboff submarine model using numerical simulations. Both studies affirm that understanding the forces acting on maneuvering submarines is critical for the tuning of PID variables, indicating a shared recognition of the importance of precise hydrodynamic parameters in achieving successful maneuverability. Le et al. [19] introduced the integration of fuzzy logic with PID control for ship autopilots, underscoring the adaptability of control systems in various environmental conditions. Fang et al. [9] expanded upon this concept by presenting a self-tuning PID controller utilizing neural networks aimed at reducing ship roll in random waves. While Fang et al. [9] primarily focused on surface vessels, their findings underscore the potential for adaptive control methods in submarine maneuvering, aligning with the adaptability emphasized by Le et al. [19]. This highlights a broader trend toward enhancing control system performance through intelligent, adaptive techniques. Gong et al. [11] further explored the theme of stability in submarine operations with their nonlinear PID controller designed to stabilize submarines during straight-line suspension underwater. Their focus on dynamic feedback control of rudder angles complements the work of [3] Carrica et al., who conducted experimental and numerical studies on zig-zag maneuvers of the KRISO container ship (KCS) in shallow water. Both studies demonstrate the necessity for advanced control strategies to improve maneuverability, emphasizing the precision required for executing complex maneuvers. The advancements in optimization techniques are highlighted in the work of Dlabač et al. [31], who designed autopilot PID controllers for ship course-keeping using particle swarm optimization (PSO). Ren et al. [26] demonstrated that CFD can effectively predict hydrodynamic performance, confirming that RANS simulations yield crucial insights into flow separation and performance. Both studies reflect a significant trend toward integrating advanced computational methods to refine PID control systems. Zhang et al. [32] examined the effects of environmental conditions on submarine stability during surfacing maneuvers, providing insights into vertical stability that complement the horizontal dynamics explored in earlier works. This holistic perspective on stability emphasizes the necessity for comprehensive PID control strategies that account for both horizontal and vertical forces during maneuvers. Lin et al. [21] studied the stability and maneuverability of a semiautonomous submarine model equipped with an autopilot system. Their findings reaffirm the effective use of PD controllers for maintaining course during critical maneuvers, reinforcing the necessity of advanced control methods in submarine operations, much like the insights provided by Kim et al. [17] in their empirical studies on zig-zag maneuvers. Liang et al. [20] proposed a multimodel switching strategy integrating PID control with machine learning algorithms, showing improved performance under varying conditions. Recent research by Jambak et al. [15] further demonstrates this trend by presenting a practical control-oriented ship model for autonomous vessels, addressing challenges in executing turning maneuvers. Delen and Kinaci [8] conducted direct CFD simulations of standard maneuvering tests for the DARPA Suboff, confirming that while the submarine exhibited high turning ability, it lacked course-keeping stability, a significant issue also explored through zig-zag and turning circle tests. This finding underscores the necessity of rigorous testing in PID tuning, particularly for maneuvers requiring quick adjustments. The connection between these findings and those of Kim et al. [17] highlights the collective emphasis on the importance of validation in refining control strategies. Hu et al. [13] explored autonomous orientation for AUVs using a light intensity gradient (LIG)-based navigation system.

The ongoing integration of advanced computational techniques, experimental validation, and adaptive control strategies highlights the critical importance of these methods for enhancing submarine maneuverability and course-keeping performance, particularly during complex maneuvers like zig-zag tests. In this chapter, a preliminary study was carried out to increase the course-keeping capabilities of a submarine form (DARPA Suboff) by starting from a few systematic maneuver tests in a fully submerged condition. For this purpose, a propulsion test was carried out in a free-running condition at $V = 2.75$ m/s. Support was taken from the PI control algorithm in determining the propulsion point (where the force balance is provided). Determining the self-propulsion point with the PI controller has made the study much faster and more cost-effective. Then, the available analysis was continued, and 10/10 zig-zag tests of the form were carried out. The determination of PID coefficients for course-keeping was carried out with the method developed by Nomoto [24]. This study significantly reduces the computational time required to determine the self-propulsion point and estimate the coefficients essential for course-keeping control. Furthermore, it distinguishes itself by applying the Nomoto method commonly utilized for surface vessels, to a submerged object, representing an innovative approach compared to existing research. The effectiveness of the coefficients in course-keeping capabilities will be examined in future studies.

## 2 Geometry and Test Cases

The DARPA Suboff model, defined as the benchmark form proposed by Groves [12], is a well-established design utilized for underwater vehicle research and open-source computational fluid dynamics (CFD) studies. The DARPA Suboff has been extensively used in various benchmarking cases, serving as a self-propulsion point for comparisons in hydrodynamic performance and control studies. The three-dimensional geometry of the submarine is shown in Fig. 1. Main particulars

**Fig. 1** The profile view of DARPA Suboff (AFF-8) with E1619 propeller

**Table 1** Main parameters of DARPA Suboff [12] and E1619 propeller

| Hull (AFF-8) | | Propeller (E1619) | |
|---|---|---|---|
| Parameters (unit) | Values | Parameters (unit) | Values |
| $\lambda$ (-) | 24.000 | D (m) | 0.262 |
| $L_{OA}$ (m) | 4.356 | P/D (-) | 1.15 |
| $L_{PP}$ (m) | 4.261 | Z (-) | 7 |
| $D_{MAX}$ (m) | 0.508 | $D_H$/D (-) | 0.226 |
| S ($m^2$) | 6.348 | $A_E/A_0$ (-) | 0.608 |
| $\Delta$ ($m^3$) | 0.706 | Rotation (-) | Clockwise |

**Table 2** Test matrix

| Cases | Rudder angle (degrees) | Velocity (m/s) | Degrees of freedom |
|---|---|---|---|
| Zig-zag maneuver | 10 | 2.75 | 3 (surge, sway, yaw) |
| Self-propulsion (straight ahead) | 0 | 2.75 | 1 (surge) |

of DARPA Suboff (AFF-8) and the general features of the E1619 propeller are given in Table 1. The study was conducted at a fundamental velocity of 2.75 m/s. Initially, the self-propulsion point was determined using PI control under straight-ahead conditions (considering only surge motion). Subsequently, in the subsequent simulations, the other two motions (sway and yaw) were activated, and 10/10 zig-zag maneuver tests were performed. Test cases are given in Table 2.

# 3 Numerical Methodology

## *3.1 Pyhsics*

The flow equations governing the submarine's behavior were discretized using the finite volume method (FVM). The unsteady Reynolds-averaged Navier-Stokes (RANS) equations were resolved using a segregated flow solver, employing an uncoupled approach that is particularly effective for incompressible flows. The solution procedure was governed by a SIMPLE-type algorithm, which efficiently manages the pressure-velocity coupling within the flow domain. In this research, the SST $k-\omega$ turbulence model was chosen for its robust performance in near-wall turbulence modeling, particularly in ship hydrodynamics simulations. The simulation time step has been adjusted according to ITTC guidelines [14]. For the

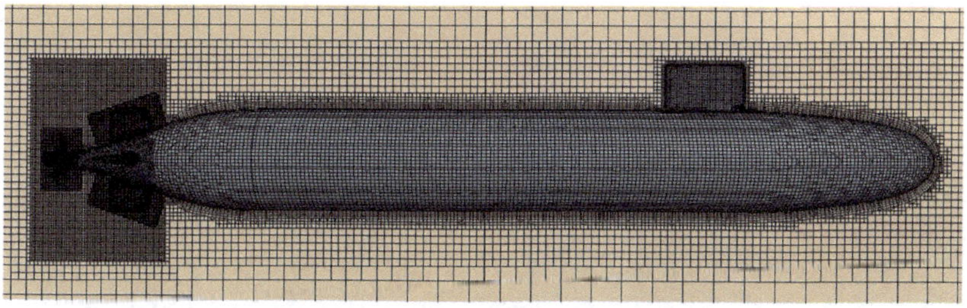

**Fig. 2** Grid structure around the DARPA Suboff

self-propulsion simulations, the propeller was modeled using the moving reference frame (MRF) method, which allows for steady rotational motion to be applied during the simulation. The propeller rotation rate (n) was adjusted to achieve a forward velocity of 2.75 m/s, matching the operational cruising speed of the submarine. In the maneuvering simulations, a zig-zag test was performed using a 10° rudder deflection. The rudder's motion was modeled using rigid body motion (RBM), allowing for rotational deflections during the maneuver.

## *3.2 Mesh Generation*

The primary grid structure consisted of triangular cells, which are well-suited for complex submarine geometries. A trimmed cell mesher was employed to produce the cell volumetric mesh, known for its high quality and minimal skewness, particularly in areas with complex geometries such as around the propeller and rudder [4]. Additional mesh refinement was applied in wake region. The grid structures of Darpa Suboff are shown in Fig. 2.

## *3.3 Boundary Conditions*

The computational domain size was meticulously defined according to the International Towing Tank Conference (ITTC) guidelines [14], ensuring that the flow characteristics were captured accurately while maintaining computational efficiency. Figure 3 illustrates the size of the computational domain. A Cartesian coordinate system was employed to define the mesh and boundary conditions.

Although the majority of the boundary surfaces were designated as inlet boundaries, no constant velocity was applied via the regions because the simulations involved self-propulsion and maneuvering tests. A pressure outlet boundary condition was applied at the downstream end of the domain, allowing the flow to leave with a specified pressure—typically atmospheric pressure—representing open-

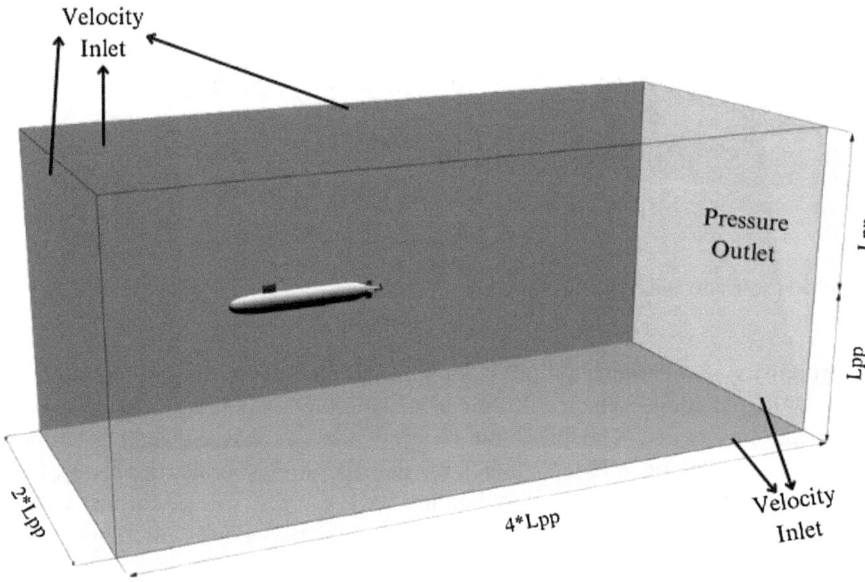


**Fig. 3** Computational domain with boundary conditions

water conditions. Additionally, a no-slip wall boundary condition was enforced on all solid surfaces, ensuring zero velocity at the walls. Figure 3 highlights the relevant boundary conditions applied in the simulations. To enable detailed analysis of the submarine's maneuvering behavior, separate regions were created for the rudders and propeller. The rotation domains for both the rudder and propeller were defined to accurately represent their dynamic interactions with the surrounding flow.

## 4 Straight-Ahead Maneuver with a PI Controller

The self-propelled point refers to the equilibrium condition at which the thrust produced by the propeller equals the hull resistance, and as a result, a constant forward speed is achieved. At this point, the net force acting on the submarine is zero. The block diagram of the controller is illustrated below in Fig. 4, demonstrating the interaction between key components of the control system. This includes the reference input, the controller itself, the plant, and the feedback loop. The controller receives the error signal, which is the difference between the desired setpoint and the actual output of the system. Based on this error, the controller computes an appropriate control signal to drive the plant, aiming to minimize the error over time. Feedback from the output is continuously provided to adjust the control actions, ensuring that the system maintains the desired performance criteria.

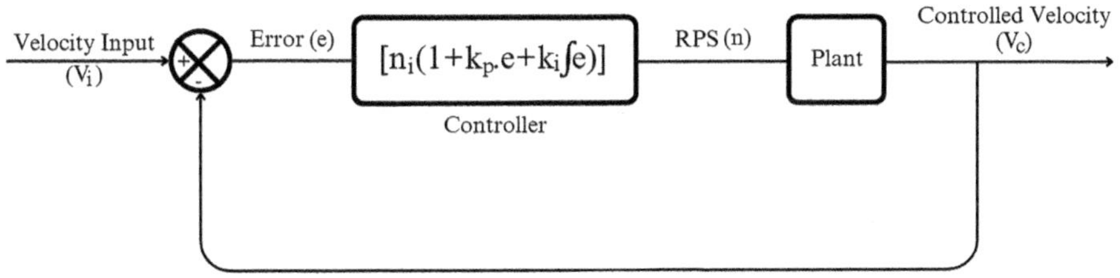


**Fig. 4** The block diagram for PI-driven self-propulsion test

A proportional-integral (PI) controller was chosen in this study due to its ability to eliminate steady-state errors by addressing immediate and accumulated discrepancies. This approach ensures a more accurate balance between thrust and resistance, enhancing stability and long-term performance under varying conditions. The parameters $K_p$ and $K_i$ required by the PI controller can be calculated using the following equations:

$$K_p = \frac{\alpha}{\rho L_{pp}^2 V^2} \tag{1}$$

$$K_i = \frac{\beta}{\rho L_{pp}^3 V^2}. \tag{2}$$

It is important to note that the values of $\alpha$ and $\beta$ do not necessarily need to be optimal; however, finding suitable values can help reduce computational costs [30]. The values $\alpha$ and $\beta$ selected as 100. The $K_p$ and $K_i$ parameters are calculated as $K_p = 7.005137 \times 10^{-4}$ and $K_i = 1.610376 \times 10^{-4}$.

The time-dependent variation of thrust and resistance during the self-propulsion simulation with a PI controller in Fig. 6. The graph illustrates how thrust gradually increases to match the required resistance, leading to the stabilization of the vessel's speed over time. Additionally in Fig. 5, the time-dependent error, velocity, and rotation rate variation during PI-driven self-propulsion simulation are depicted. The graph demonstrates that the error decreases over time, indicating that the system is gradually achieving the desired speed.

Under free-running conditions, the propeller rotation rate (n) required to achieve the target speed of 2.75 m/s was determined to be 9.315 rps, as controlled by the implemented controller. When compared with rotation rate values from existing studies at the same target speed, this result showed a high degree of compatibility. Specifically, while the mean rotation rate value presented in Table 3 is 9.335 rps, the estimated operating point in this study was calculated to be 9.315 rps. As evidenced by the data presented in Table 3, thrust and torque coefficients were computed $K_T = 0.240$ and $K_Q = 0.047$, respectively. It can be said that these results are consistent with both other studies and the average values. Moreover, Fig. 6

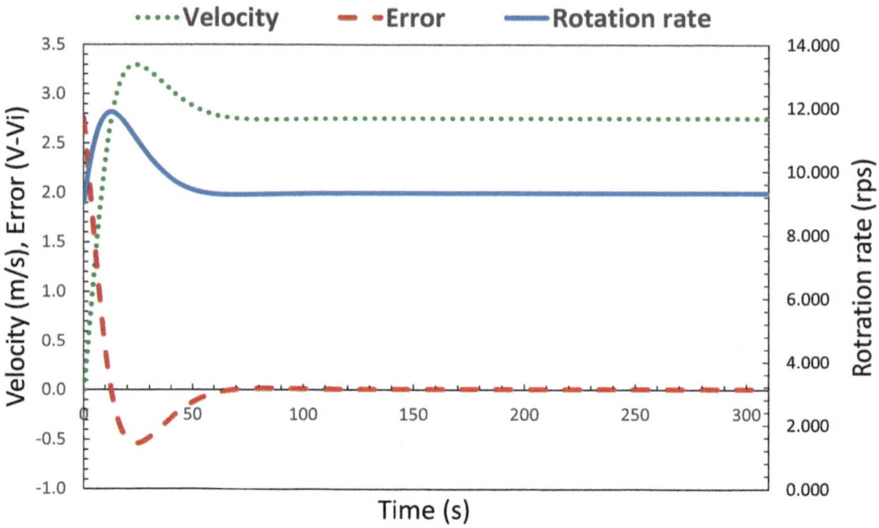


**Fig. 5** Time variation of propeller rotation rate, hull velocity, and velocity error in straight-ahead test conditions when PI-Drive is active

**Table 3** Comparison of $K_T$ and $K_Q$ values for self-propelled cases

| Self-propelled cases | Type | $n$ (rps) | $K_T$ | $K_Q$ |
|---|---|---|---|---|
| Present study | CFD | 9.315 | 0.240 | 0.046 |
| Chase and Carrica [5] | CFD | | 0.234 | 0.047 |
| Kinaci et al. [18] | CFD | | 0.231 | 0.046 |
| Sezen et al. [27] | CFD | 9.36 | 0.234 | 0.045 |
| Ozden et al. [1] | CFD | 9.258 | 0.242 | 0.046 |
| Ozden et al. [1] | EFD | 8.969 | 0.275 | 0.056 |
| Sezen et al. [28] | CFD | | 0.234 | 0.045 |
| Cosgun [7] | CFD | | 0.232 | 0.045 |
| Lungu [22] | CFD | 9.78 | 0.234 | 0.046 |
| Delen et al. [8] | CFD | 9.33 | 0.240 | 0.046 |
| Average | | 9.335 | 0.240 | 0.047 |

highlights how the PI controller adjusts the propeller rotation rate to ensure that the target velocity is maintained effectively.

## 5 Determination of PID Coefficients for Controlling the Course-Keeping Maneuver

PID controllers offer simplicity, reliability, and cost-effectiveness compared to modern control techniques like model predictive control (MPC) or AI-based systems.

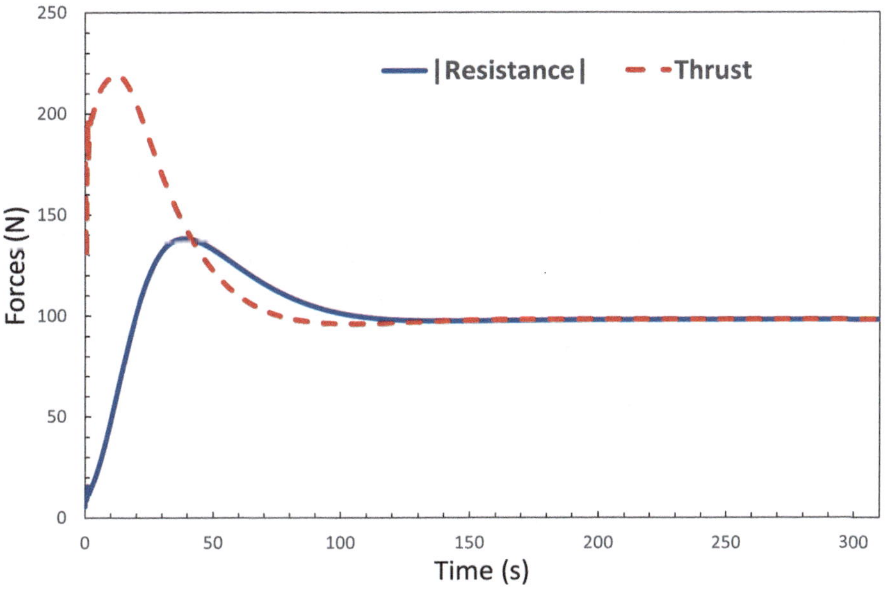


**Fig. 6** Change of forces (thrust (T) and resistance (R)) acting on the hull with PI controller

They are easy to implement and understand, require minimal computational power, and do not depend on detailed system modeling. While modern methods excel in predictive capabilities and complex scenarios, PID controllers remain the preferred choice for straightforward, feedback-driven systems due to their quick deployment and versatility. Due to these advantages, applications on PI and PID were carried out in this study.

## *5.1 Theoretical Background*

The Nomoto parameters $K$ and $T$ can be ascertained by examining data derived from ship maneuvering experiments or model assessments. Nomoto [24] and Journée and Pinkster [16] elucidated a technique for ascertaining $K$ and $T$ based on zig-zag maneuver data. This methodology is purported to yield more reliable outcomes compared to alternative approaches; however, it exhibits a degree of complexity. A more streamlined technique, attributed to Clarke [6], is delineated herein.

$$\dot{r} + \frac{1}{T}r = K\delta_0. \tag{3}$$

Upon integrating Equation 3 from $t_1$ and $t_2$, the following expression is derived:

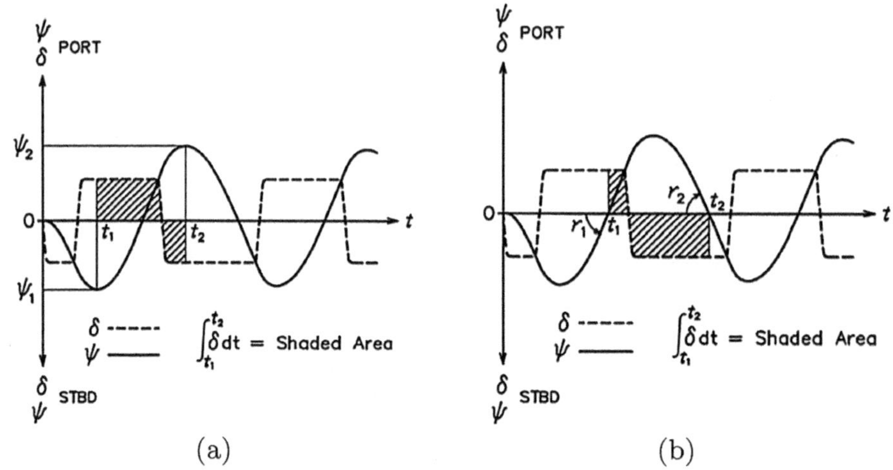


**Fig. 7** Determination of K and T parameters from a zig-zag maneuver [10]

$$T \int_{t_1}^{t_2} \dot{r}\, dt + \int_{t_1}^{t_2} r\, dt = T(r_2 - r_1) + (\psi_2 - \psi_1) = K \int_{t_1}^{t_2} \delta\, dt, \tag{4}$$

where the subscripts 1 and 2 refer to $t_1$ and $t_2$, respectively.

Subscripts 1 and 2 correspond to the time instants $t_1$ and $t_2$, respectively. If the time intervals $t_1$ and $t_2$ are selected such that $r_1 = r_2 = 0$, as illustrated in Fig. 7a, the value of $K$ can be expressed as

$$K = \frac{\psi_2 - \psi_1}{\int_{t_1}^{t_2} \delta\, dt}. \tag{5}$$

On the other hand, if $t_1$ and $t_2$ are chosen such that $\psi_1 = \psi_2 = 0$, Fig. 7b gives

$$\frac{K}{T} = \frac{r_2 - r_1}{\int_{t_1}^{t_2} \delta\, dt}. \tag{6}$$

It is imperative to acknowledge that, within the first-order Nomoto framework, nonlinear terms and cross-coupled hydrodynamic derivatives (for instance, $Y_r'$ and $\dot{N}_v$) are disregarded [10]. The control system can be expressed using the classical PD (proportional-derivative) controller formula:

$$\delta = K_p(\Psi_d - \Psi) - K_d\dot{\Psi}. \tag{7}$$

By applying this equation, the characteristics of the closed-loop system are determined using the first-degree Nomoto model. The dynamics of the control law are represented by the following closed-loop equation:

$$T\ddot{\Psi} + (1 + KK_d)\dot{\Psi} + KK_p\Psi = KK_p\Psi_d. \tag{8}$$

In this context, the damping ratio ($\zeta$) and natural frequency ($\omega_n$) are calculated as follows:

$$\zeta = \frac{1 + KK_d}{2\sqrt{TKK_p}}, \quad \omega_n = \sqrt{\frac{KK_p}{T}}. \tag{9}$$

Accordingly, the values of $K_p$ and $K_d$ can be computed as

$$K_p = \frac{T\omega_n^2}{K} \tag{10}$$

$$K_d = \frac{2T\omega_n - 1}{K}. \tag{11}$$

Additionally, Fossen [29] provides the following recommendation for the integral component:

$$\frac{K_i}{K_p} \approx \frac{\omega_n}{10}. \tag{12}$$

In this regard:

$$K_i = \frac{\omega_n^3 T}{10K}[2]. \tag{13}$$

## 5.2 *Darpa Suboff Model 10/10 Zig-Zag Maneuver*

The 10/10° zig-zag maneuver was conducted using the model. Prior to performing the zig-zag test, the initial velocity was reached by applying PI control. The results of the zig-zag maneuvers are presented in Table 4 and visualized in Fig. 8. In this context, $\Psi$ represents the heading angle, $r$ represents the yaw rate, and $\delta_{Rudder}$ represents the rudder angles. Based on the results of the zig-zag maneuver test, several important conclusions can be drawn regarding the submarine's control capability and maneuverability. The first and second overshoot angles ($\Psi_1 = 62.91°$ and $\Psi_2 = 69.48°$) are quite high. This indicates that the submarine's responses to rudder rotations need careful management. Particularly, the high overshoot angles suggest that the submarine tends to deviate significantly during sudden maneuvers, raising concerns about its heading stability and course-keeping capacity. This behavior is primarily attributed to its geometric characteristics and relatively small rudder surface area. These findings indicate that the DARPA Suboff form requires a dedicated control system to maintain course stability effectively. The initial turning

**Table 4** 10/10 zig-zag maneuver results

| Case | $\Psi_1$st OSA | $\Psi_2$nd OSA | Initial turning time $t_a$ | Reach $t_A$ | Period $T_{ZZ}$ |
|---|---|---|---|---|---|
| 10/10 | 62.91 deg | 69.48 deg | 2.64 s | 19.4 s | 38.6 s |

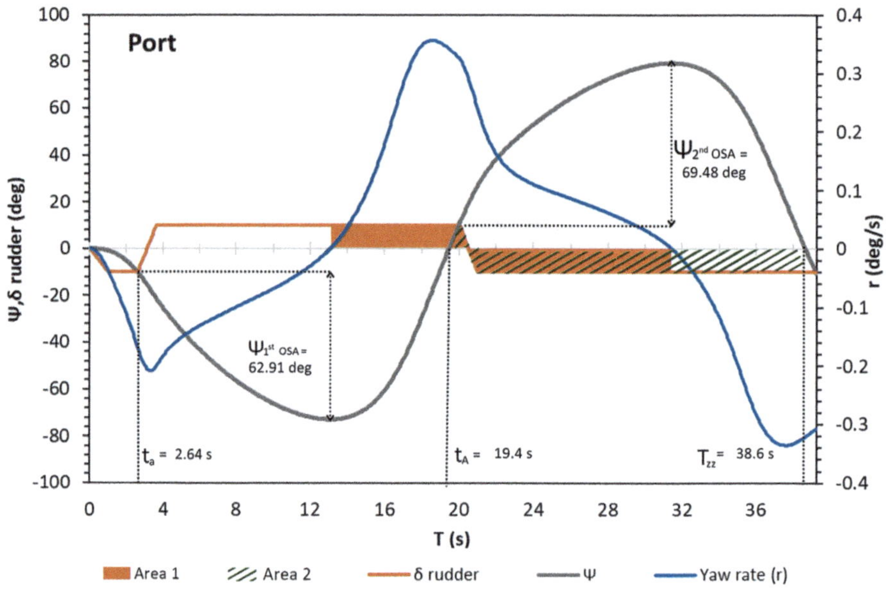


**Fig. 8** 10/10 zig-zag maneuver results

time of 2.64 seconds demonstrates that the submarine responds quickly to rudder commands, indicating a prompt reaction to sudden course changes. This is a positive feature, as it plays a critical role in the submarine's agility and maneuvering capability. However, the reaching time of 19.4 seconds indicates that it takes longer than expected for the submarine to settle at a specific angle. This suggests that the submarine takes time to establish a stable course during maneuvers, highlighting the need for fine-tuning the control system. These results indicate that the submarine's rudder and control systems require better optimization; otherwise, it may struggle with abrupt maneuvers and maintaining a stable course.

## *5.3 Determining PID Coefficients*

The Nomoto model [24] was used to derive the PID coefficients. The obtained data of heading angle ( $\Psi$), yaw rate ($r$), and rudder angle ($\delta_{Rudder}$) were processed to generate Fig. 8, which is used in the process of determining the values of $K$ and $T$. In Eq. 5, the integral $\int_{t_1}^{t_2} \delta \, dt$ represents the area under the curve in section $a$ of

**Table 5** PID parameters for course-keeping maneuvers

| Parameters | Value |
|---|---|
| $K$ | −4.3297 |
| $T$ | −5982.7 |
| $K_p$ | 13.8177 |
| $K_i$ | 0.1382 |
| $K_d$ | 276.5857 |

the figure. This integral is critical for deriving $K$, as it accounts for the cumulative effect of the rudder angle ($\delta$) over the specified time interval. When both $r_1$ and $r_2$ are set to zero, the equation simplifies, allowing us to isolate the effect of the rudder angle on the change in heading angle ($\psi$). This relationship illustrates how the effective application of rudder inputs influences the heading change without external disturbances, providing valuable insight into the submarine's maneuvering characteristics. On the other hand, in Eq. 6, if $t_1$ and $t_2$ are selected such that $\psi_1 = \psi_2 = 0$, the equation simplifies further. The integral $\int_{t_1}^{t_2} \delta\, dt$ in this case represents the area under the curve in section $b$ of the figure, and it is used to determine the ratio of $K$ to $T$, where $K$ is the maneuvering gain and $T$ is the time constant. When $\psi_1 = \psi_2 = 0$, the yaw rates ($r_1$ and $r_2$) are used to calculate the system's response to rudder inputs, providing key parameters to evaluate the submarine's dynamic stability and maneuverability. The coefficients $K$ and $T$ were computed using Eqs. 5 and 6. Once the values of $K$ and $T$ were determined, the PID controller coefficients $K_p$, $K_i$, and $K_d$ were calculated using Eqs. 10, 11, and 13. The resulting coefficients are presented in Table 5.

## 6 Conclusions

This chapter successfully investigated the applications of a CFD-based PI and PID controller to determine the propulsion characteristics and improve the course-keeping capabilities of the DARPA Suboff submarine, respectively. The self-propulsion point of AFF-8 was identified using the PI controller, with the proportional and integral gains calculated based on the model's hydrodynamic performance. It was observed that the self-propulsion characteristics found with this method are quite compatible with the literature. Thanks to this method, the man-hours and calculation costs spent on the iterative process of determining the self-propulsion characteristics have decreased considerably. The subsequent 10/10 zig-zag maneuvers were simulated to gather data on yaw, heading, and rudder angles, enabling the calculation of the Nomoto model's $K$ and $T$ parameters. These values were then used to determine the PID coefficients, which were applied to improve the submarine's maneuvering response. The implementation of the PI controller allowed for precise determination of the self-propulsion point, optimizing the balance between thrust and resistance.

The findings of this study highlight the effectiveness of using advanced control strategies such as PID and PI controllers for submarine course-keeping. By integrating CFD simulations with real-time control mechanisms, the research demonstrated the potential for improving submarine handling in challenging operational scenarios. The effectiveness of the coefficients in course-keeping capabilities will be examined in future studies. Future research should focus on automating vessels through advanced control systems, such as PID-based course-keeping and velocity control. These findings provide a solid foundation for further advancements in marine robotics, particularly in improving the autonomy of underwater vehicles.

**Acknowledgments** This work was supported by the Research Fund of Istanbul Technical University under the project number MYL-2024-45566 and is part of the first author's MSc thesis in progress at Istanbul Technical University. Computing resources used in this work were provided by the National Center for High-Performance Computing of Turkey (UHeM) under grant number 1017352023.

## References

1. Arıkan Özden, Y., Ozden, M., Demir, E., Kurdoğlu, S.: Experimental and numerical investigation of Darpa Suboff submarine propelled with INSEAN E1619 propeller for self-propulsion. J. Ship Res. **63** (2019). https://doi.org/10.5957/JOSR.09180084
2. Assani, N., Pavić, I.: Srdan Vukša, Marina Laušić analysis of the Nomoto ship model response to course ... ICTS 2020 Portoroz (2020)
3. Carrica, P.M., Mofidi, A., Eloot, K., Delefortrie, G.: Direct simulation and experimental study of zigzag maneuver of KCS in shallow water. Ocean Eng. **112**, 117–133 (1 2016). https://doi.org/10.1016/j.oceaneng.2015.12.008
4. CD-adapco, S.: STAR CCM+ User Guide Version 12.04, vol. 62 CD-Adapco, New York, NY (2017)
5. Chase, N., Carrica, P.M.: Submarine propeller computations and application to self-propulsion of Darpa Suboff. Ocean Eng. **60**, 68–80 (2013). https://doi.org/10.1016/j.oceaneng.2012.12.029, https://www.sciencedirect.com/science/article/pii/S0029801812004489
6. Clarke, D.: The Foundations of Steering and Manoeuvring. School of Marine Science and Technology, University of Newcastle upon Tyne (2003)
7. Çoşgun, T.: Numerical self-propulsion assessment of a generic submarine model at various forward speeds. J. Eta Maritime Sci. **9**, 192–199 (2021). https://doi.org/10.4274/jems.2021.24654
8. Delen, C., Kinaci, O.K.: Direct CFD simulations of standard maneuvering tests for Darpa Suboff. Ocean Eng. **276** (2023). https://doi.org/10.1016/j.oceaneng.2023.114202
9. Fang, M.C., Zhuo, Y.Z., Lee, Z.Y.: The application of the self-tuning neural network PID controller on the ship roll reduction in random waves. Ocean Eng. **37**, 529–538 (5 2010). https://doi.org/10.1016/j.oceaneng.2010.02.013
10. Gokarn, R.P.: A Study of Ship Manoeuvrability. Springer Nature, Singapore (2024). https://doi.org/10.1007/978-981-97-0625-9
11. Gong, C., Zhu, X., Wang, F., Jiang, P., Yu, W.: Research on submarine straight-line track control underwater based on nonlinear proportion differential. Math. Prob. Eng. **2016** (2016). https://doi.org/10.1155/2016/8432764
12. Groves, N.C., Huang, T.T., Chang, M.S.: Geometric characteristics of DARPA SUBOFF models (DTRC Model Nos. 5470 and 5471). David Taylor Research Center (1989)

13. Hu, P., Liu, W., Yang, J., Yu, X., Xu, L., Guo, L.: Underwater autonomous orientation using submarine light intensity gradient. Mechatronics **98** (2024). https://doi.org/10.1016/j.mechatronics.2023.103134
14. International Towing Tank Conference: ITTC-recommended procedures and guidelines: practical guidelines for ship CFD applications (2011), table of Contents: 1. Overview, 2. Pre-Processing, 2.1 Problem Characterization, 2.1.1 Resistance, 2.1.2 Wall Function, 2.1.3 Surface Roughness, 2.1.4 Incident Waves, 2.1.5 Motions, 2.1.6 Flow Features, 2.1.7 Region of Influence, 2.2 Geometry Creation and Modification
15. Jambak, A.I., Bayezit, I., Kinaci, O.K., Fidan, B.: Control oriented modeling and cascade control of container ships. In: 2023 29th International Conference on Information, Communication and Automation Technologies, ICAT 2023 – Proceedings. Institute of Electrical and Electronics Engineers Inc. (2023). https://doi.org/10.1109/ICAT57854.2023.10171320
16. Journée, J., Pinkster, J.: Introduction in Ship Hydromechanics. Delft University of Technology, The Netherlands (2002). Lecture MT159, Draft edition
17. Kim, D.H., Kim, Y., Baek, H.M., Choi, Y.M., Kim, Y.J., Park, H., Yoon, H.K., Shin, J.H., Lee, J., Chae, E.J., ho Shin, Y., Kim, E.S.: Experimental study of the hydrodynamic maneuvering coefficients for a BB2 generic submarine using the planar motion mechanism. Ocean Eng. **271** (2023). https://doi.org/10.1016/j.oceaneng.2022.113428
18. Kinaci, O.K., Gokce, M.K., Alkan, A.D., Kukner, A.: On self-propulsion assessment of marine vehicles. Brodogradnja **69**, 29–51 (2018). https://doi.org/10.21278/brod69403
19. Le, M.D., Nguyen, S.H., Nguyen, L.A.: Study on a new and effective fuzzy PID ship autopilot. Artif. Life Robot. **8**, 197–201 (2004). https://doi.org/10.1007/s10015-004-0313-9
20. Liang, L., Shi, Y., Mou, J.: Submarine multi-model switching control under full working condition based on machine learning. J. Shanghai Jiaotong Univ. (Sci.) **27**, 402–410 (6 2022). https://doi.org/10.1007/s12204-021-2284-6
21. Lin, Y.H., Lin, Y.T., Chiu, Y.J.: Stability and manoeuvrability simulation of a semi-autonomous submarine free-running model Suboff with an autopilot system. Appl. Sci. (Switzerland) **11**, 1–33 (2021). https://doi.org/10.3390/app11010410
22. Lungu, A.: A des-based study of the flow around the self-propelled Darpa Suboff working in deep immersion and beneath the free-surface. Ocean Eng. **244**, 110358 (2022). https://doi.org/10.1016/j.oceaneng.2021.110358, https://www.sciencedirect.com/science/article/pii/S0029801821016541
23. Moradi, M.H., Katebi, M.R.: Predictive PID control for ship autopilot design (2001). www.elsevier.comllocate/ifac
24. Nomoto, K.: A model for ship maneuvering. Int. Shipbuild. Prog. **13**(145), 21–30 (1966)
25. Pan, Y.C., Zhang, H.X., Zhou, Q.D.: Numerical prediction of submarine hydrodynamic coefficients using CFD simulation. J. Hydrodyn. **24**, 840–847 (2012). https://doi.org/10.1016/S1001-6058(11)60311-9
26. Ren, J., Xu, D., Xu, J.: Rans simulation for the maneuvering and control of a Suboff submarine model. Fluid Dyn. Mater. Process. **16**, 561–572 (2020). https://doi.org/10.32604/fdmp.2020.09791
27. Sezen, S., Dogrul, A., Delen, C., Bal, S.: Investigation of self-propulsion of Darpa Suboff by rans method. Ocean Eng. **150**, 258–271 (2018). https://doi.org/10.1016/j.oceaneng.2017.12.051, https://www.sciencedirect.com/science/article/pii/S0029801817307898
28. Sezen, S., Delen, C., Dogrul, A., Atlar, M.: An investigation of scale effects on the self-propulsion characteristics of a submarine. Appl. Ocean Res. **113**, 102728 (2021). https://doi.org/10.1016/j.apor.2021.102728, https://www.sciencedirect.com/science/article/pii/S0141118721002054
29. Skjetne, R., Smogeli, Ø.N., Fossen, T.I.: Modeling, identification, and adaptive maneuvering of cybership II: a complete design with experiments. IFAC Proce. Vol. **37**, 203–208 (2004). https://api.semanticscholar.org/CorpusID:113957884
30. Song, S., Demirel, Y.K., Atlar, M.: Penalty of hull and propeller fouling on ship self-propulsion performance. Appl. Ocean Res. **94** (2020). https://doi.org/10.1016/j.apor.2019.102006

31. Tatijana Dlabač, Martin Ćalasan, M.K., Marvučić, N.: PSO-based PID controller design for ship course-keeping autopilot. Brodogradnja **70**, 1–15 (2019)
32. Zhang, Z., Guo, L., Wei, P., Wang, X., Feng, D.: Numerical simulation of submarine surfacing motion in regular waves. Iran. J. Sci. Technol. Trans. Mech. Eng. **44**, 359–372 (2020). https://doi.org/10.1007/s40997-018-0259-5

# Index




B. U. Töreyin et al. (eds.), *8th EAI International Conference on Robotic Sensor Networks*, EAI/Springer Innovations in Communication and Computing,
https://doi.org/10.1007/978-3-031-92143-8

GPSR Compliance

*The European Union's (EU) General Product Safety Regulation (GPSR) is a set of rules that requires consumer products to be safe and our obligations to ensure this.*

*If you have any concerns about our products, you can contact us on ProductSafety@springernature.com*

In case Publisher is established outside the EU, the EU authorized representative is:

Springer Nature Customer Service Center GmbH
Europaplatz 3
69115 Heidelberg, Germany

**Batch number: 09699116**

Printed by Printforce, the Netherlands